MARINE

A Guided Tour of a Marine Expeditionary Unit

Tom Clancy's
Bestselling Novels Include:

The Hunt for Red October

Red Storm Rising

Patriot Games

The Cardinal of the Kremlin

Clear and Present Danger

The Sum of All Fears

Without Remorse

Debt of Honor

Nonfiction:

Submarine:
A Guided Tour Inside a Nuclear Warship

Armored Cav:
A Guided Tour of an Armored Cavalry Regiment

Fighter Wing:
A Guided Tour of an Air Force Combat Wing

Created by Tom Clancy and Steve Pieczenik:

Tom Clancy's Op-Center

Tom Clancy's Op-Center: Mirror Image

Tom Clancy's Op-Center: Games of State

TOM CLANCY

MARINE

A Guided Tour of a Marine Expeditionary Unit

B

BERKLEY BOOKS, NEW YORK

Most Berkley Books are available at special quantity discounts for bulk
purchases for sales promotions, premiums, fund-raising or educational use.
Special books, or book excerpts, can also be created to fit specific needs.
For information write to Special Markets at the address listed below.

MARINE

A Berkley Book / published by arrangement with
Jack Ryan Limited Partnership

PRINTING HISTORY
Berkley trade paperback edition / November 1996

The Putnam Berkley World Wide Web site address is
http://www.berkley.com/berkley

ISBN: 0-425-15454-8

BERKLEY®
Berkley Books are published by
The Berkley Publishing Group, 200 Madison Avenue,
New York, New York 10016.
BERKLEY and the "B" design are trademarks belonging to
Berkley Publishing Corporation.

PRINTED IN THE UNITED STATES OF AMERICA

10   9   8   7   6   5   4   3   2   1

Contents

For Air Force Captain Scott O'Grady.
A downed and doomed "zoomie" whose faith in his God,
his country, his service, and himself,
along with the help of a few Marines, brought him
home to us. God bless him, and
the members of the 24th MEU (SOC)
who made us all proud to be Americans once again.

Acknowledgments

It is now time for the best part of book writing: thanking those who helped make it possible. We start with my longtime partner, researcher, and friend, John D. Gresham. Once again, he traveled across the landscape, from Fort Worth, Texas, to Rota, Spain, gathering the stories and digging out the facts that make this book special. Perhaps most important of all, he kept the promises to our partners in industry and the military, which are the things that make books like this possible. Again, we have also been given the gift of wisdom and experience from series editor Professor Martin H. Greenberg. Laura Alpher is again to be complimented for her wonderful portfolio of drawings, which have added so much to this book. Tony Koltz and Mike Markowitz also need to be recognized for their continuing support that was so critical and welcome. Thanks again goes to Cindi Woodrum, Diana Patin, and Roselind Greenberg for their support in backing us up as always.

A book like this would be impossible to produce without the support of senior service personnel in leadership positions, and this one is no exception. Our first thanks go to General Charles "Chuck" Krulak, the 31st Commandant of the Marine Corps. Thanks also to his hardworking PAO, Major Betsey Arends. Another group, less well known but equally important, that was vital to our efforts consisted of the members of the various USMC public affairs offices (PAOs) and protocol organizations that handled our numerous requests for visits and information. Tops on our list were Brigadier General Terry Murray, Lieutenant Colonel Patricia Messer, and Lieutenant Mike Neuman of the Headquarters PAO. Along with them, Major General Paul Wilkerson, Captain Whitney Mason, Lieutenant Scott Gordon, and many others worked hard to get their stories across. Down at Quantico, Colonel Mick Nance and Gunner Bill Wright made our visits both memorable and livable in the incredible heat of 1995. At NAVSEA, Captain George Brown, Barbara A. Jyachosky, Sue Fili, Captain Manrin Gauthier, Captain Stan Harris, Colonel Al DeSantis, George Pickins, Paul Smith, and Gene Shoults told the shipping story. Over at the intelligence agencies, once again there was Jeff Harris and Major Pat Wilkerson at NRO, Russ Eggnor's photo shop at CHINFO, Lieutenant Colonel Jim Vosler and Penny Chesnut at DMA, and Dwight Williams at DARO. Many other helpful Marines studded the landscape to pass on their wisdom to us. Thanks to you all.

It is out at the units that you get the real story, though, and this year was a treasure chest of experiences and new friends. At the 26th MEU (SOC), there was the incredible Colonel Jim Battaglini, who is a national asset, along with such memorable personalities as Colonel "Fletch" Fergeson, Sergeant Major Bill Creech, Gunny Sergeant Tim Schearer, and Major Dennis Arnellio. Over at BLT 2/6, there was Lieutenant Colonel John Allen, an officer and Virginia gentleman. HMM-264 was led by the crusty and wise Lieutenant Colonel "Peso" Kerrick, and MSSG-26 by the capable Lieutenant Colonel Donald K. Cooper. Thanks also to Brigadier General Marty Berndt and Lieutenant Colonel Chris Gunter for sharing their adventures from 1995. And for all the other Marines at all the bases, we say, "Ooh-rah!" and many thanks for guarding the walls of freedom.

Out in the fleet, there were many wonderful folks as well. Special thanks to Captain C. C. Buchanan, who made PHIBRON 4 a great place to work and learn. Captains Ray Duffey and Stan Greenawalt as well as their incredible crew made USS *Wasp* our home-away-from-home. Captain John M. Carter of USS *Shreveport* and Commander T. E. McKnight of USS *Whidbey Island* are to be thanked as well for letting us break bread and share time with them and their crews. And out in the Med, Commander Mike John, Lieutenant Commander Bill Fennick, Ensign Dan Hetledge, and many others made our trip to Spain special.

Again, thanks are due to our various industrial partners, without whom all the information on the various aircraft, weapons, and systems would never have come to light. At the aircraft manufacturers there was Barbara Anderson, Robert Linder, Lon Nordeen, Gary Hakinson, Mary Ann Brett, and David Wessing of McDonnell Douglas; Joe Stout, Karen Hagar, Jeff Rhodes, James Higginbotham, and Doug McCurrah of Lockheed Martin; Russ Rummnay, Pat Rever, and Paige Eaton at Bell Textron; and finally, Bill Tuttle and Foster Morgan of Sikorsky. We also made and renewed many friendships at the various missile, armament, and system manufacturers including: the incomparable Vicki Fendlason and Tony Geishanuser at Texas Instruments; Larry Ernst at General Atomics; Glenn Hillen, Bill West, Kearny Bothwell, and Cheryl Wiencek at Hughes; Tommy Wilson, Adrien Poirier, Edward Ludford, Dave McClain, and Dennis Hughes at Loral; Eric O'Berg and William D. Eves at Delco; Jim McIngvale, Steve Davis, and many others at Litton Ingalls; Karl G. Oskoian at General Dynamics; Madeleine Orr Geiser and Bill Highlander at United Defense; Lee Westfield and Ms. Kathleen Louder at Right Away Foods; Rhonda Restau at Oregon Freeze Dry; Paige Sutkamp at the Wornick Company; Russ Logan at Beretta; Art Dalton and Brian Berger at Colt; Ronney Barrett at Barrett Firearms, and, last but certainly not least, Ed Rodemsky of Trimble, who again kept us up to date on the GPS system.

Again, we give thanks for all of our help in New York, especially Robert Gottlieb, Debra Goldstein, and Matt Bialer at William Morris. At Berkley Books, our appreciation again goes out to our editor, John Talbot, as well as David Shanks, Patti Benford, and Kim Waltemyer. For retiring friends like Jim Myatt and Robin Higgins, thanks for all you did and gave to the Corps and the country. Thanks also to our press pals, including Gidget Fuentes, Lisa Burgess, and Chris Plant. And for all the folks who took us on adventures, thanks for teaching the ignorant how things work for real. For our friends and loved ones, we have to once again thank you. For being there when we can't. God's blessings and goodwill upon you all.

Foreword

On January 5th, 1991, a third night of fitful sleep gave way to another day of incredibly tense living for U.S. Ambassador Bishop and the 281 personnel trapped with him in Somalia's capital city of Mogadishu. Included were officials from thirty nations, 12 diplomatic heads of mission, and 39 Soviets. After a message for help and two aborted rescue attempts by other nations, those remaining in the war-torn country, uncertain of their future, joined ranks and hunkered down inside the besieged and soon-to-be-overrun American Embassy compound.

Aboard the USS *Trenton* (LPD-14), 466 nautical miles away, two CH-53E Super Stallion helicopters with forty-six Marines and 9 Navy SEALs lifted off the flight deck into the Arabian night. Their mission—to evacuate the American Embassy in Mogadishu. After flying for seventeen hours, and two midair refuelings, the helos flew over the unsuspecting city at a twenty-five-foot altitude and landed in the compound at 0710—just as the rebels were scaling the walls. Within minutes Marines had secured the embassy. Shortly thereafter, the two helicopters departed with the first 61 evacuees. Less than twenty-four hours later, all 281 personnel had been successfully evacuated. The Amphibious Readiness Group with its embarked Marine Expeditionary Unit (Special Operations Capable)—ARG/MEU (SOC)—welcomed back its tired but successful warriors and quietly steamed back over the horizon.

Four years later and four seas away, a fatigued Air Force captain entered the sixth day of his fight for survival in rugged northern Bosnia. At home, a nation awaited news of her first native son shot down while supporting United Nations and NATO operations in this conflict. Out of sight, eighty-seven nautical miles away aboard the USS *Kearsarge* (LHD-3), another MEU (SOC) launched its Tactical Rescue of Aircraft and Personnel (TRAP) force. In the predawn darkness of June 8th, 1995, with less than a two-hour notice, forty-three Marines boarded two helicopters and launched into the Adriatic dawn. Joined by Cobra helicopter gunships and Harrier jump jets, they flew east over the missile-infested mountain to recover a tired, but relieved, Captain Scott O'Grady from the grasp of the pursuing Serbs.

Within twenty-four hours the rescued pilot was en route back to his home station at Aviono, and ultimately to the White House. Back aboard ship, the Marines cleaned their weapons and maintained their helicopters and equipment. They then rested as the ships sailed quietly over the horizon toward another readiness training exercise, all part of their scheduled 180-day tour of duty afloat. In both of these sagas, the individual of the hour was the United States Marine. For over 220 years, Marines have served at the end of America's operational reach—on freedom's far frontiers. These Marines are the backbone of the ARG/MEU (SOC) team, our regional commanders' force of choice for both forward presence and crisis response. When American interests are threatened abroad, Marines are on scene answering the call.

Marines and MEU (SOC)s are *not* special operations forces. They are general purpose forces who have successfully completed several months of intense specialized training, education and evaluation. Then they deploy forward with ARGs at their country's bidding—often in harm's way. They are America's warrior class: there when needed and prepared to "do what must be done." They seek only to serve their nation, and they enjoy the strong camaraderie born of shared sacrifice and hardship.

Marines have been doing this with rare consistency and success for more than 220 years. Since their inception in November 1775, when our Founding Fathers "...resolved, that two Battalions of Marines be raised...[and]...that particular care be taken that no person be appointed or enlisted into said Battalions, but such as are good seamen or so acquainted with maritime affairs as are able to serve to advantage by sea," Marines have continually demonstrated their readiness and utility. On their inaugural amphibious raid in the Caribbean in March 1776, Marines captured British cannon and powder to support the Continental Army. Since then they have been our nation's premier naval expeditionary warfighters, ever capable of executing a wide range of crucial missions "from the sea." On numerous occasions the Navy/Marine team has responded quickly and successfully to Presidential, Congressional, or military orders with such wide latitudes as "attack, take, and destroy as you may find," "perform duties as may be directed," or "render appropriate assistance." A 220-year legacy of readiness, teamwork, and courage is the result. Generations of Marines have repeatedly proven the veracity of both the Marine Corps motto of "Semper Fidelis" ("Always Faithful") and the reputation earned at Iwo Jima, where "uncommon valor was a common virtue."

Beginning with the "Banana Wars" in Haiti, Santo Domingo, and Nicaragua in the 1920s and 1930s, the Marine Corps bred a new generation of lean, battle-hardened fighters who were as proficient at amphibious landings and long-range jungle patrols as they were at urban warfare and quelling civil disturbances. Their stock-in-trade was readiness, versatility, and a deadly earnestness in fulfilling any assigned mission. These Marines got there fast, and with surprise, and came from the sea. They traveled light, fought hard, and lasted long. This reputation was not lost on either actual or potential adversaries. From these "interwar" experiences came the doctrine and training that would propel the Marine Corps for over forty years. The 1933 *Tentative Manual for Landing Operations* and the 1939 *Small Wars Manual* were the result. With the evolution of these operational practices in places like China and the Caribbean came the concept that the United States Marine Corps played a unique role in America's national defense. Besides being amphibious, Marines emerged as America's premier force-in-readiness.

As World War II dawned and our Corps grew by over fivefold, the legacy of ready and versatile soldiers of the sea was emblazoned on yet another age of American youth. Lieutenant Colonel Merritt "Red Mike" Edson's 1st Raider Battalion conducted its August 7th, 1942, landing on Tulagi with Marines steeped in this training and tradition. So too did the 1st Parachute Battalion that same day on Gavutu. Lieutenant Colonel Evans Carlson's 2nd Raider Battalion raided Makin Island a day later with Marines forged in the same fire. Each of these new units harnessed the raw energy of "basic Marine," and emboldened them with special and focused training, unit cohesion, and clarity of purpose. These units were special only because they consisted of special warriors: Marines capable of and willing to achieve extraordinary tasks because they had unquestioning confidence in themselves, their leaders, and their training.

These hard-earned lessons of the mid-20th century sustained Marine Corps training through Vietnam and well into the 1970s. With a prolonged investment in jungle and counter-guerrilla warfare as well as mountain and arctic warfare, the Marine Corps gradually refined a growing body of special operations capabilities. This included helicopter-borne reinforcement operations like "Sparrow Hawk" and "Bald Eagle," amphibious and riverine raids, snipers and discriminate shooters, and non-combatant evacuation operations (NEO) and TRAPs. What may have been missing in doctrinal cohesion was more than made up for with battle-tested tactical proficiency and well-honed operational procedures. Regardless of whether conducting long-range deep reconnaissance patrols or direct action missions like sniping, Marines had a well-earned reputation as fighters with courage, savvy, and skill.

After Vietnam, the U.S. military refocused on the Cold War, and the Marine Corps returned to its historic role as the nation's amphibious force-in-readiness. In the Pacific, Marines evacuated Saigon and Phnom Penh, boarded the *Mayaguez,* and rescued hurricane victims. From the Caribbean to the Mediterranean, Marine Amphibious Units (MAUs) executed NEOs and peace-keeping operations in Cyprus, Grenada, and Beirut. Around the globe, MAUs planned for and rehearsed countless other contingencies. From 1983 through early 1985, these lessons were codified with the activation of the new Marine Amphibious Unit/Special Operations Capable—MAU (SOC). This two-thousand-Marine unit was built around a Marine infantry Battalion Landing Team (BLT) as the Ground Combat Element (GCE), a composite helicopter squadron as the Aviation Combat Element (ACE), and a MAU Service Support Group (MSSG) as Combat Service Support Element (CSSE). This triad, along with the parent MAU Command Element (CE), represented the "pointy end of the spear" in America's foreign policy.

The six designated MAU (SOC)s, three on each coast, were trained, evaluated, and certified to perform eighteen critical and discrete missions. Several were amphibious in nature, such as the Marine Corps' time-tested amphibious raid. Others were contingency-response missions like evacuations and rescues. Several more were combat-related maritime special-operations missions. These included security operations, reinforcement operations, specialized demolitions operations, and military operations in urban terrain. Others included "stability" missions such as civic-action operations that provided dental, medical, and/or engineering support and mobile training teams that taught basic weapons, maneuver, and maintenance skills. Intelligence, counterintelligence, and tactical-deception operations completed another mission subset.

An integral part of the MAU (SOC) concept was the simultaneous development of the Maritime Special Purpose Force (MSPF). This internally sourced and task-organized, highly trained rapid-response force could participate in all of the above missions: specifically TRAP, demolitions, and operations. However, its essential role was conducting in-extremis hostage rescues. The MSPF, like its parent MAU, was never intended to be a special force. Instead, it was designed to provide Marines with special training and mission-essential equipment, keeping them ready and able to conduct the nation's bidding in circumstances requiring rapid response and quick thinking.

Since the formalization of the SOC program over a decade ago, and with a name change from amphibious (MAU) to expeditionary (MEU) to better reflect its adaptive nature and fast response focus, the ARG/MEU (SOC) continues to carve a unique and vital niche in America's defense establishment. The stark fact is that any MEU (SOC) can execute any one of its eighteen missions within six hours of an alert. They are trained and prefer to execute all their missions at night, or in limited visibility from over the horizon with tightly controlled communications. These operating characteristics put the MEU (SOC) at the cutting edge of night-flying and night-shooting technology. With the now-proven rapid-response-planning sequence, and years of exhaustive development of standing operating procedures and execution checklists, the MEU (SOC) program remains at the cutting edge of Marine combat training and preparation.

For almost forty-one years it was my honor and privilege to be a United States Marine. For much of that period, I was closely involved with the execution and refinement of the MEU (SOC) skills and initiatives just outlined. Through it all, some of my proudest moments were reserved for those many gallant warriors who selflessly answered their nation's frequent and clarion call to "send in the Marines." The history of the MEU (SOC) program has been written in their sweat and blood. It is a history that once again offers proof that special men with special training, forged in the fires of discipline and sacrifice, and operating as a team, can routinely achieve uncommon success when accomplishing even the most challenging missions.

Tom Clancy's engaging work on the MEU (SOC) captures much of this history and spirit. It provides the reader a lens through which to see today's Marines, and to experience their training, their challenges, and the intense confidence and camaraderie that continues to bind them. I commend it to your reading. It reaffirms my long-held belief that Marines are truly America's warriors, "...the few and the proud." Again for a brief moment, it has been my honor to reflect on the history of the courageous accomplishments of our Marine warriors. To all the Marines and sailors who have made our nation's Marine Expeditionary Forces truly special operations capable, take care of yourselves, take care of each other and—Semper Fidelis!

Al Gray, Marine
General, United States Marine Corps
29th Commandant of the Marine Corps

Introduction:
Marine—Part of the American Soul

et me pose a question to you. Do we actually have to *learn* who the men and women of the United States Marine Corps are? Or is it just an inbred part of our identity as Americans, like baseball and apple pie? Well, no, not really. Nevertheless, the Marines are older than baseball, much older in fact. It's generally accepted that America's birthday is July 4th, 1776, with the signing of the Declaration of Independence by the Continental Congress in Philadelphia. Interestingly though, the Marines were there first. *Their* institutional birthday is November 10th, 1775, predating the birth of the United States by fully eight months. Thus, the history of America *is* the history of the Marine Corps, and they have always been there for us.

It is perhaps the vision of Marines storming ashore onto a hostile beach that is the most enduring image of the Corps. Their amphibious tradition began in the Revolutionary War with the successful assault on Nassau in the Bahamas (we gave it back). Since then the Corps and its members have been at the crossroads of American and world history. Later, our first overseas assertion of national power was in the Mediterranean to fight the Barbary Pirates—Marine Lieutenant Presley O'Bannon on the "Shores of Tripoli," successfully attacking Derna and winning the Mameluke sword, which is still part of the uniform today. Marines also helped to raise the Bear flag in California. Marines even captured John Brown at Harpers Ferry, while under the command of two native Virginia Army officers, Colonel Robert E. Lee and Captain J.E.B. Stuart. When World War I came, Marines so impressed the French in 1918 that the forest they captured (Belleau Wood) was renamed in their memory. In World War II, Marines engaged in America's first major ground actions when we took the offensive against Japan on the steaming island of Guadalcanal in Operation Watchtower. During the Korean War, Marines anchored the stop line around Pusan, and then blew the Korean War wide open with their dramatic landing at Inchon. Almost everywhere our country has gone in the last twenty-two decades, the United States Marine Corps was the team that knocked on the door—or just kicked it in! Marines have even led us into outer space. The first American to orbit the earth—Lieutenant Colonel John H. Glenn, Jr. (now the senior senator from Ohio)—was a Marine aviator. They do get around.

The Marines have a global reputation. Whether it's fear or respect—probably a little of both—people around the world know exactly who the U.S. Marines are. At the Royal Tournament in London back in 1990, I saw the U.S. Marine Corps Band welcomed so warmly as to make me wonder if the British thought it was *theirs.* Clearly the Marines have a highly developed sense of public relations, but all that does is make people aware of who they are and what they've done. The Army's 82nd Airborne, the proud "All American" division with its bloused pants and jump wings, calls itself "America's Honor Guard," but look outside the White House and you find Marines. Probably there is no more easily recognized symbol of our country anywhere in the world—aside from the Stars and Stripes itself—than a Marine in dress uniform. What does it mean? It means the Marines *are* America. The Corps is an organization in which legend and fact intertwine to the point that you have to believe it all, because it really is true, ought to be, or soon will be. As recently as this last summer, in the science-fiction movie *Independence Day,* who saved the world from destruction? A Marine fighter pilot (ably played by actor Will Smith), of course.

The United States Marine Corps is America's national SWAT team. When there is trouble, they usually get there first. Their lifelong partnership with the U.S. Navy sees to that, since almost every nation in the world is accessible from the sea, and the Marines can appear like a genie from a bottle, deployed by helicopter from ships well beyond the horizon, projecting force within minutes of the President's phone call. Why? Lots of reasons. To rescue American citizens. To render disaster assistance. To stabilize a dangerous situation. To begin the invasion of a country to be liberated from tyranny. To do almost *anything,* because the Marine Corps by its nature is both a sharp and flexible instrument of national policy, with a lot of weight and power behind it.

Weighted? Flexible? These are terms that may not seem applicable to the "devil dogs" of the Corps. You would be wrong to assume this though. The Marine Corps is a package deal. Under the Marine Air-Ground Task Force (MAGTF) structure that every unit of the Corps fights from, you get almost every kind of combat power that can be imagined. Mostly you get riflemen—because *every* Marine is a rifleman. Tankers, artillerymen, helicopter and fixed-wing aviators, all one integrated MAGTF force package whose members all wear the same uniform, attend the same schools, pass the same standard tests, and talk the same language. Their Navy brethren are kind enough to provide transport, logistics, and medical corpsmen—and heavier air and fire support if any is needed. As a result, person-for-person, the United States Marine Corps may be the most dangerous group on the planet.

Weighted? Flexible? But how about smart? Somewhere in their history, the members of the Corps seem to have gotten a reputation for being simple-minded "jarheads." Let me tell you here that this is a *major* misconception. Marines have been among the most innovative of the world's military forces. Consider the following: there have been five major tactical innovations in twentieth-century ground combat. They are:

• *Panzerblitz* (**Armored Assault**): The use of heavy mounted formations was systematized by Hans Guderian of the German Reichwehr in the early 1930s. From this came the development of the large armored formations that were the spearheads of the campaigns in Europe in World War II. Since that time, armored units have been the cutting edge of the world's ground forces.

• **Airborne Assault:** The idea of dropping light infantry by parachute into an enemy's rear actually dates back to Ben Franklin in the late eighteenth century—he proposed using balloons to lift the troops. The idea was then resurrected in 1918 by General Billy Mitchell, though the Germans were the first to use them in combat against France and the low countries in 1940. Later, airborne assaults would be executed by all the major powers in World War II.

But, what about the other three?

• **Amphibious Assault:** This particular concept resulted from the British disaster at Gallipoli during World War I. After the Great War, two Marine colonels studied that campaign, diagnosed its failures and found in them both a formula for success and a mission for the Corps. Also called "Combined Operations" (by the British), Amphibious Assault almost overnight became a recipe for success. The Marine Corps wrote the cookbook.

• **Close Air Support:** The use of aircraft to support ground troops is another Marine innovation, practiced and perfected in the "Banana Wars" of the 1920s, and brought to the point that no American fighting man feels entirely clothed without aircraft overhead—preferably piloted by Marines.

• **Airmobile (Helicopter) Assault:** The technological perfection of the airborne assault, this concept was first used by Marines following Korea (they called it "vertical envelopment") to deliver riflemen and their support units in cohesive packages to decisive points behind the enemy front lines. It's both safer and more effective than falling from the sky in a parachute. With the addition of supporting attack helicopters, airmobile units are among the most mobile and well-armed in the world.

In short, for tactical innovation, the score for this century is U.S. Marine Corps 3–the World 2. All this from the smallest of the uniformed services in terms of size and budget. And some would tell us that Marines are dumb? Like a fox.

In this book, I'm going to take you on a tour of the most "Marine" unit left in the Corps today: the Marine Expeditionary Unit–Special Operations Capable (MEU [SOC]). In the seven MEU (SOC)s currently in existence, the Corps has placed the bulk of its amphibious and airmobile assault capability, and packaged them into battalion-sized MAGTFs that are forward deployed into trouble areas of the world. In this way, national leaders and regional commanders have a "kick-in-the-door" (the Marine leadership likes to call it "Forced Entry") capability that is right where it needs to be. We'll be looking at the 26th MEU (SOC), which is one of three such units in the East Coast rotation. Along the way, I think that you will be able to get a feel for the people and equipment that make up the 26th, and the Corps in general. You should, when finished, have a much better understanding of why I believe in Marines: their missions, their people, and their traditions. America's "911 Force."

Marine 101: Ethos

From the halls of Montezuma to the shores of Tripoli,
We will fight our country's battles in the air, on land and sea.
First to fight for right and freedom, and to keep our honor clean,
We are proud to claim the title of United States Marines.

—Marine Corps Hymn

"**M**arine." Say the word to any American, and you can count on a strong reaction. The word brings a vivid image to the mind of every American listener—perhaps John Wayne in *The Sands of Iwo Jima* or Jack Nicholson in *A Few Good Men.* Outside the United States, there are equally strong reactions, both positive and negative. Like other American icons such as Harley Davidson, Disney, and FedEx, the United States Marine Corps (USMC) is known as an institution that works. When the world throws problems at an American President, it is often Marines who are sent to make them right.

This book will focus on one of the basic building blocks of today's Marine Corps, the Marine Expeditionary Unit—Special Operations Capable, or MEU (SOC). It is a rapid-response unit, patrolling a dangerous world while waiting for the President of the United States to get a "911" call for armed intervention. Currently, the USMC maintains seven MEU (SOC)s: three on each coast, and one on Okinawa. Two or three of these units are deployed aboard ship into forward areas at any one time. Each MEU (SOC) is a self-contained naval/air/ground task force, capable of putting a reinforced Marine rifle battalion (over one thousand men) ashore. For decades, MEUs have provided U.S. Presidents with the ability to project power from the sea. MEUs (they were then known as Marine Amphibious Units or MAUs) led the way into Grenada and Beirut in 1983, and were among the first forces sent to Saudi Arabia when the 1990 Persian Gulf Crisis erupted. They were there when the first peacekeeping and relief forces went into Somalia in 1992, and were there again for the evacuation two years later. And MEUs are out there right now as you read this, training and staying ready, just in case they are needed.

This book will take you inside one of these units, and through it, inside the USMC as a whole. As you meet the people in the MEU and examine their equipment, I think you will learn why they represent an irreplaceable asset for the United States, an asset that's even more important today than it was just five years ago. You will come to understand how they work, their dedication and the personal sacrifices they make. For these are truly the people who stand guard on the walls of freedom, while the rest of us sleep safely in our homes.

Marines practice at
Camp Lejeune, N.C.
Regular exercises keep
these Sea Marines some
of the best combat infantry
in the world today.
JOHN D. GRESHAM

The Marine Corps Edge: Ethos

In my earlier books *Armored Cav* and *Fighter Wing*, the first chapter was devoted to an examination of critical technologies that give a particular service its combat edge. But in this book, things have to be a bit different. This is because most of the Marine Corps technology base is shared with the other three services. In fact, except for amphibious vehicles and vertical/short-takeoff-and-landing aircraft design (VSTOL), virtually every piece of equipment Marines use was developed by, and even bought for the Army, Navy, or Air Force. From rifles and uniforms to bombs and guided missiles, the Marines know how to get the most out of a Department of Defense dollar.

You might ask why we even have a Marine Corps, if all they do is use other folks' equipment and wear their clothes. Well, the answer is that Marines are more than the sum of their equipment. They are something special. They take the pieces that are given to them, arrange them in unique and innovative ways...and throw in their own distinctive magic. There is more to military units than hardware. There is the character of the unit's personnel: their strengths, experience, and knowledge, their ability to get along and work together amid the horrors of the battlefield. There is an almost undefinable quality. That quality is the Marine Corps' secret weapon. Their edge. That quality is their *ethos*.

Ethos is the disposition, character, or attitude of a particular group of people that sets it apart from others. It is, in short, a trademark set of values that guides that group towards its goals. The Corps has such an ethos, and it is *unique*. And it explains, among other things, why the Marines' reputation may well frighten potential opponents more than the actual violence Marines can generate in combat. Now, you may be thinking that I've gone off the deep end, comparing an abstract concept like ethos to hard-core technologies like armored vehicles or stealth fighters, but the "force-multiplier" effect on the battlefield is similar—an overmatch between our forces and those of an opponent. Trying to quantify such a concept is a little like trying to grab smoke in midair. To say that it is "X" percent training or "Y" percent doctrine is to trivialize what makes Marines such superb warriors. It is also probably inaccurate. Therefore, I think it is quite appropriate to explore what makes a Marine, any Marine, different from an Army tanker or an Air Force fighter pilot.

Though most Marines are unable to fully explain this mystical power, the Marine ethos is a combination of many different shared values and experiences. And it comes from what all Marines have in common, much like the brothers and sisters of a large family. In fact, this is how they refer to each other: as brother and sister Marines. Marines are unique among American service personnel in that they all must pass the same tests, no matter whether they are officers or enlisted personnel. This is in stark contrast to the other services, which rigidly separate their officers and enlisted personnel, maintaining separate career tracks, professional responsibilities, and even standards of performance and behavior to which they are held. In the Corps, everyone is a Marine!

This means that the leadership of the Corps works hard to give every Marine a common set of core skills, capabilities, and values to draw upon when they face the emotional crucible of combat. For example, once a year every Marine from the guards on American embassy gates to the Commandant of the Corps has to pass a physical fitness test (running and various other exercises), or be drummed out. In addition, every Marine always has to be fully qualified as a rifleman with the M16A2 5.56mm combat rifle; and officers also have to be fully qualified with the M9 9mm pistol. You might consider such standards petty, but when the call of "Enemy sappers on the wire!" is shouted, you want everyone from cooks to fighter pilots armed and ready to fight, shoulder to shoulder. This is the Marine way of doing things, and it has been for over 220 years.

Along with common standards and skills, every Marine shares a common heritage. This is more than just textbook history, for the Corps leadership believes that Marines need to know they are part of a team with a past, a present, and a future. What they do today is based upon the lessons of the past, just as the future should be based on a firm foundation of present experience. For Marines, their rich past is a living, ever-present reality. The Marines, alone among the services, require basic recruits and officers candidates to study their history as soon as they enter training. They all learn the important milestones that have defined the character of the Marine Corps and its ethos.

There is much to study in the Marines' twenty-two decades of existence, but a few defining moments stand out. These milestones—some predate the creation of the United States itself—are the historical structure which holds that ethos together. Let's take a look at them.

The Beginning: Tun Tavern, 1775

If you want to understand the Marine Corps ethos, it helps to start at the beginning. Created on November 10th, 1775, by the Second Continental Congress, the Corps served the new Continental Navy in the role Royal Marines had traditionally filled on board ships of the Royal Navy. Royal Marines were (and are) tough soldiers who suppressed mutiny and enforced discipline among the "press-ganged" (in effect, kidnapped) ships' crews, manned heavy cannons, and gave the ship's captain a unit of professional soldiers for boarding enemy vessels or landing on an enemy shore. These missions were rooted in the history of the Royal Navy, and the leaders of the Continental Congress felt their new Navy should also have Marines.

Four weeks after their legislative creation, the first Marine unit was formed in Philadelphia, at an inn called the Tun Tavern. The beginnings were modest: just one hundred Rhode Island recruits commanded by a young captain named Samuel Nicholas, a Philadelphia Quaker and innkeeper. These early recruits were all volunteers (beginning a tradition that continues in today's Corps). They fought their first action in March of 1775. Embarked on eight small ships, they sailed to the Bahamas and captured a British fort near Nassau, seizing gunpowder and supplies. Later, during the Revolutionary War, Marines fought several engagements in their distinctive green coats, such as helping George Washington to cross the Delaware River, and assisting John Paul Jones on the *Bonhomme Richard* to capture the British frigate *Serapis* during their famous sea fight.

From these humble beginnings came the start of the traditions that make up the Marine Corps that we know today. Its ranks are filled primarily with volunteers, and its missions are joint (i.e., in concert with other services like the Navy) and expeditionary in character. But perhaps most important is that when duty first called, Marines were among the first organized forces of the new nation to be committed to combat. This tradition of being "first to fight" is the first characteristic that their history brings to the ethos of the Corps.

The Halls of Montezuma...and the Shores of Tripoli

For a time following the Revolutionary War, the Marines were disestablished. But they were reborn with the revival of the United States Navy and its "big frigates" like the USS *Constitution* and USS *Constellation*. Once again, Marines went aboard to support the Navy in missions to protect American shipping and interests. As the 18th century came to a close, the interests of the United States assumed a more global character, and the Navy and Marines had to protect them.

During this period the Marine Corps conducted a series of operations, known as the War against the Barbary Pirates, that defined its role for the next two centuries. Four outlaw states along the coast of North Africa (the "Barbary Coast")—Algeria, Tunis, Morocco, and Tripoli—drew their primary source of income from capturing and ransoming merchant ships and their crews transiting the Mediterranean. For a time, the U.S. Government paid the ransoms, as other nations had done for years. But by 1803, the American and British governments had tired of this, and sent squadrons of combat vessels to suppress these maritime outlaws. Over four hundred Marines and other soldiers were committed to the effort, which inspired the line "*to the shores of Tripoli*"[1] in the Marine Corps Hymn. Their early achievements included the destruction of the captured American frigate *Philadelphia*. Later, in 1805, an expedition against Tripoli included eight Marines and a force of Arab mercenaries, which marched across six hundred miles of desert to storm the town of Derna. The

---

1 There are two ports on the Mediterranean named Tripoli. The one in the song is Tripoli, Libya, not to be confused with Tripoli, Lebanon.

war against the Barbary States was America's first overseas military operation, and Marines were in the thick of the action.

By the 1840s, the young United States of America had started to flex its muscles, coveting the tempting, sparsely populated, and vast Mexican territories of the Southwest. President James Polk, deciding to make this dream real, organized the conquest of Texas and California. Following the annexation of Texas in July of 1845, he dispatched Marine First Lieutenant Archibald Gillespie on a covert mission to the U.S. consul at Monterey, California, with special instructions for the takeover of that Mexican territory. Gillespie joined the famous explorer John C. Fremont, who led the California rebellion a year later.[2]

Meanwhile, the United States had declared war on Mexico. General Winfield Scott's invasion force included a battalion of about three hundred Marines led by Brevet Captain Alvin Edson. Landing at the Mexican port of Vera Cruz in March, 1847 aboard specially designed landing boats (the first purpose-built landing craft), they helped take the port in a matter of just two weeks. They also undertook a series of coastal raids to pin down other Mexican forces along the coast. Later, reinforced by additional Marines, the combined Army/Marine force marched on the Mexican capital, taking part in the final assault of the Battle of Chapultepec (September 13th, 1847).[3] The victory at the fortress of Chapultepec, the famous "*Halls of Montezuma*," led to the capture of Mexico City, and itself became a part of Marine Corps folklore. The scarlet stripes Marines wear on their dress pants are said to be in remembrance of the blood shed in the Mexican War.

While Marines took part in other actions, from quelling labor unrest to fighting in the War of 1812 and the American Civil War, it was these two conflicts just mentioned that defined the roles and missions of the Corps in its first century. Most notably, Marines fought alongside their Army and Navy brothers-in-arms, a precursor of the joint warfare so typical of today's military operations. The ethos had been born and was taking form.

Martial Tradition: The Music of John Philip Sousa

You do not need to be in the military to know that every organization has its own character or culture; for human groups spontaneously create culture. At IBM, it was conservative suits, John D. Watson's motto "THINK" on every desk, and a silly company song. Within other organizations, like the Jesuits or the Baltimore Orioles, those who belong to them are empowered by their culture, which is articulated in their traditions, rituals, and collective memories. Employees or members of an organization use the symbols of their culture to identify their roles and missions in the world.

Music forms an important part of Marine tradition. Though the Corps formed a band in the 1800s to play at ceremonial functions around Washington, D.C., it was

---

2 In 1846 American settlers poured into California, where they soon outgunned the tiny Mexican garrisons. Fremont led these settlers in the "Bear Flag Revolt," which briefly established California as a fictitious independent country. This is why the state flag is still inscribed "California Republic." Later, during the Mexican War, the Navy's Pacific Squadron under Commodore John D. Sloat seized the towns along the California coast, with the help of Fremont's local forces. California officially joined the United States in 1850 as the thirty-first state.

3 They did so alongside the "Brave Rifles" of the 3rd Cavalry Regiment, the predecessors of the modern 3rd Armored Cavalry Regiment which we explored in *Armored Cav: A Guided Tour of an Armored Cavalry Regiment.*

pretty much like other military bands of the period (i.e., loud and probably out of tune) until 1880, when Colonel Charles McCawley (the 8th Commandant) appointed composer and musician John Philip Sousa to lead the Marine Corps Band. Sousa created and popularized the Corps' martial music tradition. And in so doing, he revolutionized marching music and the bands that played it. He also composed a body of music that is at the core of Marine Corps tradition today. His compositions included "Semper Fidelis" (1888), "Washington Post March" (1889), "King Cotton" (1897), and the most popular of all, "The Stars and Stripes Forever" (1897). For a dozen years he led the Marine Band, taking it on tours all over the country and the world. The effects were both deep and lasting.

Since Sousa and his music were as popular in his time as Glenn Miller or the Beatles have been in ours, his band's performances were the 19th century equivalent of a recruiting commercial for young men of the period. More than that, in the age of global imperialism, the band's bright uniforms, the precision of their drills, and the inspiring qualities of their music left a positive impression of the Marine Corps in the public mind. Perhaps Sousa's most lasting contribution to the Corps, however, was forging the Marines' relationship with the President of the United States. As the Chief Executive's personal band, the Marine Band often played at the White House and other official functions. And by the time Sousa left to form his own private band in 1892, his music and service had forever bound the Presidency to the Marine Corps. You see this when the President flies in his Marine helicopter, when you walk up to an American embassy guarded by Marines, or when you note that wherever the Navy has nuclear weapons, there are Marines guarding them. Sousa's music was the link that forged that special relationship.

World War I: The Corps Is Forged

For over a century, the Marine Corps was a tiny portion of the American military structure. Before World War I, its strength of 511 officers and 13,213 enlisted men made it just a fraction of the strength of the Army and Navy. America's entry into World War I meant that the country's modest peacetime military was going to expand exponentially in a short time. In support of this effort, the Marine Corps

Marines storm ashore a Japanese-held island during World War II.
OFFICIAL U.S. NAVY PHOTO

expanded rapidly. With the addition of new training facilities at Parris Island, South Carolina, the Marines grew to a wartime high of 2,462 officers and 72,639 enlisted ranks. This included a small number (277) of the first women Marines, recruited to free up men for combat. The war also saw 130 Marine aviators, a new kind of warrior. For World War I, the Marines deployed the largest formations in their history, brigades of up to 8,500 men, to fight on the Western Front. Rapidly pushed into that cauldron, they fought at Belleau Wood, Soissons, and St. Mihiel, and in the Meuse-Argonne Offensive. These victories came at a high cost, with the Marine Brigade suffering 11,968 casualties, with 2,461 killed. Following the Great War, Marines participated in the occupation of Germany, keeping watch on the Rhine until July 1919, when they finally returned home. Following a victory parade in front of President Wilson with the rest of the U.S. 2nd Division, they were demobilized.

For all of its costs, World War I left the Marine Corps with positive results: For the first time, the Corps was allowed to form and operate combat units as large as those of the Army. They demonstrated that their unique training and indoctrination produced a more effective and aggressive combat infantryman than the other armies on the Western front. They experimented with new ideas, like integrating women into the Corps and using airpower to support Marines on the ground. Virtually every kind of challenge that faces Marines today was discovered and dealt with during the Great War—for example, the horror of poison gas. But most importantly, the Marines had been allowed to grow large, and had shown the country what a larger Corps could achieve. This would make it easier for the Marines to expand to meet the challenge of their defining moment, the Pacific campaign of World War II.

The "Small Wars"

With the Great War won, the Corps returned to its peacetime routine of duty aboard ships, and peacekeeping missions in China and the Philippines. This was an era of "small wars," with interventions mostly in Central American and Caribbean countries in support of American foreign policy. This "gunboat diplomacy" was a typically American mix of corporate greed (dominating the regional economy) and noble intentions (rescuing local populations from despotism or anarchy). At the cutting edge of these interventions were Marines, leading the way and taking most of the casualties.

Even before World War I, the Marines took part in putting down Filipino rebels and quelling the Boxer Rebellion in China, both in 1899. During the Taft and Wilson Administrations, Marines carried out interventions in Nicaragua (1912 to 1913), Haiti (1915 to 1934), and the Dominican Republic (1916 to 1924), pacifying the Panama Canal Zone (1901 to 1914) and Cuba (1912 to 1924), and at Vera Cruz, Mexico (1914). Through these actions, the Marines became experts in what is now called "counterinsurgency" warfare. They even wrote a book, *The Small Wars Manual* (1939), which is considered a military classic, much admired but little read outside the Corps.

The small wars established the Marines as leaders in unconventional warfare—thus continuing a tradition of special missions and operations that date back to the war with the Barbary States in the early 19th century. This tradition gave the Corps a base of experience that allowed it to conduct similar missions in World War II, as well as into the postwar era and today. In fact, ignorance of the lessons in *The Small Wars Manual* contributed to the failure of U.S. policy in Vietnam and various Third World insurgencies over the years. These lessons included the importance of providing security to native populations ("civic action"), and the need to target the enemy's weakness (in finance and logistics) rather than his strength (small-unit combat in difficult terrain). Despite that failure, Marines still have the corporate knowledge of such operations, and are using it today in the training and operations of the MEU (SOC)s around the world. And they still read and use *The Small Wars Manual*. I know. They gave me a freshly printed copy.

1942: First to Fight

The years leading up to World War II saw Marines at the edge of the developing conflict. Marines were caught between warring Chinese and Japanese forces in 1932 Shanghai when fighting broke out there. Other incidents involving Marines in China followed. When World War II finally engulfed the United States in 1941, the Corps was in the heat of the fighting from the start. Over a hundred Marines died in the attack on Pearl Harbor, and thousands more would fall in the weeks and months ahead. Marine units initially served as base garrisons defending remote outposts. The tiny Marine force on Guam surrendered on December 10th, and the Midway garrison was bombarded by a pair of Japanese destroyers. All around East Asia, small forces of Marines fought for their lives in the early days of the Great Pacific War, usually without enough men or equipment to be more than speed bumps for the onrushing Japanese forces.

One exception was the tiny atoll of Wake, where a Marine island defense battalion with a handful of fighter planes held off repeated Japanese assaults before they were overwhelmed on December 23rd, 1941. For over two weeks, the defenders of Wake Island held off a vastly superior force of Japanese ships and troops, inspiring the whole nation with their plucky spirit and sacrifice. Unfortunately, Navy leaders at Pearl Harbor, struggling to protect what was left of the shattered Pacific Fleet, canceled a relief mission, allowing the island and its defenders to fall without support. Wake damaged the long-standing trust between the Corps and the Navy, a memory that still rankles Marines and shames sailors.

The Navy would soon have a chance to square things with their Marine brethren. The spring of 1942 saw the Navy and Marines reversing the Japanese tide of expansion at the Battle of Midway. Navy carrier aviators wiped out their Japanese opponents, and this time stayed to support the Marines. As a result, Midway held out against determined air attacks, but Marine aviators defending the island were decimated while flying obsolete Navy "hand-me-down" aircraft. The leaders of the Corps vowed that the next time Marines had to fight, they would have proper equipment, aircraft, and Navy support. They would not have long to wait.

Down in the Solomon Islands, Allied intelligence found the Japanese constructing an airfield on the island of Guadalcanal which threatened Allied supply lines to Australia and had to be neutralized. Luckily, the prewar expansion of the Corps had begun to pay off, and there now was a division-sized force in the Pacific to do the job. In August of 1942, the 1st Marine Division splashed ashore onto the

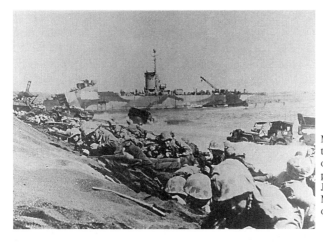

Marines take cover on the beachhead at Iwo Jima on February 19th, 1945, prior to moving inland. Iwo Jima was the largest all-Marine amphibious operation of World War II.

OFFICIAL U.S. NAVY PHOTO

beaches of Guadalcanal and nearby Tulagi and seized the airfield, beginning one of the most vicious campaigns of World War II. For the next six months, Allied and Japanese ground, naval, and air forces fought a battle of annihilation in the jungles, skies, and seas around Guadalcanal. When it was over, the Marines had played a key role in winning a decisive but costly victory, with 2,799 Marines wounded and 1,152 killed. Marine aviation helped drive the Japanese from the skies over Guadalcanal. This also took its toll, with some 127 Marine aviators wounded, 55 killed, and 85 missing in action. But when it was done, the Marines and their Navy partners had turned the tide of battle in less than a year following Pearl Harbor. For the Marines, it validated their claim of "first to fight." They were the first Allied ground force to take the offensive against Axis forces in World War II, a point they still take pride in today.

The Central Pacific: Winning the Bases with Vision

As early as the early 1900s, following their victory over Czarist Russia, the Japanese had dreamed of expanding their empire into China, Southeast Asia, and the Pacific islands. This dream did not go unnoticed. Even before World War I, the U.S. and Great Britain had prepared contingency war plans against Japan, the American version being the famous War Plan Orange. The U.S. plan was based upon a long march across the Central Pacific, with the navies of the two nations eventually slugging it out in one huge decisive battle. Capturing and holding the island bases that would be needed was the task of the Marine Corps, which had been studying problems of amphibious warfare for decades.

Among the many Marines busy thinking about amphibious operations was Commandant Lejeune, who declared in 1922 that it was vital to have "a Marine Corps force adequate to conduct offensive land operations against hostile naval bases." By the 1930s the Corps had made great strides, including the publication of a *Tentative Landing Operations Manual*, which became the bible for early amphibious exercises. Marines worked to master new technologies that would allow them to carry out new missions. Landing craft, naval gunfire control equipment, and command radio equipment were key to the new job. Marines appear to have been the first aviators in the world to perfect precision delivery for aircraft bombs when they developed dive bombing. German officers observed Marine bombing demonstrations in the 1930s, which led to the adoption of the technique by Stuka dive bombers of the Luftwaffe.

It was in the Central Pacific, though, that the Marine Corps forged the amphibious assault doctrine that became its enduring tradition. In an "island hopping" campaign, the Marines and Navy conducted a series of landings to take the bases originally designated in War Plan Orange. The drive across the Central Pacific began in the fall of 1943 at Tarawa Atoll in the Gilbert Islands. Although almost everything possible went wrong (incorrect tidal projections, poor communications and naval gunfire support, etc.), the main island of Betio was taken in seventy-six bloody hours. Despite the heavy cost in Marine and Navy casualties (1,113 killed and 2,290 wounded), much was learned, and the lessons paid for in blood at Betio saved lives later on other islands. Following Tarawa, the Marine/Navy team took the atolls of the Marshall Islands in a swift campaign in early 1944. Capturing Kwajalein and Eniwetok Atolls, they bypassed other Japanese-held islands in the chain.

The next campaign was to be the decisive battle that strategists on both sides had planned for almost half a century, the drive into the Marianas and the resulting naval battle of the Philippine Sea. Led by the legendary General Holland M. "Howlin Mad" Smith in the spring of 1944, the joint Marine/Army amphibious force took Saipan, Guam, and Tinian in just a matter of weeks. This gave the Americans bases to launch B-29 strategic bombers against Japan, less than 1,500 nm/2,750 km away. From these bases, the war was taken to the Japanese homeland in an intense firebombing and mine-laying campaign. It also provided the bases that launched the atomic bomb strikes that ended the war.

More than the other services, the Marine Corps clearly saw its mission in World War II, and developed appropriate technologies and skills to accomplish its critical task of amphibious assault. This is in sharp contrast to the Army Air Force, which saw strategic bombing alone as the key to victory,[4] and the Navy, which thought their battleships' guns would win the war.[5] The Corps understood that war is a joint operation—if we were going to win, all the services were needed—and this vision has continued into the postwar era. Always innovators, since the end of World War II the Marines have been leaders in the developments of helicopters, air-cushioned landing craft, and other technologies.

Iwo Jima: The Defining Moment

Following the mainly Army-led invasion of the Philippines in late 1944, the Marines were ready to begin their drive towards the home islands of Japan. They would need to be ready, because the next battle in the Pacific would be the toughest yet: Iwo Jima. A tiny pork-chop-shaped island (just eight square miles) in the Bonin chain just 670 nm/1,225 km from Japan, it was a vital link in the drive on the

---

4 Airpower zealots believed that daylight precision bombing, using the top-secret Norden bombsight, would quickly cripple the German war effort. It turned out that the accuracy of bombing had been overestimated by several orders of magnitude, and most of the bombs dropped on Germany were wasted.

5 Navy submarines sank far more enemy shipping in WWII than the cherished and vastly expensive battleships, despite the handicap of scandalously unreliable torpedoes during the first year of the war.

The defining moment of the Corps: the raising of the American flag by Marines atop Mount Suribachi on February 23rd, 1945. At this moment, then–Secretary of the Navy James Forrestal is said to have uttered, "...the raising of. that flag means a Marine Corps for the next five hundred years...."

OFFICIAL U.S. MARINE CORPS PHOTO VIA ASSOCIATED PRESS

Japanese homeland. In February 1945, 71,245 Marines hit the volcanic ash beaches of Iwo Jima. There were 21,000 Japanese dug in, determined to fight to the death. They did. Over the next month, Marine and Navy casualties were almost 27,000, and virtually every Japanese on the island was killed. The American lives were not wasted, however, for the island runways began to save the lives of B-29 crewmen even before the fighting ended.

These dry facts are all well and good. But beyond them is a deeper reality: The battle of Iwo Jima was *the* single defining moment in the history of the Corps. Iwo Jima was an impregnable fortress if there ever was one, far tougher than anything along Hitler's vaunted "Atlantic Wall." The Japanese had spent over a year fortifying the island, including over eleven miles of tunnels which were dug mostly with hand tools! Japanese leaders clearly understood that its loss would put even American P-51 Mustang fighters within range of the home islands. For the Marines, Iwo Jima was going to be the last all-Marine invasion of the war, and they needed to take it, both for what it meant to the war effort and for what it meant to the image of the Corps. Iwo Jima was *their* island, and they meant to take it whatever it cost. They got their wish.

From the moment that they hit the black sand beaches, Iwo Jima was the hell of every Marine's nightmare, with death and horror behind every rock and in every hole. But tough as the Japanese and their island fortress was, the Marines and their Navy support offshore were tougher. Yard by yard, rock by rock, the Marines cleared the island. And in so doing, they wrote a unique page in American military history. In front of the full view of the wartime press, Marine units advanced against suicidal Japanese forces, taking everything that was thrown at them. Twenty-four Marines won the Medal of Honor on Iwo,[6] more than at any other battle in history.

---

6 Twelve of these awards were posthumous. Several Navy medics serving with Marine units on Iwo Jima were also awarded the Medal.

Lastly, Iwo Jima gave the Corps and America its most famous and enduring World War II image, the raising of the American flag atop Mount Suribachi. When the flag was raised at 10:20 A.M. on February 23rd, 1945, Secretary of the Navy James Forrestal, watching from one of the offshore ships, turned to General Holland "Howlin' Mad" Smith and said, "General, the raising of that flag means a Marine Corps for the next five hundred years." Now memorialized with the magnificent memorial overlooking the Potomac River in Rosslyn, Virginia, this was the defining moment of the Marine Corps. More than any other aspect of the Marine ethos, the indomitable spirit of the Corps at Iwo Jima tells who they are. It says there is nothing Marines won't try if ordered to do so, and no cost they will not pay to accomplish that mission. Later, at Chosin Reservoir, Khe Sanh, and the Beirut barracks, Marines remembered the sprit of Iwo Jima, dug in, and accomplished their missions, no matter what was asked of them. *That* is the Marine ethos defined.

"Send the Marines"

Following the war, the Corps endured the same downsizing as the other military services. The hollow shells of just two divisions remained: the 1st at Camp Pendleton, California, and the 2nd at Camp Lejeune, North Carolina. The year 1950 saw a rapid Marine response to the outbreak of war in Korea. Once President Truman committed ground troops to help the beleaguered South Koreans, Marines were among the first reinforcements to arrive. Unfortunately, after the brilliant landing at Inchon (September 15th, 1950) by Marine and Army forces and the drive to the Yalu River, the Marines settled down to a miserable routine of trench warfare. They spent the next twenty-two months fighting as "leg" infantry alongside the other UN forces. This misuse of the Marines' unique amphibious capabilities made a deep impression on the leadership of the Corps, who determined it would never happen again. Their response to the problems of Korea was a new organizational doctrine, the Marine Air-Ground Task Force (MAGTF). The idea was to keep the air, land, and logistical elements of Marine units together, as an integrated team. In this way, Marines on the ground would not have to depend upon the Air Force for close air support (CAS) or the Army for supplies. They would be able to shape their own tactics and doctrine. Half a century later, Marines always go to fight in MAGTFs.

The election of Dwight D. Eisenhower as President brought a new respect for the capabilities of the Marine Corps. Eisenhower and his successors began a tradition of sending highly mobile MAGTFs to trouble spots around the world for pacification, peacekeeping, or plain old-fashioned gunboat diplomacy. Some, like the landing operation in Beirut in 1958, were highly successful. Others, like the 1965 Dominican Republic operation, were widely viewed as repressive mistakes. Direct U.S. military intervention in Vietnam in 1964 began as a series of landings designed to prop up the government of South Vietnam. Marines served in Vietnam from the first to the very last in 1975, usually assigned to the I Corps area in the northern sector of South Vietnam.

For the Corps, the tendency of Presidents to "send the Marines" simply affirms their "first to fight" reputation, as well as the inherent flexibility of the MAGTF concept. Willingness to move first and fast, and being ready to do so, is part of the Marine ethos—when you want something done right, give the job to the Corps!

Ribbon Creek: Remaking the Corps

The postwar years were busy for the Marines, as they were often called upon to support U.S. interests overseas. But with the coming of the Cold War, the Corps sought to make itself ready for its part in America's defense mission. Thus, Marines endured atomic battlefield tests in Nevada and began to absorb new equipment and tactics. All of this came from a general view that the Corps was remaking itself into a high-technology force that was ready to fight on the nuclear battlefield. Then came the tragedy at Ribbon Creek. In 1956, a drunken drill instructor at the recruiting depot at Parris Island, South Carolina, marched a group of seventy-four recruits into a tidal swamp called Ribbon Creek. Six of them died. The tragedy led to a total reform of Marine recruit training.

Ribbon Creek brought on a strong Congressional and public reaction. This came from genuine concern for the welfare of individual Marines and the Corps as a whole. Clearly, Americans wanted the Corps to be a reflection of their values and ideals. Several hundred instructors were relieved of duty as a result of investigations into their conduct in training Marines. In addition, Ribbon Creek led to a profound transformation in the way the Corps viewed and trained its recruits. The shift reinforced the attitude that all Marines are brothers or sisters to their fellow Marines. Even today, the memory of Ribbon Creek influences the way new recruits are handled—not with kid gloves, but with respect for their safety and dignity. This too is part of the Marine ethos: to take care of their brother and sister Marines.

From Desert One to Desert Storm

November 1975 found the Corps celebrating its two hundredth birthday...and again fighting for its life in Congress. This time the issues were manpower, and the question of the Marines' capability to fight on modern battlefields. The 1980s and 1990s provided ample proof that they were capable. Meanwhile, two events in this period would have a fundamental effect on the Corps. The first was the failed embassy hostage rescue in Iran, in which Marine helicopter pilots took part. A result of this disaster was a *really* hard look at joint warfare, leading to the Goldwater-Nichols reform act of 1986. The second was the creation of the Rapid Deployment Joint Task Force (RDJTF) for use in the Middle East. First commanded by Lieutenant General Paul X. Kelley (the future 28th Commandant), it was not very rapid, could not deploy, and was not much of a force. But it was a step on the road to the creation of the U.S. Central Command—the force that would later emerge victorious in Desert Storm.

The election of President Reagan in 1981 led to renewed growth for the Marine Corps, as it did for the other services. Programs like the CH-53E Super Stallion transport helicopter and the AV-8B Harrier II fighter bomber, starved for funding during the Carter Administration, now were fully funded for production. Navy amphibious shipping, which had dropped to only sixty-seven units, was built

A Marine Corps AV-8B Harrier II flies over the oil fires of Kuwait following the 1991 Persian Gulf War. A two-division Marine Expeditionary Force liberated most of Kuwait, including the capital city.

up as well. The next few years were good ones for the Corps, with a steady influx of new equipment, personnel, and doctrine. Key among these was the development of the Maritime Prepositioning Force, groups of prepositioned ships loaded with equipment and supplies to support a Marine Expeditionary Brigade of 16,500 men in the field for a month. Based at three locations around the world, MPS allows a rapid response to an emerging crisis by a Marine force with serious teeth. The other major development was the creation of the MEU (SOC). Created by General Alfred Gray (the future 29th Commandant), the MEU (SOC) was a response to the terrorism of the 1980s and the need to deal with fast-breaking situations in hours, not days or weeks. It was this force that the Marines took into the last days of the Cold War and the beginning of the New World Order of the 1990s.

When Iraq invaded Kuwait, the 7th Marine Expeditionary Brigade (MEB) immediately deployed to Saudi Arabia from its home at Twenty-nine Palms, California. There, at the port of Al Jubayl, they joined up with the supplies and equipment of the MPS dispatched from Diego Garcia. Some of these supplies even helped to sustain an early arriving brigade from the 82nd Airborne Division. By the start of the ground war, the Marine force ashore had grown to two full divisions, an air wing with over 450 aircraft, and two combat service support groups, totaling over seventy thousand Marines and sailors.

When the ground war began on February 24th, 1991, two divisions of Marines drove north into Kuwait, while other units of the Corps were busy offshore in the Persian Gulf. The combined 4th and 5th Marine Expeditionary Brigades, with seventeen thousand men embarked in thirty-one amphibious ships, threatened an assault on the Kuwaiti coast. This had the effect of freezing seven Iraqi divisions in place, guarding against an invasion that never came. Meanwhile, elements of the 4th MEB, while waiting to play their part in Desert Storm, carried out a daring rescue of the American embassy in Mogadishu, Somalia, in January 1991. Using in-flight refueling, CH-53Es evacuated the entire embassy staff and other civilians from that war-torn city.

It was a very busy time! It still is. Since 1991, Marines have gone wherever American interests were on the line—peacekeeping in Somalia, disaster relief in Florida, riots in California, or rescuing a downed pilot in Bosnia. The key to the Marines' flexibility is a strong sense of their chosen roles and assigned missions. By clearly understanding who they are, where they have been, what they have done, and what they are capable of doing in the future, Marines will remain America's premier shock troops, the "first to fight." This is the Marine Ethos.

Forward From the Sea: The Marine Corps Mission

Book IV of Julius Caesar's *War Commentaries* describes his amphibious invasion of Britain with two Roman Legions in 55 B.C. The details would be familiar to any Marine who has ever hit a beach. Though Marines are capable of conducting many other missions, storming ashore is the role most associated with the Corps. These include guarding U.S. embassies and diplomatic personnel overseas, helicopter transportation for the President and senior Administration leaders, and security for "special" (i.e., nuclear) weapons and their storage sites. Marines have also excelled in peacekeeping operations in Somalia and Haiti, riot control in Los Angeles, security operations at the 1984 Los Angeles Olympic Games, and disaster relief almost anywhere an earthquake, hurricane, or other natural catastrophe strikes. Finally, just ask Marines themselves what they think their mission is. I can guarantee that they will give you an icy look, square their shoulders, and respectfully tell you that they are riflemen, first and last, whatever their actual job specialty. Keep that in mind as we look at the *real* roles and missions of the Corps that follow.

All the missions we mentioned above are surely important, but going over enemy shorelines and winning battles is what defines the mission of the Marine Corps today. Following the end of the Cold War and Desert Storm in the early 1990s, the Navy published a white paper called *From the Sea*, which redefined American seapower for the 21st century. *From the Sea*, and a revised edition called *Forward...From the Sea*, have been controversial. The Navy has backed away from its traditional "blue water" combat role,[7] now that the Soviet naval threat has, in effect, disappeared. With no real blue-water threat on the horizon, Navy leaders see the roles and missions of the sea services increasingly tied to operations in the "littoral" or coastal zones of the world. Littoral zones in the Middle East, Indian Ocean, and Asia probably offer the highest likelihood of conflict in the coming years. The majority of the world's population centers are within these areas, along with vast industrial, energy, and mineral resources. Since the end of World War II, most of the U.S. Navy's operations have taken place in littoral areas, notably the Persian Gulf, the Gulf of Tonkin, and the Mediterranean Sea.

So ingrained is the mentality of the sea service towards battles in the open ocean, some naval analysts have questioned the Navy's turn to the coastline.

---

7 Note, "Blue water" is generally understood to mean naval operations on, above, or under the ocean, beyond the continental shelf, which is nominally defined as the 200-fathom (1200-foot/600-meter) depth-contour line. "Blue water" may also be viewed as ocean areas beyond the direct influence of enemy land-based airpower.

However, whether or not you accept the doctrine in *Forward from the Sea*, the Marines see it as yet another validation of their basic mission as America's seaborne assault force. Now, after 220 years, that mission is finally part of the official U.S. Navy doctrine. It has even survived a recent Department of Defense (DoD) Roles and Missions commission, which left much of the Marine force structure virtually untouched after months of examination.

Clearly, the Marine Corps' first mission is the maintenance of the three active division-aircraft wing teams as rapid-reaction forces for trouble spots around the world. These forces can support other Allied forces already in place, or open a new flank from the sea. This is exactly what happened in Korea in the 1950s, Vietnam in the 1960s and 1970s, Desert Shield in 1990, and Desert Storm in 1991. In each case, Marines added mass to joint operations with the U.S. Army. While this mission may not be the favorite of the leadership at Marine Corps Headquarters, given the decreasing size of the Army, it is vital. The three active Marine divisions represent almost 25% of U.S. ground forces today. This means that any major overseas deployment will probably include one or more of these powerful division-air wing teams.

Another mission is for large, regimental-sized Marine units (up to fifteen thousand Marines) to fall upon prepositioned equipment stocks in land-based depots (such as in Norway) or stowed on ships of the Maritime Prepositioned Squadrons (MPSRONs) based in the Mediterranean Sea, at Agana Harbor at Guam, and at Diego Garcia in the Indian Ocean. These stocks include all of the arms, equipment, and supplies necessary to sustain the unit in the field for a month. The advantage of this scheme is speed, because the only thing that has to be delivered are the Marines, who would fly in aboard aircraft from the Air Mobility Command (AMC), the Civil Reserve Air Fleet (CRAF), and chartered airliners.

Once the Marines begin to arrive in-theater, the prepositioned stocks from the ships would be unloaded, distributed to the units, and then deployed. This means that a combat ready force of Marines can be on duty in a crisis area in a matter of days, as happened during Operation Desert Shield. The whole scheme requires a friendly nation willing to host the prepositioned stocks on its soil (like Kuwait or Norway), or a port facility capable of rapidly unloading the heavy ships of the MPSRON. Thus, while the prepositioning concept has worked on the occasions that it has been used (1990, 1994, and 1995 in the Persian Gulf), there are no guarantees that future conflicts will occur in places with such convenient facilities. Without friendly ports and airfields, prepositioning falls apart. Luckily, the Marines and their Navy brethren can deal with such problems. They just fall back on old tradition, head over the beach, and take what they need.

Marines are old-fashioned shock troops, still able to come from the sea and win the first battles of a war. Despite the huge cuts in force structure between 1990 and 1995, the Marine Corps lost only about 11% of its strength, primarily because its missions were well understood and appreciated by Congress, which controls the money. Much of the amphibious capability that the Corps built up in the 1980s has been retained, and the capability to "kick in the door" is still an option for U.S. policymakers. The Navy retains enough sealift to move and land about 1.25 Marine divisions, though not all at once, or at the same place. As a result, considerable time may be required to assemble a substantial landing force of amphibious ships and

Marines, if it is possible at all. To overcome this problem, the Navy and Marine Corps have developed a strategy of rotating small, forward-based Amphibious Ready Groups (ARGs) into potential trouble areas. In this way, one or more battalion-sized landing forces (each about 1,500 to 2,200 Marines) can arrive anywhere in the world within a matter of days, sometimes even hours.

Each of these Battalion Landing Teams (BLTs), with a helicopter squadron and support group, forms a MEU (SOC). The MEU (SOC) does for amphibious warfare what the Aircraft Carrier Battle Group (CVBG) has done for airpower in naval warfare. It provides U.S. policymakers with options to threaten an enemy's coast, take or destroy a vital target such as a port or airfield, and conduct raids or rescue operations. Only the Marines, with support from the Navy, can keep a landing force hovering for months off a hostile coast, and then strike at a moment's notice.

Unlike the heavy armored forces of the U.S. Army, Marine units are infantry formations, whose feet provide their mobility once they hit the ground. Well armed with personal weapons, they tend to be lightly equipped with heavy artillery and armored vehicles. Their offensive potential and mobility, once they are on dry land, requires reinforcement with additional artillery, armor, and transportation. For the afloat MEU (SOC) units, though, such augmentation is unlikely. Their strategy is based on stealth, maneuver, and deception. Once they're on location, they either accomplish their goal quickly and leave, or dig in and hold until relieved by other friendly forces.

While Marine units may be somewhat less mobile than their Army counterparts after they hit dry land, they have many ways to reach a particular strip of coastline. They can ride helicopters, crawl ashore in armored amphibious vehicles, or land from conventional or air-cushioned landing craft. And a MEU (SOC) can put units ashore using all of these options simultaneously, if weather and seastate conditions are favorable. The enemy can even be hit in many, widely separated places all at once, if that is desirable. Such operational mobility is paralyzing to an enemy, and will often enable the Marines to achieve surprise.

Marines have developed their own form of "maneuver warfare." They leap across or over the water to gain operational mobility, hit an enemy at weak points, and confuse and befuddle his command structure. Whenever possible, they try to avoid stand-up fights, preferring to shock an enemy into running or surrendering. The key to all this is intensive training and practice down to the squad level. This requires intelligence and initiative on the part of every Marine, from the commanding officer to the most junior private. Far from the image of "dumb jarheads," today's seaborne Marines are among the most intelligent, motivated, and positive young people you will ever meet. They have to be, because there just are not enough of them to go around.

Marine resources are stretched pretty thin these days. As an example, between October 1993 and October 1994, Marine deployments included:

- **October 1993–**The 22nd MEU (SOC), based around the USS *Guadalcanal* with the USS *America* Carrier Battle Group in support, moved from the Adriatic, where it supported operations off former Yugoslavia, to Somalia, where it landed forces to enforce UN peacekeeping and famine relief efforts.

- **April 1994**—The 11th MEU (SOC), based around the USS *Peleliu*, deployed from operations off Somalia to the waters offshore from Mombassa to support famine relief in the Rwandan Civil War and evacuation of non-combatants.
- **August 1994**—The 15th MEU (SOC), based around the USS *Tripoli*, deployed from supporting operations in Mombassa for humanitarian relief at Entebbe, Uganda, and in Rwanda for victims of the Rwandan Civil War.
- **October 1994**—The 15th MEU (SOC), again based around the USS *Tripoli*, steamed with the USS *George Washington* Carrier Battle Group from the Adriatic to the Persian Gulf to assist in deterrence against Iraq, which had moved two elite Republican Guards armored divisions into the Basra area.

These movements represented almost half of the MEU deployments that year. What does that mean? If you are a deployed Sea Marine, you have a better than even chance of seeing some sort of crisis. This is the lot of today's deployed Marines.

Tools of the Trade: Marine Units

The Marine Corps is the only branch of the armed services whose size and structure are spelled out in the United States Code, by Public Law 416 of the 82nd Congress (1952),[8] which states that the Corps shall be composed, at a minimum, of three division-sized ground units and three Marine Air Wings (MAWs). The 1st and 2nd Divisions each have about eighteen thousand Marines. But the 3rd Division, split between Hawaii and Okinawa, is down below ten thousand. Each MAW has about 250 aircraft (fighters, attack aircraft, helicopters, etc.). Along with the combat units, there are service and support units to provide supply and equipment maintenance. Backing the whole Marine Corps is a large segment of the U.S. Navy (nicknamed the 'Gator Navy) with the duty of transporting Marines and supporting them in their assigned missions.

The total active-duty strength (as of 1996) of 174,000 Marines is parceled out to the three divisions and three MAWs, as well as various supporting units. The Marine Reserve includes an additional 108,500 people (approximately), spread among units around the country. Reserve units are used to augment active units when they deploy. Each division includes an artillery regiment and two or three Regimental Landing Teams (RLTs), each of which contains several Battalion Landing Teams (BLTs). Each RLT usually has three BLTs under its command, each with about one thousand Marines. But RLTs provide the BLTs used to make MEUs, so a Marine division commander is usually short a battalion or two. In addition, other units are frequently detached to support peacekeeping and humanitarian operations.

Thus, to view the Corps as three monolithic division-sized blocks on the battlefield is not realistic. Desert Storm saw the largest Marine ground force that can

---

8 This was codified in law so that no future President or Secretary of Defense could disband the Corps or reduce it to a token force merely through an executive order or administrative reorganization. President Harry S Truman was not fond of the Marine Corps. The Congress, during those years, was not fond of President Truman.

Marines board a CH-53E Super Stallion helicopter aboard USS *Wasp* (LHD-1). Marines always operate in Marine Air Ground Task Forces (MAGTFs), frequently in joint operations with the Navy and other services.   *JOHN D. GRESHAM*

probably be assembled in one place, when General Walt Boomer commanded the two divisions of the 1st Marine Expeditionary Force (MEF). In addition, almost every Marine unit in the world was gutted to deliver that force into battle. In fact, if the late Kim Il Sung had wanted to take South Korea, his last, best chance was probably in January 1991, when most of the deployable U.S. forces were facing Iraq!

The basic building block of Marine operations is the BLT, which is a rifle battalion of over 900 men, with attached units bringing it up to a total of 1,200 to 1,300 Marines. The BLT is probably the smallest unit the Corps would deploy into a crisis area. Commanded by a lieutenant colonel (O-5), it is a task-oriented team that can attach or detach units, as the mission requires. For example, the basic BLT, with three Marine rifle companies, might gain a platoon of four M1A1 tanks or a company of wheeled Light Armored Vehicles (LAVs) to beef up its combat muscle. The BLT normally has a reconnaissance platoon and a sniper platoon added to provide intelligence to the commander and his staff. Amphibious tractor or rubber boat companies might also be attached, depending upon the assigned mission. Marine units tend to be tailored for specific mission requirements, and are supremely flexible in both organization and equipment.

The Marine Air-Ground Task Force (MAGTF)

Whatever a Marine unit is tasked to do, it would operate as part of a Marine Air-Ground Task Force or MAGTF. The MAGTF is the basic working task unit of the Marines, a concept that has been at the core of operations by the Corps for over half a century. It combines an infantry-heavy ground component—anything from one BLT to several divisions—with supporting artillery and other heavy weapons. Attached is an air component—anything from a reinforced squadron of helicopters and attack fighters to several full Marine Air Wings (MAWs). The entire MAGTF has a logistical service and support component to provide supplies and maintenance. All of this is melded into a single team commanded by a senior Marine officer, anything from a colonel to a lieutenant general.

MAGTFs come in a variety of shapes and sizes, depending on how big a commitment the President of the United States cares to make. For example, during the early stages of Operation Desert Shield following the Iraqi invasion of Kuwait in August 1990, the Marine Corps deployed the 7th Marine Expeditionary Brigade (MEB) based at Twenty-nine Palms, California. The 7th MEB had four infantry battalions, a light armored infantry battalion, a brigade service support group, and a reinforced Marine Air Group (MAG). By November 1990, the force had quadrupled in size, and come under the headquarters of the 1st Marine Expeditionary Force (MEF), which included all of the 1st Marine Division from Camp Pendleton, California, the 3rd Marine Air Wing from El Toro, California. the 1st Force Service Support Group (FSSG), and other reinforcements from active and Reserve Marine units around the world. By the start of the ground war in February 1991, the 1st MEF mustered over seventy thousand Marines. Throughout the Persian Gulf deployment, the Marine force was a fully integrated MAGTF, with all of the necessary components to enter combat. In this particular case, the Marines of the 1st MEF were under the command of Lieutenant General Boomer, who reported to General Schwarzkopf, Commander-in-Chief of the United States Central Command (CENTCOM). Another seventeen thousand Marines of the 4th and 5th MEBs were afloat in the Gulf, and came under the command of the Navy's 7th Fleet.

While the division-sized MEFs have made most of the headlines for the Corps in the last five years, it is the smaller, battalion-sized MEUs that do most of the day-in, day-out work. Their rapid mobility aboard the ships of their Amphibious Ready Groups (ARGs) and their ability to rapidly adapt to assigned missions make them popular among Washington politicians. This explains why, in a time of severe budget restrictions, funding for a 7th unit of the *Wasp*-class, a multipurpose amphibious assault ship, sailed through Congress with hardly a notice.

America needs the capabilities of the Marines and their MEUs; they buy time and provide options that airborne divisions and heavy bombers just cannot provide. Marines of the 24th MEU (SOC) were able to stand by on just twenty minutes notice, for over a week, to rescue Air Force Captain Scott O'Grady from Bosnia following his shootdown by a surface-to-air missile (SAM). Presence is important. In the mind of a potential aggressor, the idea of 1,500 Marines sitting off his coast has a calming effect. It may make him stop, think, and decide, "Well...not today." No dictator, warlord, or international thug wants 1,500 heavily armed, well-trained, and uninvited guests dropping by suddenly to adjust his attitude. That, in the end, is why we need sea-based Marines.

In chapters that follow, I'll try to give you a feel for the "nuts and bolts" of a MEU (SOC), its people, equipment, and organization. We'll have a chance to talk with the Corps' top Marine, and get to know how a young person becomes one of the "brothers and sisters." In addition, you'll see the equipment that is used by the sea Marines, as well as spend some time with one of the MEU (SOC)s that helps maintain forward presence for the United States. By the time we are done, I think you will have a feeling for how the Marines do their vital jobs, and why they can proudly bark their motto, "Semper Fi!" ("Always Faithful!"), when asked how things are in *their* world.

Warrior Prince of the Corps:
An Interview with General Charles Krulak

The Marine Corps reserves a special trust for the officers that pass through its ranks. Each of them is charged with responsibilities and obligations that frequently exceed those of their counterparts in the other military services. Every few years, one of these officers, after a lifetime of commitment and devotion to the Corps and its personnel, takes on a trust that goes beyond even that and is given a title unlike that of any other military officer—Commandant. Just the name sounds like responsibility incarnate. It is. The position of Commandant of the Marine Corps has traditionally been awarded to a leader of unique qualifications; and to look at the list of those who have held the job is to see the history, direction, and ethos of the Corps embodied. There are peaks and valleys on the list, as there are in the history of any great organization; many of those on the list were not even generals—no Marine of that rank even existed until after the close of the American Civil War. Yet every one of them has reflected the culture and direction of the Corps, for it is their leadership that should, and does, set the pace for the Marines during their tenure and beyond.

Now, it should be said that the Marines have been blessed during the past few years with what has to be considered both inspired and timely leadership. Since the early 1980s in particular, the Corps has known a string of truly great Commandants, each of them with significant gifts and strengths that have made the Corps into the highly ready and capable force that it is today. They took a force demoralized and battered by the experiences of the Vietnam War, and made it into an organization that Americans and our allies trust and our enemies fear. The road back actually started in the 1970s when the 26th Commandant, General Louis H. Wilson, told the Corps to look to itself to solve the problems that Vietnam created. Then there was General Paul X. Kelley, the 28th Commandant, who rebuilt the material capabilities of the Corps during the early 1980s. To General Kelley, himself the first commander of the Rapid Deployment Joint Task Force (the precursor of the present Central Command, CENTCOM), fell the job of getting the resources to buy the equipment, munitions, and supplies that the Marine Corps eventually used to succeed in the Persian Gulf, Somalia, and elsewhere. Following him was General Alfred M. Gray, the 29th Commandant of the Corps. Known as "the Warfighter" and possessing a voice like the ghost of "Chesty" Puller (gravelly with a distinct Southern drawl), General Gray will always be remembered as the Commandant who reestablished the concept of combat as the core mission of the Marine Corps. He did this through a renewed emphasis on warfighting basics and professional military education and a program of new manuals. The wisdom of this was proven by the performance of Marines in the field, especially in the Persian Gulf during 1990 and 1991. He further demonstrated his creative powers by conceiving and designing the unit we will

LEFT: General Charles Krulak, the 31st Commandant of the U.S. Marine Corps.
OFFICIAL U.S. MARINE CORPS PHOTO

RIGHT: General Krulak preparing to go on field maneuvers at Camp Lejeune, N.C.
OFFICIAL U.S. MARINE CORPS PHOTO

explore later, the Marine Expeditionary Unit–Special Operations Capable, MEU (SOC). Following General Gray came General Carl E. Mundy, Jr., the 30th Commandant of the Corps. General Mundy's greatest achievement was reestablishing the preeminence of the Marines in joint operations and holding the active strength of the Corps at 174,000 (216,000 with reserves) during the so-called "Bottom-Up Review" in 1993. The latter effort was particularly impressive as the other services suffered substantially deeper cuts, proportionately, during this era of the massive Federal budget reductions that have been a hallmark of the 1990s.

In early 1995, with General Mundy's tenure as Commandant was coming to an end, there was great speculation among members of the Corps over who would be his replacement. There were many excellent candidates, but within the Marines there was a favorite, a man whose name was whispered with a voice of hope and respect. Then in February 1995, on the fiftieth anniversary of the Battle of Iwo Jima, the announcement came down from the White House that this man, General Charles "Chuck" Krulak, the son of one of the Corps' most famous Marines, was to be nominated. It was a job, some would say, which he was not only qualified for, but born for. A warrior prince of the Marine Corps was arriving to take up the post that his father had come so tantalizingly close to holding some three decades earlier. The story of these two men, the most outstanding father and son combination in Marine Corps history, is worth looking at more closely, and thus we shall.

Father and Son: The Krulaks

In 1934, when Victor "Brute" Krulak graduated from the Naval Academy to become a Marine (the nickname was from his days as a coxswain at Annapolis), it is doubtful that he ever considered the family odyssey he was beginning. A veteran of prewar service as a China Marine and battle in World War II, Korea, and Vietnam, he is in many ways a living symbol of the Corps—much like his former commander and mentor, the legendary General Lemuel C. Shepherd, Jr. It was Brute Krulak who took the photos of Japanese landing barges in China and urged the creation of the early U.S. landing craft that would be so important to Marine amphibious operations in the Second World War. He would also personally command the first unit of amphibian tractors, as well as write influential reports on tactics and doctrine that are still important today. Later in World War II, the elder Krulak led Marines in raids and assaults on numerous Japanese-held islands. After the war, as a full colonel, he was influential in the development of the first vertical assault experiments using helicopters. He played a key role in shaping the National Security Act of 1947, which established the Marine Corps as a separate service. And he was instrumental in the creation of Public Law 416, which established the size of the Marine Corps as not less than three combat divisions and air wings and which accorded the Commandant of the Marine Corps coequal status with the members of the Joint Chiefs of Staff when considering matters that directly concern the Marine Corps.

The senior Krulak's service continued into the 1960s, by which time he was generally considered one of the nation's leading experts on the new science of guerrilla warfare. In early 1964, now a Major General (two stars), he helped plan raids into North Vietnam, even before active United States involvement started. He was later promoted to Lieutenant General and was placed in command of the Fleet Marine Forces, Pacific (FMFPAC), where he commanded the Marines in Vietnam through much of that ill-starred effort. He also came tortuously close to the post of Commandant of the Marine Corps: He was actually promised the post when the appointment was given to another Marine officer. Since that time, he has gone on to write his own book on the Corps, *First to Fight*, which many see as a classic work on the Marines and warfare. Yet despite his many personal achievements, it will be, perhaps, the achievements of his son that will go down as his single most lasting contribution to the Marine Corps. For when Chuck Krulak entered the Marines, Brute Krulak gave the institution a unique gift, a warrior around whom they could rally in a time of need. Let's talk to Chuck Krulak about it.

Tom Clancy: When did you first decide you wanted to be a Marine?

General Krulak: I decided that I wanted to be a Marine between the ages of eight and ten. It was the period when my dad was in Korea and immediately after his return. During that time he was involved in the fight to save the Marine Corps, which resulted in the amendment to the National Security Act of 1947. I couldn't help being impressed by his efforts and by those of other senior officers and politicians [involved in the legislative fight that ensued] who came in and out of our house at that tremendously important time in the history of the Marine Corps. These men were involved in great efforts the results of which are reflected in the Corps we have today.

Tom Clancy: Did you have a sense of just who your father was, and how important he was in the history of the Marine Corps?

General Krulak: I didn't understand at the time. I knew that during the National Security Act struggle—during the second session of the 82nd Congress—he was doing something important; he was gone a lot, and my mother would just tell me that he was doing "important work." It was not until much later, however, that I realized how critical and pivotal these events were for the Marine Corps.

Tom Clancy: Let's talk about your career. It began at the Naval Academy, what years were you there?

General Krulak: 1960 to 1964. My class [of '64] had some very special people in it. The current Commander-in-Chief, U.S. Pacific Command [Admiral Joe Preuher] was my classmate, as was Secretary of the Navy John Dalton, plus a whole group of Navy admirals that are currently on active duty. It was a very special class in the sense of what has transpired for all of us and the Naval Service since graduation.

Following graduation from the Naval Academy and commissioning as a 2nd Lieutenant, General Krulak went to serve in South Vietnam. The experience was a defining moment for the young Krulak, and is best described by his own words.

Tom Clancy: What units did you serve with in Vietnam?

General Krulak: Immediately after attending the Basic School, I joined the 1st Marine Division and served as a platoon commander with Golf [G] Company, 2nd Battalion, 1st Marines (2/1). Shortly after arriving at Camp Pendleton, the war broke out, and I was deployed with 2/1. I spent thirteen months in Vietnam, and was the commander of Golf Company by the time I returned.

Tom Clancy: Many people came away with vivid impressions of their experiences from Southeast Asia. What were yours? How does your own Vietnam experience effect what you do now?

General Krulak: I think that is a great question because the war has definitely had an impact on much of today's military leadership. I believe that we have a group of idealists at the senior levels of the military today—whether you are talking about a now-retired Colin Powell, a John Shalikashvili, or a Tony Zinni [the current commander of the I MEF]. I'm talking about the folks who came out of that war, particularly those who did multiple tours, who believed that there must be a more clearly defined reason for going to war again. Then, we went through the immediate aftermath of the war, where you had the race-relation and manpower problems which made the choice of a continued military career difficult. For an officer to lead a platoon or a company in Vietnam, come home and face the institutional issues of the day, and still progress up the promotion ladder required a deep-seated set of values—a commitment that next time, things were going to be done differently. No matter how it is described, I think that somewhere in each of us is a touchstone that we reach back to when things are tough. It helps us remember what happened in Vietnam and the promises we made that we would do things differently next time. I remember that even during the war we saw that things could be done better. We all made it through some tough times by sticking to our guns and to our ethos.

Tom Clancy: You did two tours in Vietnam. Would you please talk about them?

General Krulak: I went to Vietnam with the 1st Marine Division on Operation Harvest Moon in early 1965 [one of the earliest Marine operations in Vietnam], and stayed through 1966. I went back in 1969 with the 3rd Marine Division and spent all of my time in the Northern (I Corps) sector. That second tour went through the "Vietnamization" and drawdown phases of the war. What means more to me than a list of operations, dates, and places is what my experiences in Vietnam meant in terms of shaping me as an officer, and that goes back to what I learned about the Corps and Marines during that time. The most important lesson that came out of Vietnam for me was that Marines take care of Marines. I saw this over and over again; young Marines with wives and children back home who had every right to just look out for number one and make it back alive, but they all, to a man, would lay down their own life for the lives of their fellow Marines. Regardless of what was happening with the war on a macro sense, what I was experiencing on the ground with my Marines had a profound effect on me.

Following Vietnam, Chuck Krulak entered the normal career track of a Marine officer, which is to say that he did a variety of things. Some were normal "Marine" jobs, while others had what is now called a "joint" flavor.

Tom Clancy: After Vietnam, where did you serve?

General Krulak: I went back to the Naval Academy as an instructor, and I took with me the deep personal sense of our ethos. It dawned on me that this time [as a midshipman at Annapolis] is a critical period for new Navy and Marine officers. It is where their values—which mean more than anything—will be developed, so I took that responsibility very seriously.

From there I went out and commanded the Marine Barracks at Naval Air Station [NAS], North Island, California, and then went to study at the Army Command and General Staff College [Fort Leavenworth, Kansas]. From there I reported to the 3rd Marine Division, and served as the operations officer [S-3] of 2nd Battalion, 9th Marine Regiment [2/9]. Then I moved to Headquarters Marine Corps in Washington, D.C., and worked in the Manpower Division, followed by a tour as a student at the National War College [Fort McNair, Washington, D.C.]. After that, I was sent to Hawaii and served as the Fleet Marine Forces Pacific plans officer, then executive officer of the 3rd Marine Regiment, and as commanding officer of the 3rd Battalion, 3rd Marines [3/3]. I then moved over to head the prepositioning ship program for the staff of the 1st Marine Expeditionary Brigade [MEB], and eventually became the brigade operations officer [G-3].

I then moved back to northern Virginia and spent one of the most challenging years of my life as the military assistant to Mr. Don Latham, the then Undersecretary of Defense for C^3I [Command, Control, Communications, and Intelligence]. I was the oversight officer for the battle management system on the Strategic Defense Initiative. It was incredible to be involved in a system that was definitely on the cutting edge.

From there I went to the White House. I was there during the last year of the Reagan Administration [1988] and the first year of the Bush Administration [1989] as Deputy Director of the White House office. Following my White

House tour, I was the assistant division commander of the 2nd Marine Division at Camp Lejeune, North Carolina. Then I took over command of the 2nd Force Service Support Group.

Like almost half of the Marine Corps, during the fall and winter of 1990 and early 1991, the junior Krulak, now himself a general, found himself in the sands of Saudi Arabia. Though trained as an infantry Marine, he wound up in a different position—serving as a logistician. As commander of the 2nd Force Service Support Group [FSSG], it was his job to keep over ninety thousand Marines involved in Desert Shield and Desert Storm supplied and fed.

Tom Clancy: Tell us about your work in Southwest Asia during Desert Shield and Desert Storm as head of the 2nd FSSG.

General Krulak: Initially MARCENT [Marine Component, United States Central Command] was going to rotate forces into the theater. But that idea was vetoed. Instead of replacing the existing in-theater forces, we reinforced them to about twice their original strength. So the 2nd Marine Division, the 2nd Marine Aircraft Wing, and my combat support organization fell in on top of the 1st MEF [composed of the West Coast-based 1st Marine Division, 3rd Marine Aircraft Wing, and 1st FSSG]. From the two combat service support units [1st and 2nd FSSG], we formed a direct support command that supported the Marine warfighters up front and a general support command to run the ports and bring the supplies into the theater. Then–Brigadier General Jim Brabham commanded the general support unit, and I commanded the direct support unit, which was the 2nd FSSG. My mission was to provide support for two full Marine Divisions, a heavily reinforced Marine Aircraft Wing [West Coast-based 3rd MAW, augmented by the East Coast-based 2nd MAW], the Army's "Tiger" Brigade [from the 2nd Armored Division], and my own 2nd FSSG troops.

After the war, I became MARCENT Forward and commanded the forward-deployed CENTCOM Marine forces responsible for the reconstitution of the Maritime Prepositioning Ships. This was a daunting task that required retrieving equipment spread over one hundred miles across the desert, transporting that equipment to the port of Al Jubail in Saudi Arabia, then back-loading the equipment on the ships.

Tom Clancy: Since you've been an infantry officer for much of your career, could you tell us your thoughts about the effects of logistics on the success of the war?

General Krulak: My impression is that the success of the Marines in Southwest Asia will go down in history as a victory based on logistics. That's a tough thing to say for a career infantryman. From the very first days of Desert Shield, this was an operation heavily dependent on logistics. Just five days after Iraq invaded Kuwait, the five-ship, Maritime Prepositioning Squadron [MPSRON] 2 was ordered to Saudi Arabia. The following day, three ships of MPSRON 3 were ordered to the region. From receipt of the mission on August 7th until the final off-load on September 7th, the MPSRONs provided supplies and equipment to support over 53,000 Marines and sailors for thirty days. Not only did this effort completely validate the requirement for MPS, it formed the foundation of the tremendous logistic effort to follow.

Regarding the ground war itself, you hear and read a lot about the minefield breaching and "left hook" in the war, as well you should; but to get the Marines prepared to make that assault was a logistical nightmare. The only reason that the logistic part of the war has not gotten its due is that the ground portion was so successful. The difficulty inherent in supporting and sustaining that large a force was tremendous; but it was never really an issue, because the war went so fast. The follow-on support requirements just went away. Had it lasted, it would have been something else. The lessons learned are many and varied. It took about two thousand short tons a day to keep MARCENT in ammunition alone during the ground war.

To me the Gulf War proved the accuracy of the maxim "Amateurs study tactics, and professionals study logistics." This is one of the great things about the MAGTF. It has its own tactical and logistical capability. You get everything with one call. Whether you order up a full MEF [a Marine Division/Aircraft Wing/FSSG], or an MEU [SOC], these units have their own logistics base that does not absorb itself as it conducts operations. They carry what they need with them, so that field operations can be sustained for a period of time [usually fifteen to thirty days] without the need for immediate resupply or reinforcement. That's the "expeditionary" part of the Marine Corps today. We have the offshore resources on the amphibious and MPS ships to sustain forces on the land without anyone's permission to base forces ashore.

Following the war and his own homecoming, Brigadier General Krulak began a task as painful as it was important, the drawdown and restructuring of the Marine Corps in the post–Cold War world. Under the so-called Base Force concept, all of the military services were to be downsized, with excess and redundant units and capabilities eliminated. General Krulak's job was to design and supervise this effort for the Marine Corps, without actually destroying it, or its vital capabilities.

Tom Clancy: You came out of Desert Storm, and then what happened?

General Krulak: When General Mundy assumed duty as the 30th Commandant, he assigned me as the head of the Personnel Management Division at Headquarters in Washington. I no sooner took over when he held an off-site meeting with all of his three-stars [Lieutenant Generals]. Out of that meeting came the decision to put together the Force Structure Planning Group [FSPG] to actually develop the plan to take the Corps down to the mandated [Base Force] level of 159,000 personnel. Essentially, we were tasked to take our existing Corps and build a new Corps. So the study group spent the next year working that issue and then, under the direction of General Mundy and with his personal involvement, selling our plan to Congress and the rest of the military services. The key was that, as the FSPG looked at the National Military Strategy and the Marine Corps' role, we determined that we could not meet the needs of this nation at 159,000. Our work showed that the number we actually needed was 177,000, of which we got to keep 174,000 active-duty Marines—a number that was validated by the Department of Defense Bottom-Up Review.

I was then promoted to lieutenant general in October of '92 and went to Quantico to command the Marine Corps Combat Development Command. During my two years there, we as a Corps were formalizing and institutionalizing the combat development process, which was the brainchild of General Gray. From there, I moved back to Hawaii and took over my father's last command, Marine Forces Pacific.

Following in his father's footsteps and commanding the Marine Forces of the Pacific was an honor for Chuck Krulak. But more was to come for the young three-star, as we will soon hear.

Tom Clancy: When you learned that you were being considered for the post of 31st Commandant of the Marine Corps, what went through you mind?

General Krulak: My very first thought was, "Am I up to the job?" I questioned whether I was the right man for the job because there were such great people in the running. General Mundy and Secretary [of the Navy] Dalton interviewed every three- and four-star general in the Marine Corps and all were qualified to lead the Corps. We have great generals, and Secretary Dalton made certain that everyone got his day in court. His personal efforts during this process are unmatched in the history of the Navy Secretaries. My second thought was about my wife Zandi, and the pressures that would fall on her. My third thought was that I had a great job as Commander, Marine Forces Pacific, and whatever happened I was going to continue to be challenged.

Tom Clancy: During this time, was there any thought on your part about how close your own father came to being appointed Commandant of the Marine Corps?

General Krulak: No. That was on his mind, though, because the reality is that he came a lot closer than most Marines know he did. He had, in fact, been told that he had the job, and then he didn't get it. So his concern was that history would repeat itself, and I just told him, "Quit worrying about it, because *I'm* not worrying about it." It was not an issue with me personally. I was not looking for the job. In my opinion, the last thing you want in an organization with this type of deep ethos of service is someone who actually wants or is posturing to be the Commandant. That's an ego issue and the wrong motivation. The job is so hard, so demanding, that if any service chief isn't doing it for what I call the "right thing," then he's going to have a real problem.

Tom Clancy: The day comes and you receive word that the President has nominated you to be the 31st Commandant of the Marine Corps. What did it feel like?

General Krulak: It was a phenomenal experience. I found out while circling in a plane about five thousand feet above Mt. Suribachi on Iwo Jima. General Mundy, his wife, my wife, and I were headed to Iwo to commemorate the celebration of the fiftieth anniversary of the invasion of the island. A radio operator handed General Mundy a small yellow message form. He looked at it, and then pulled my wife over to look at it. She looked at it, and started to cry. He then gave it to me, and it said, "The President of the United States has today signed and forwarded to Congress your nomination to be the 31st Commandant of the Marine Corps." It was an unbelievable feeling. Every emotion you could possibly think of came over me. You name it: from exhilaration to, "Oh, my God, what is happening?"...to relief...to fear.

The actual announcement was unforgettable. We were on top of Mt. Suribachi—virtually on top of some of the most glorious pages of Marine Corps history—when Secretary Dalton made the announcement to the assembled dignitaries, not least of whom were the survivors of that great battle. I was being told by the Secretary of the Navy that I was becoming Commandant at the exact same place where fifty years earlier Navy Secretary James Forrestal had looked over at General Holland M. "Howlin' Mad" Smith and, upon seeing the flag raised at the top of Mount Suribachi, said, "The raising of that flag...means a Marine Corps for the next five hundred years."

My feelings were overpowering. There is a family connection here, because Holland M. Smith was *my* godfather. Now, half a century later, I'm standing where my godfather once stood and Secretary Dalton is telling the godson of that man that he would be the Commandant who would take the Marine Corps into the 21st century. It was a very emotional moment. I thought of my dad immediately. He and my mom were so excited and happy for me. I am convinced it meant more to them then it did to me.

Tom Clancy: Are you yet aware just how important this matter of your becoming Commandant was to the Marines out in the Corps?

General Krulak: No. I often say that they could have picked any of a number of officers to do the job. There were so many great generals who could have done it. I tend to believe that the commandancy makes the officer, not the other way around.

Tom Clancy: During the 1980s and 1990s, the Marine Corps seems to have been blessed with a string of truly great Commandants. Could you give us your thoughts on some of them?

General Krulak: You really need to go back into the 1970s when you talk about the string of great Commandants. That's where we began implementing policies that gave us the quality manpower to operate the equipment and conduct the operations that made us so successful in the 1980s.

General Louis H. Wilson [26th Commandant of the Marine Corps].
General Wilson inherited a Corps riddled with the personnel problems associated with the post-Vietnam era [racial tension, high desertion and discipline rates, recruiting problems, etc.] and tackled these issues with the same ferocity he demonstrated in combat. He literally turned the manpower tide for the Corps. He was determined to improve the quality of the personnel in the Corps to the point where he vowed to go down to "just two Marines if those two are the kind that we want." I call that the "Wilsonian Doctrine," and it began a revolution that is responsible for the quality of Marines we have in the Corps today.

General Robert H. Barrow [27th Commandant of the Marine Corps].
General Barrow expanded on General Wilson's manpower initiatives. He continued to tighten the quality screws; and in 1983 over ninety percent of new recruits were high school graduates. He also launched his own "war on drugs" and issued the policy that put an end to the Corps' tolerance of problem drinkers. The percentage of substance abusers fell from 48% in 1980 to less than 10% by 1985, and the Corps became known as a quality institution sought out by some of the best young men and women our country had to offer.

General Paul X. Kelley [28th Commandant of the Marine Corps].
General Kelley's vision of what we were going to need for equipment and his willingness to fight tooth and nail to obtain the funds to modernize the Corps are his great legacy. While we often talk about the warfighting ethos that we took to the desert in Southwest Asia, we should never forget that he was the Commandant who gave us the means and the implements to fight and win on the battlefield. General Kelley is an unsung hero of the Corps. Ironically, some fifteen years later, one of my biggest challenges is equipment modernization, but it's the equipment he fought for during his tenure as Commandant that *I* must now fight to replace.

General Alfred M. Gray [29th Commandant of the Corps].
General Gray gave the Marine Corps a brilliant mind that saw beyond the immediate moment. He saw a need to totally revamp the way we think, train, and educate ourselves. He cultivated our maneuver warfare mind-set, so that when we went into Desert Shield/Desert Storm, we didn't see the minefields that we faced as insurmountable obstacles; we just searched for the gaps, breached them, and went on. He gave us the doctrine to do that job, and more since then. A great, great man, and a real thinker. Everyone who looked at him saw this rough, tough son-of-a-gun; but he was, and is, smart as a whip.

General Carl E. Mundy [30th Commandant of the Corps].
General Mundy was a kind, wonderful man, but he knew how to fight. Some wondered if he was going to be able to defend the life of the Corps in the post-Cold War drawdown, and he proved to be a bulldog. His leadership in the battle for an end-strength 174,000 Marines was remarkable. General Mundy will also be remembered for his great moral courage and deep love of Corps and country. He articulated the ethos of our Corps as well as any Commandant. General Mundy and his wife Linda brought real meaning to the Marine Corps family and to the concept that Marines take care of their own.

When General Krulak took command in mid-1995, he inherited a Marine Corps whose strength had been for the most part preserved, but which was facing many new challenges: aging equipment, personnel issues, and basic questions about the role of the Corps in the run-up to the 21st century. Grabbing the bull by the horns, he rapidly took control and began to exert his own unique ideas onto the structure of the Marines. He published his now-famous Commandant's Planning Guidance, so that every Marine in the Corps would know what the new boss had planned for them. He also opened up new channels for direct communications of ideas, including Internet access directly to himself. Let's hear his thoughts on this.

General Charles "Chuck" Krulak (right) with the author during a recent visit to the Commandant's office in the Pentagon.
JOHN D. GRESHAM

Tom Clancy: What has been your philosophy in these early days (summer and fall of 1995) of your tenure as Commandant?

General Krulak: I felt that I had one year from the start of my tenure as Commandant to set the course and speed for what I believed needed to be done. The remaining three years are to be for follow-through. We now have major projects and initiatives started and have generated momentum. During the next three years, we will continue to give course and speed corrections to the things that we see as important. I tried to get us going, with some clear-cut, definitive goals to make sure that everybody involved knew our plan and was prepared to step out and act. That's what the *Commandant's Planning Guidance* was all about. To let everybody know what my philosophy was and is and then get on board and charge!

Tom Clancy: Okay, let's talk about some of the things you are working on within the Corps. First, let's hear what you think of the state of the force that you have inherited. Currently, your authorized end-strength is 174,000 active duty personnel. Will you be able to hold onto that?

General Krulak: I think that it [Marine Corps end-strength] will be under attack almost immediately. In fact, it already is. The Administration [of President Bill Clinton] is locked into the force levels defined by the Bottom-Up Review of 1993; but we have major budget problems in the Department of Defense. Part of the problem is that DoD has more infrastructure [bases and facilities] than there's money to support that infrastructure. I am concerned that there will be pressure to make each of the services smaller, both by reducing personnel and infrastructure, and utilizing the money saved to modernize the armed forces. For the nation, a drawdown of the Marine Corps would be a terrible mistake. The Marine Corps was never a Cold War force. Our mission did not change with the end of the Cold War era, so there is no need for other major changes in the Marine Corps specifically in response to the demise of the Soviet Union. Where we can assist this nation as the other services adjust to the post-Cold War period is to be this country's "risk-balance" force. We provide to the nation the ability to take a risk—in this case allowing the rest of the military services to draw down quickly while still having an organization that is ready to respond. We are the most ready when the nation is the least ready, and you don't want to reduce the only force that provides this nation the capability to react while at the same time assuming the risks associated with the rapid post-Cold War drawdown.

Tom Clancy: There has been some envy on the part of the other services at your success at holding on to a relatively high percentage of your Cold War end-strength. Will you please tell us your perceptions of drawdown process with regard to the Marine Corps?

General Krulak: What General Mundy and the Marine Corps did right was create the Force Structure Planning Group that I spoke of earlier and build a plan that made sense. It was a tremendously rigorous effort to analyze the national military strategy and then balance our capabilities against that strategy. From this we came up with the requirement for a Marine Corps with a personnel base of 177,000 active-duty personnel, of which we actually kept 174,000. Now, when people say that we did not cut our strength, they fail to look at the facts. They fail to see that we went from 198,000 active-duty Marines to 174,000. We cut 50% of our tanks and 33% of our tactical aviation strength. We lost a third of our artillery, as well as all six of our Marine Expeditionary Brigade Headquarters units and a quarter of our combat service support units.

What is really critical is that most of our cuts had to come out of our muscle—our combat power—because as a service, we were already very lean. When we did identify our requirement for 177,000, a hard number with no fluff, we still had to cut. That's why at this point, I'm determined to keep our end-strength at 174,000. Having said that, we can't get stuck on a number, because our challenge today is to determine what we need to fight and win the battles of the 21st century. That's my problem: to get to the 21st century, making the best use of technology and our remaining personnel base, while still giving the nation what it needs.

One of the biggest challenges faced by General Krulak is maintaining the flow of new Marine recruits into the Corps. The combination of public perception regarding the drawdown of the military as well as a limited pool of recruiting dollars has made this task ever more difficult. Let's hear the Commandant's thoughts on this tough problem.

Tom Clancy: Talk a little about the raw material of the Marine Corps—the recruits—and the recruiters and the recruiting process. What are your thoughts on the recruiting problems facing the Corps as you continue to search for qualified men and women?

General Krulak: First of all, my respect and love for recruiters knows no bounds. As the former head of the Personnel Management and Personnel Procurement Divisions at Headquarters Marine Corps, recruiting was one of my responsibilities, so I have a very good sense of the recruiting process. We have great recruiters and they're doing a tremendous job.

Nevertheless, we have a couple of problems. First, not all of the American people know that we're hiring. They see the military cutting back, they read about the reductions-in-force, and wonder why they should allow their sons or daughters to join the Corps. They just don't see any career possibilities or longevity in the service today. We can tell from our various youth-attitude surveys that America's youth doesn't know we are hiring.

So, the first thing I need to do is to enhance our recruitment advertising. That takes dollars. But at the same time, we need to reach our target market with our message. That message is embodied in our new commercial called *Transformation*.

Transformation symbolizes what the Marine Corps does for this nation: We take America's youth, what you called "raw material," and we transform them into Marines. We instill in them our core values —honor, courage, and commitment. We teach them to be the leaders of tomorrow's Corps and the leaders of their communities and country the day after tomorrow. We recognize that we are recruiting a different kind of American today. They're coming from a different society, with different values than those that have been the hallmark of the Corps' value system. We transform them, and that transformation lasts forever. That's important for our nation and our nation's youth. But they have to know we will do that for them, and that is where advertising comes into play.

I won't sacrifice quality for quantity, and I believe the "Wilsonian Doctrine" was the right approach. Like General Wilson, we will willingly sacrifice numbers to get the very best of our youth. Then we will transform them forever...into Marines and, more importantly, productive citizens of this great nation of ours.

Tom Clancy: On to another personnel matter. Could you talk a little about the changing roles for young women in the Marine Corps?

General Krulak: Our women make tremendous contributions to the Corps. I had 201 women under my command during Desert Shield and Storm and I would not have been combat-effective without them. To a Marine, they were superb. As the Commandant, however, I am tasked to train, equip, and provide fighting forces to the regional commanders-in-chief. I have to consider this as we select and procure the right equipment and train the right people to do the job the nation expects of us. It is also my responsibility to ensure the we maximize the effective utilization of those resources. I do not believe that I am maximizing the utilization of the limited resources of the Marine Corps by putting women at the point of a rifle platoon or in units that engage in direct ground combat.

One of the hallmarks of the 1990s has been that as U.S. forces have gotten smaller, they have also gotten busier. Higher operational tempos (Optempos) have resulted in some notable difficulties, even for the Marines. General Krulak has been forced to deal with some unique problems in the areas of morale, as well as some surprising quality-of-life issues. Let's hear what he has to say.

Tom Clancy: Morale always seems to be an issue in the military. Can you talk some about the challenges this presents for you?

General Krulak: First of all, I have not encountered the kinds of morale problems in 1995 that we had in the past. The morale problems of 1995 are minuscule compared with those, for example, of the 1970s. Nevertheless, the first thing I am doing for morale is to show the individual Marine that their Commandant cares about them, as individuals. So when members of Congress asked what they could do for me as I made my in-calls, I asked for an additional ten to twenty million dollars for things like rain gear and boots instead of dollars for additional amphibious shipping, aircraft, and vehicles. When they said, "What are you talking about?" I said, "What I want to give my Marines is field equipment that is of newer design than the Korean War!" I think they thought I was somewhat "off-the-wall," but the bottom line is that the first thing that Marines saw from this Commandant was new boots, rain gear, and the new load-bearing equipment system and backpack.

With the exception of the woodland camouflage pattern, the field jacket that we use today is the same design that our Marines wore in the 1950s. You have sportsmen walking around in state-of-the-art boots, and our current boots are terrible! Our rain suits are made of rubber, which does not breathe, so the Marine wearing it is as wet on the inside as he is outside. Today if a Marine's sleeping bag gets wet, it weighs over 40 lb/18 kg, and we use tents designed back in World War II!

Everyone talks about being worried about the "quality of life," but what they fail to understand is that the bulk of an infantryman's life is spent in the field. In the fleet, many Marines spend more time in the field than they do at home. For our forward-deployed Marines, like those in the MEU (SOC)s, this is especially true. We spend money on new barracks and other base facilities, but don't buy our Marines the basic clothing and equipment they need to survive and be comfortable in the field.

Secondly, there is the matter of my style of leadership. I don't have many pretenses, so the last thing I want when I stop somewhere is a lot of preparation and fanfare over my visits. Now, there are people who disagree with this philosophy and feel that a visit from the Commandant is cause for a major outpouring of effort, but the onus ends up on the enlisted Marines who need to be spending their time being Marines, not preparing for my visit. So I try to fly in unannounced; and that precludes excess preparation work and allows me to see my Marines as they are. I want them to know that their Commandant is coming to see them. They now know that, and I learn a lot from my visits talking with the troops.

In my discussions with Marines, I hear the usual about what I call external morale issues [barracks, recreational facilities, etc.] and we are already working on these things. What I focus on, however, are the more deep-seated, internal things, such as pride in the organization and making sure that our leaders have what John A. Lejeune [the 13th Commandant of the Corps] called "a self-sacrificing love for the Marine Corps." Those are things that we can always positively influence, regardless of the budget, so those are the areas we need to concentrate on. I have to tell you that Marines, whatever their rank, will respond to this type of approach. They have to know that they will all be treated fairly, and that nobody above them is going to harm them by ruining their career automatically because of a mistake.

At the same time, though, that Marine has to know that we will not tolerate lying, cheating, or stealing. As warfighters we must understand the dimensions of physical courage. There are no greater supporters of peace than those who are sworn to risk their lives when war occurs. However, our profession also demands moral courage—the strength of character and the integrity to do what is right. Acts of moral torpitude have no place in the Marine Corps.

While the new Commandant has strong ties to the history and traditions of the Marine Corps, he has a keen appreciation of the usefulness of modern technology to help his Marines. In particular, he has used electronic mail and the Internet to open up direct lines of communications with his Corps. Let him tell you about it.

Tom Clancy: One of your major initiatives has been to open the lines of communication with Marines of all ranks. To accomplish this, you even obtained an address on the Internet. Could you please talk about your new communications systems with your Marines?

General Krulak: It is phenomenal! Some of our best ideas and initiatives originate with the lance corporal and corporals who work, live, eat, and sleep Marine Corps twenty-four hours a day. I don't think a Commandant can effectively lead the Corps without input from Marines. So with this E-Mail and Internet access, they can send their ideas directly to me, and they do. I get messages from corporals to gunnery sergeants, with suggestions telling us how we can do things better in the Marine Corps. I want to focus their suggestions, so I have asked them to think about and address three questions: "What are we doing now that we shouldn't be doing?" "What aren't we doing now that we should be doing?" and "What are we doing now that we could be doing better and how?"

They are answering those questions, and we have made major changes in the Corps today based on their input. We have, or are, considering changes in training, in the promotion system, and in our performance evaluation system. The changes are driven by lance corporals through colonels dialing up and dropping me a note with an idea. You have to see the quality of what they are saying to appreciate just how intelligent they are and how much they care about improving the Corps. It is truly motivating!

One of the most difficult tasks facing General Krulak and the Marine Corps as they head into the 21st century is the need to modernize their equipment during a time when there is very little money and even less support around Washington, D.C., to do so. With their modernization budget (for new and replacement equipment, as well as upgrades/conversions) slashed to almost historic lows, the challenges for the Commandant are immense. Let's hear his thoughts on this.

Tom Clancy: Could you talk now about the Marine Corps modernization budget? Obviously you're making due with an absurdly small amount, compared with the other services. What is the outlook?

General Krulak: Let me preface my comments with some background about our budget. Procurement and modernization for some equipment [Marine aviation, amphibious shipping, and landing craft] are funded by the Department of the Navy. The shortfall you are referencing has to do with what we call "green dollars," or dollars earmarked specifically for Marine Corps procurement. With that as background, the Marine Corps needs a "green" modernization budget between $1 billion and $1.2 billion. That is one of my biggest challenges. We had $474 million in FY-95; and that's less than half of the historic average; and that means we have had to sacrifice either readiness or modernization, because we can't have both at that level. If I don't get that budget up to the necessary level, we'll be in real trouble.

In fact, we're in trouble right now. We have 5-ton trucks that are almost twenty years old. You don't drive a car that is that old, but we'll be sending Marines into combat in those vehicles. Our amphibious assault vehicles [the AAV-7s] are twenty to twenty-five years old. There are problems on the aviation-dollar side as well—we are flying CH-46 medium-lift helicopters that are headed into their fourth decade of service as we are sitting here! We have some real modernization problems that we need to come to grips with as a service and a nation.

Tom Clancy: With that introduction, I'd like to run some of the key modernization programs by you, and get some of your comments on them. Tell me about the V-22.

General Krulak: The V-22 is critical to the nation and Marine Corps. We're going to get it, and get it quicker than anyone thinks we will. Once other services realize the capability that tilt-rotor technology brings, I believe that they will join us in procuring this aircraft. It has all the capabilities of a helicopter in terms of vertical flight, but has the speed and distance more akin to a fixed-wing aircraft. Imagine how useful this aircraft would have been in places like Somalia or Burundi, or might be in Bosnia. We're currently programmed to get the first squadron of V-22s in 2001, but I would like to be able to buy two or three squadrons a year [twenty-four to thirty-six airframes], as opposed to the current planned buy rate of fourteen airframes per year. Again, I believe that once people understand and appreciate how incredibly capable this aircraft is, the buy will be accelerated.

Tom Clancy: How about the Harrier re-manufacture?

General Krulak: The re-manufactured Harrier will be our "bridge" aircraft until the Joint Advanced Strike Technology [JAST]/Joint Strike Fighter [JSF] program gets us to ASTOVL (Advanced Short Takeoff, Vertical Landing—a variant of the JSF). With the updated AV-8B Harrier II Plus, we have an extremely good aircraft that has remarkable capabilities compared to earlier versions of the plane. In fact, thanks to the re-manufacture program, it is virtually a new airplane. It is not, however, the plane we want for the 21st century. That's the ASTOVL strike fighter. Our goal is for the Marine Corps to "neck down" to just one single strike aircraft, the ASTOVL version of JSF. Combine that with the capability of the V-22, the heavy-lift CH-53E, our light attack and utility helicopters, and our support aircraft, and we will have a Marine aircraft wing that brings incredible capability to the combatant commander.

There will be tremendous savings when we pool all the Marines working on or flying in a number of different airframes, and put them all into one of our extremely capable, hard-charging air wings with fewer types of airframes. We will realize significant economies of production, as well as operations and maintenance. When you are talking modernization, you have to think beyond today or tomorrow, and think about the day after tomorrow. That is how we approach everything as Marines. Everyone is excited about the AV-8B Harrier II Plus. Yet while I believe that is great, and it may be doing what we want today, it is a bridge to the ASTOVL strike fighter of the future.

Tom Clancy: Tell me about the Advanced Amphibious Assault Vehicle (AAAV).

General Krulak: The AAAV is as critical to our future success as the V-22. Seventy percent of the world's population lives within 300 miles [480 kilometers] of a coastline in the littorals. The end of the Cold War ushered in a new era of global instability where regional strife will dominate. While we can't predict exactly where a crisis will occur, there is a good chance it will require a response originating from the sea. If we, as a nation, are going to have forward-deployed forces effective at managing instability around the world, we need the AAAV that operates rapidly in the water from a good standoff distance [up to 25 nm/46 km] as well as on dry land. It will be able to carry Marines, weapons, and equipment under armor with a full nuclear, chemical, and biological (NBC) over-pressure protection system. It will also give us the ability to engage enemy armor with superior mobility and firepower. It will give us tremendous flexibility in a variety of combat environments and conditions.

Ship-to-shore delivery is not an end unto itself, but a beginning, because you still have to maneuver and fight when you get on dry land. Right now, we don't have a system for moving Marines under armor that can keep up with the M1A1 tank. You can't have an effective mechanized force if your personnel can't keep up with your tanks and reconnaissance vehicles. The AAAV will give us that capability.

Tom Clancy: What about the Predator and Javelin systems?

General Krulak: We need a solid fire-and-forget anti-armor capability, and these two systems will get us to the future. Like the AV-8B Harrier II Plus though, I see Predator and Javelin as "bridge" systems, to get us the follow-on generations of truly "brilliant" fire and forget anti-armor technology.

Tom Clancy: How does the Lightweight 155mm Howitzer (LW 155) fit into the future?

General Krulak: We really need a true lightweight 155mm howitzer. The current M198 towed howitzer is just too heavy. The LW 155 will give the MAGTF commander greater operational and tactical flexibility in executing his mission. It maintains the current thirty-kilometer range and lethality; but the increased mobility will significantly improve artillery ship-to-shore movement and increase the survivability, responsiveness, and efficiency of artillery units supporting ground operations. We need this system, and will be selecting a contractor to do the job soon.

Tom Clancy: You talk a lot about technology. Do you envision a role for GPS (Global Positioning System) in the future?[9]

General Krulak: I would like see a GPS receiver on every Marine before the end of my commandancy, but I think one per squad leader is more realistic. This will solve so many of the problems that the ground-maneuver forces have had in the past. It will greatly simplify yet improve our ability to determine where our units are and where the enemy is—the basic battlefield picture.

Tom Clancy: Communications in combat are always a concern. What do you see on the horizon in this area?

General Krulak: I want the individual Marine to be fully integrated from a communication standpoint with all echelons above and below. Take a laptop computer tied into a GPS receiver and you have a real-time picture showing all the locations of friends, foes, etc. With a touch of his finger on the computer screen, this "digitized Marine" will have the capability to call in fire on the enemy with deadly accuracy every time.

    The technology is there. What we need to consider is the impact that it will have on how we fight. You give a system like that to every squad leader and you're looking at a completely different battlefield scenario. So the challenge is to take advantage of and field such technologies that will change the existing paradigm of warfare as we know it. In Desert Storm we said, "If you can see a target on the battlefield, then you can kill it." Ten years from now, however, I think we'll be saying, "If you can sense the target, you can kill it!" We need to start thinking seriously about the impact that will have. We need to consider how it will influence the sizes and types of formations on the battlefield. We need to look at how we are going to survive on that battlefield, a battlefield where sensing an enemy is death to them.

    Another challenge facing General Krulak and the Marines, as well as his Navy counterpart, the Chief of Naval Operations, is the need to complete the upgrade of the Navy's fleet of amphibious ships. With the job only half done (about eighteen of the planned thirty-six ships having been delivered by the end of 1995), let's hear the Commandant's thoughts on finishing the job.

Tom Clancy: Let's talk about the U.S. Navy—your other half. Right now the Navy is planning to finish building a fleet of thirty-six state-of-the-art amphibious-warfare ships (LHAs/LHDs/LSDs/LPDs) formed into twelve Amphibious Ready Groups (ARGs) that will replace the current fleet of almost fifty such ships you currently have. Are these thirty-six ships/twelve ARGs enough to meet your requirements, and are they the right ships for the jobs?

---

9 For a description of the Global Positioning System, see **Navigation,** starting on page 91.

General Krulak: We need to be able to lift three Marine Expeditionary Brigades [MEBs are task-organized and can range in size from twelve thousand to sixteen thousand Marines]. Thirty-six ships can't do that. Congress realizes the need for increased amphibious lift and has put additional resources to this requirement. I believe the need for adequate amphibious lift will become even more apparent in the early 21st century, when eight out of the ten top economies in the world will be found on the rims of Pacific and Indian Oceans. In this scenario, forward-deployed amphibious and naval expeditionary forces will be critical to our ability to manage instability in those geographic areas. I think the ARG concept with a MEU (SOC) embarked meets our needs today, but we will need a different capability in 2005 and 2010, when we are trying to protect our national interests in the littorals of places like the Indian and Pacific Oceans. If you think that twenty B-2A stealth bombers with sixteen guided bombs each comprises a presence, virtual or otherwise, you don't know the Asian people. If you want the people of the Asian rim to feel the presence of American forces, let them see and touch the gray-painted side of a U.S. warship. The U.S. can't survive in the Pacific and Asian regions if all we have to offer is a regional Commander-in-Chief [CinC] flying in on a VC-20 Gulfstream VIP jet to hold a press conference saying that U.S. forces are there, when the truth is that they are a month or more away!

Now, how do you cover areas as vast as that? You cover them with Marines afloat on Navy ships—ships like the recently commissioned USS *Carter Hall* [LSD-50]. This is a Landing Ship Dock almost nine hundred feet long; not some old LST. I say give those up and use thirty-six warships, amphibious ships of the line! Let us design and configure them, and build the MEU (SOC) of the 21st century. You'll still send out an ARG, but with three of the most phenomenally capable amphibious ships in the world. Each might have one "mini-MEU (SOC)" on board, so that they can cover the vast distances that we will be required to oversee in the 21st century. They'll use things like video teleconferencing data links for command and control, and will only come together when they have to concentrate and apply their full power to a contingency.

So, what I see in the building program of today is the possibility of thirty-six miniature ARGs, each one composed of just one ship with a mini-MEU (SOC) on board.

Tom Clancy: Could you tell us a bit about how you feel about the current amphibious shipbuilding programs?

General Krulak: On Amphibious Assault Ships. The *Wasp*-class [LHD-1] ships provide us with great capability. In particular, the possibility of upgrading the command and control technology on those vessels so we can effectively interface with virtually any other command and control system makes them into an extremely capable system. You can run exactly the kinds of operations that I described previously with a split-ship ARG, disaster or humanitarian relief, or use it as the headquarters of a Joint Task Force [JTF]. We really need that seventh one [LHD-7]; and there may very well be a press to build an eighth ship as we approach the 21st century and have to counter the kinds of instability that I see happening. The desire to maintain stability will be so great you may actually see a slow growth of forces from their current levels.

The USS *Whidbey Island/Harpers Ferry*-class [LSD-41/49] Dock Landing Ships are also doing their jobs well. Like the LHDs, you may also see additional units being built in the early 21st century, if near-term worldwide instabilities continue to grow.

In addition, there are LPD-17-class assault ships. This is the near-term "big ticket" item for us. Just last year it was only a large paper ship. Today it is well on its way to becoming a reality. They are planning to build a total of twelve within the first decade of the 21st century. The first one is due in 2001. We need to just build that first one, get it out to sea, and then determine what the follow-on units will look like. I don't want production of that first ship to be slowed and priced out of being built by adding more and more systems. I can almost guarantee that the follow-on units will be different from the first one, but we need to get that first one off the line. Also, I want shipyards to be building them at a rapid pace—the quicker we get them the better. But I want that first ship!

On Landing Craft. The Landing Craft, Air Cushioned [LCAC] gives us tremendous capability, though I would like to see a smaller version built. We could outfit these with fire-and-forget support weapons that could go in with the AAAVs and be relatively immune to the inshore mine threat. The older conventional landing craft are eventually going to go away. For now, though, we need them for delivery of follow-on equipment to the beach.

Tom Clancy: What is the status of the Maritime Prepositioning Program?

General Krulak: The MPSRONs have been winners, and the program is healthy today. Just as with everything else, though, we must look at what we will need from MPSRON in the 21st century. That is one of the tasks in my planning guidance: to see if the current MPSRON is really the way to go in the future.

I have a feeling that things may be a bit different. Once upon a time, we (the U.S. and Great Britain) managed instability in the world through a system of coaling stations that were used to refuel the warships of the day. Perhaps we need to look at Admiral Bill Owens's [the recently retired Vice Chairman of the Joint Chiefs of Staff] mobile-base concept. This is a modular self-propelled floating air/logistics base, which can be moved to a crisis area and would provide the ability to anchor a large support and logistics base right off the coast of a combat zone. We need, however, to look at the trade-off of such a concept with the current MPSRON scheme, which is mainly to haul equipment and supplies, not serve as a core base for their operations.

One of the other significant challenges faced by General Krulak is the matter of the extremely high operational tempos (Optempos) that have faced the U.S. military in general, and the U.S. Marine Corps in particular. With future Optempos projected to increase in the years ahead, his thoughts are insightful.

Tom Clancy: Let's talk a bit about the Optempos that the U.S. military in general, and the Marine Corps in particular, has been sustaining over the last few years, particularly in light of the recent cutbacks. Could you talk a little about this and its effects on the Corps?

General Krulak: Marines operate. Marines deploy. It's what we do. It's what the nation needs for us, with our Navy shipmates, to do. Optempo is going to have long term impacts on personnel and is hastening the modernization problem that is rapidly coming upon us. In regard to the latter, we are using our equipment up—going beyond the fatigue and service lives of equipment—earlier than planned. In addition, we are beginning to see maintenance problems as a result of delayed or deferred maintenance. Funding is a problem too. We are going to see lost opportunities to stretch out the life of our equipment without the proper moneys at the right time. This level of Optempos costs us training time and limits our options on how and when we repair things, which in turn gets us out of sync with the planned funding for that repair.

There is also the human cost of high Optempos. You already have problems with the families, wear and tear on the people. What is amazing, though, is that the individual Marines are loving the work, because that's what they came in to the Corps to do in the first place. The wives and families struggle, but the Marines love to work hard! It's a strange dichotomy for us to balance in the future.

The crown jewels of the Marine Corps today are its force of seven MEU (SOC)s. These compact, highly mobile forces are the key to maintaining the United States' capability to "kick in the door" to a hostile coastline, should it be required. General Krulak's thoughts on the future of these forces are important, because they represent the last remaining vestige of our once-robust amphibious capabilities.

Tom Clancy: The MEU (SOC)s. You have seven now, but will that be enough in the future?

General Krulak: I think seven are enough to do the job today, though beyond 2005 to 2010 it will not be. What we will need to do is optimize the number of MEU (SOC)s on the various amphibious platforms that we do have. For example, if you have a V-22 that can carry twenty to twenty-five combat-loaded Marines, compared to the eight to twelve carried by the current CH-46 Sea Knight, you increase your capability to deal with the threat. In addition, you may be able to off-load some of the V-22s onto the LPD-17s, and build the mini-MEU (SOC)s that we talked about earlier.

We have to get "outside of the box" in our thinking. We need to package the the MEU (SOC) with the capability to do the mission we are tasked to do, but do so in the minimum possible space aboard the ships. I mentioned earlier the "digitized Marine" squad leader who can call down accurate killing fire on anybody in a matter of seconds. We have to consider what kind of capability that kind of Marine brings to our warfighting ability. I don't know what the implications are today, but I do know that I had better find the answer if the Marine Corps is going to remain relevant in the 21st century.

In my planning guidance I directed the establishment of the Warfighting Lab at Quantico to look at these types of issues. As we develop various concepts of how we should fight or train or equip Marines, they will be tested under a concept called Sea Dragon. Because of new technologies that will be available to the Marines and sailors of the 21st century, in ten years you will see a MAGTF that has much greater capability and can cover more ground than the current MEU (SOC). The size of these units may be dictated more by technology and the capabilities of the individual ships than anything else. The question is just what systems do we really need on the modern battlefield for an expeditionary MAGTF. Do we need an M1 tank or perhaps a more mobile vehicle armed with fire-and-forget anti-armor missiles? Do we need a light tracked vehicle or a derivative of the current wheeled Light Armored Vehicle [LAV]?

These are the questions the Warfighting Lab and Sea Dragon will address. We are looking forward into the 2010 time frame and checking into a number of other things—equipment, combat support, all kinds of things. Do you think that the United States Marine Corps will look the same in ten years as it does today? I don't think so!

As we closed out our chat with the 31st Commandant of the Marine Corps, General Krulak shared some of his visions of the future, both on the roles and missions of the service as well as the ethos of the Corps in general.

Tom Clancy: Could you talk about the Marine Corps in ten to twenty years in terms of its mission?

General Krulak: I see us as the premier crisis-response force in the world. And I define crisis response as everything from major regional contingencies to disaster relief. Some military forces are so specialized they are like a window washer who only washes square or round windows. I'm telling you that we do windows! You tell me what you want done, I will configure a force for your needs. We are the most flexible military force in the world today. When you tie us to the capabilities of our sister service, the U.S. Navy, we offer a completely unique set of capabilities.

Tom Clancy: Do you feel good about what you see in the Marine Corps today and in the future?

General Krulak: Absolutely. The capabilities resident in the Marine Corps have been found to be of use and value to the nation. It's interesting that we are not doing things much different today from what we did during the Cold War and before Desert Storm. We are doing it a little more frequently, but we have not changed our philosophy much; and in the future we are going to become even more valuable. The Marine Corps that I inherited has always done just two things for this country. First, we make Marines; and they are a different type of person in their souls and their minds. Secondly, we win battles. We don't necessarily win major wars by ourselves; that is the job of the U.S. Army. We have, however, been the ones winning the early battles. If we ever stop doing either one of those things, we are finished. Therefore, all of my focus is on making Marines and winning battles. The United States of America needs the Marines.

By the time you read this, General Krulak will be at least halfway through his four-year tour as Commandant of the Marine Corps. His goals and visions will have been scrutinized, the first hard results of his initiatives will have been seen, and his programs will be showing signs of life. Yet, it is perhaps his own persona and character that will be the defining aspect of his commandancy. He has brought the Corps back to its roots, showing a hereditary line back to the qualities that have always made the Marines special to the United States. He truly is a warrior prince of the Marine Corps, and will be an important force as they enter the 21st century. In spite of the shortage of funds and the cutbacks that have been at the core of recent Marine Corps history, there will always be Marines. Trust the son of Brute Krulak to keep that promise.

Marine human material was not one whit better than that of the human society from which it came. But it had been hammered into form in a different forge.

—*This Kind of War,* T.R. Fehrenbach

In early 1996 the United States Marines were a small, elite corps of only 195,000 men and women. Every one of these, whether officer or enlisted, shares a common experience as a Marine. They face similar physical and mental challenges, and they must pass the same tests of skill and endurance. Becoming a Marine is an achievement like winning an Olympic medal. No matter what else you may do in life, once you pin on the emblem at the end of Boot Camp, you are a Marine for life. Over the years, the Corps has had its share of members it would like to forget; Lee Harvey Oswald and the idiots who raped a young girl in Okinawa in 1995 come to mind. On the other hand, former Marines such as Art Buchwald, Ed McMahon, Jim Lehrer, and Senator John Glenn exemplify many different kinds of real success.

What kind of person does the Corps want to recruit? The answer to this question determines the kind of Marines we send around the world as America's representatives and, often, our first warriors in a conflict. Does the Marine leadership want automatons who mindlessly follow the orders of a superior? Or do they want a Corps of restless, intelligent young people, asking questions and exploring new solutions to old problems? Today's recruits have to be both physically fit and mentally agile, able to work well on a team, but also able to stay cool on their own in stressful situations. Just how you find such people every year is the subject of this chapter.

The Big Green Machine: The Corps Today

They serve in every country in the world where the United States has diplomatic relations, and probably a few where we don't! Their career specialties include everything from senior managers and leaders to pilots, machinists, and computer technicians. The first thing you notice when you enter their world is that as a group they are physically fit, with the sort of "hard bodies" you might find working out at your local gym. This is a product of training, as well as the yearly requirement for every Marine (including the Commandant) to pass a rigorous physical examination called the Physical Fitness Test (PFT). Composed of a timed three-mile run combined with measured sit-ups and chin-ups on a bar, the PFT is one of the requirements that determines whether someone is still a Marine. Every day, rain or shine, at lunchtime along the riverfront park near the Pentagon, you see men and women in sweat suits running. Running hard. A lot of them are Marines. If you sit in an office all day and live on a diet of donuts and coffee, you won't pass the PFT, and failure to pass it results in an invitation to leave the Corps. This may seem harsh, but it means that Marines are on average the most physically fit personnel in the military services. Every Marine is also required to maintain proficiency with the M16A2 5.56mm combat rifle and other assigned weapons. For staff NCOs and officers this also includes proficiency with the M9 9mm pistol. Failure to maintain weapons qualification is also cause for dismissal. For some 220 years, every Marine has been qualified as a rifleman, and this is not about to change in today's Corps.

Another striking thing you notice about the Marine Corps is the surprisingly low proportion of officers, compared with other services. Traditionally the Corps has entrusted greater responsibility to enlisted personnel than other services, and it shows in the telling "nose to tail" (officer-to-enlisted-personnel) ratio in each. While the Navy ratio is about 6 to 1, the Army about 5 to 1, and the Air Force a costly 4 to 1, the Marines have some 8.7 enlisted personnel for every officer. Beyond the benefits that such a ratio has on the morale and self-esteem of enlisted personnel, there are other noticeable effects. Person for person, the Marine Corps is remarkably inexpensive to operate and maintain, since enlisted personnel cost less in salary and benefits than an equivalent number of officers. As a result, the Corps assigns many leadership and supervisory responsibilities to non-commissioned officers. This means that enlisted Marines take orders from sergeants who at one time were just like them, raw recruits headed to Boot Camp.

Marines also have a sense of their personal identity and position in the world. Ask any Marine, and he or she will be able to trace the chain of command all the way from himself or herself right up to the President of the United States. This is not simply a trick, like dogs walking on their hind legs. It is an indication that every Marine is confident of his or her place in the world. And that shows in confident behavior. More important, Marines learn that they are trusted to make good decisions, follow orders, and accomplish tasks in the best way available. If you have worked for a big corporation, with numbing layers of middle management over your head and no sense of personal empowerment, you can appreciate the refreshing clarity that Marines feel about their individual positions and missions.

In *Submarine*, *Armored Cav*, and *Fighter Wing*, I took you along the career paths of officers. This chapter will be different: It will trace the career path for the real backbone of the Marines, the NCOs (non-commissioned officers). Specifically, you'll see how a young man or woman rises through the ranks to reach the legendary rank of gunnery sergeant, or "gunny." The title harks back to the days of wooden ships, when Marines loaded and fired the Navy's cannon. Today, gunnery sergeants are the institutional "glue" that holds the Corps together, maintaining the traditions and making it clear to new recruits and officers that the gunnies really run the Corps. So follow us on the road to Gunny and learn what a career in the Corps is all about.

Prospecting for Gold: Recruiting for the Corps

The raw material for making Marines is provided by your local Marine Corps Recruiting Station. These nondescript little offices, many on the second floors of strip malls across America, are where the Corps puts its own out to find and deliver new Marine recruits for training. To find out more, I spent a Saturday morning at the Recruiting Station in Fairfax County, Virginia. Located just west of Washington, D.C., the station covers much of Northern Virginia. This is a tough place for recruiters. With a median family income of just over $70,000 per year, it is among the most affluent suburban regions in America. That makes recruiting Marines difficult. Very difficult. Running the Fairfax station is Gunnery Sergeant James Hazzard, along with Staff Sergeant Warren Foster and Staff Sergeant Ray Price. Their backgrounds range from artillery operations to helicopter maintenance. Gunny Hazzard also supervises another recruiting annex with two more Staff Sergeants in Sterling, Virginia, covering Loudoun County all the way out to the West Virginia state line. His territory extends from the high-tech headquarters of the U.S. Intelligence community (CIA, NRO, etc.) in Langley and Chantilly to the horse farms and cornfields of Leesburg.

It is a big territory, with an expanding population and economic base. The demographics combine a solidly white, conservative Protestant majority with a cross-section of almost every imaginable ethnic, racial, and religious group. Something like 70% of the high school graduates in the area go directly into college after graduation. Such young people are unlikely to see the benefits of an enlisted career in the Marines. Even within the various ethnic communities of the area, recruiting is tough. For example, within the Asian American community, tradition dictates that parents want the oldest son to go to school, return to run the family business, and eventually become the head of the family. An old Confucian proverb says: "Good iron is not used for nails, good men are not used for soldiers." That attitude makes it tough for a recruiter who is looking for a few good men.

Marine Corps Recruiting Command has set a relatively modest "mission" (the term "quota" is out of favor) of two per month for each recruiter assigned to the Fairfax station. That's 120 recruits a year for two small offices with only five personnel. An office's recruiting mission is based on the number of qualified military applicants (QMAs) historically recruited from an area. The top-scoring Marine recruiter of 1995, based in the small Midwestern town of Quincy, Illinois, averaged 5.5 enlistments per month, so you can see the problems of the Fairfax recruiters.

How does Gunny Hazard's team recruit enlisted Marines in a place like northern Virginia? Well, for starters, they have the best walking billboards in the world, themselves. As a "brand name," the Marine Corps usually enjoys a strong, positive public image. When you see a story in the media about the Marines these days, it is usually favorable. The rescue of Air Force Captain Scott O'Grady from Bosnia, the evacuation of UN peacekeeping forces from Somalia, and helping liberate Kuwait City from the Iraqis are typical Marines stories seen on the nightly network news. With that in mind, every Marine recruiter is encouraged to wear his or her dress uniform in every possible situation—out on appointments, visiting schools, or just when they are out buying groceries or picking up the dry cleaning.

Often, future recruits will just walk up and ask to talk to them about what it is like to be a Marine.

Another tool is television. While the Marines have the smallest advertising budget per capita of any of the services, they spend it wisely. Their television ads are Peabody Award winners, designed to leave a lasting and positive impression on a carefully targeted audience of high school- and college-age men and women. Each ad is designed to have a useful life of about four years, and it is run in key time slots designed to maximize its visibility. "Do you have the mettle to be a Marine?" was a classic example.

Much of the recruiting advertising budget is spent on sports broadcasts during football season (early in the school year), and basketball playoffs (during the decision-making period before graduation). A new ad, *Transformation*, was first aired on October 9th, 1995, during *Monday Night Football*. Using sophisticated computer animation and "morphing," it symbolized the mental and physical challenges overcome in transforming a young civilian into a Marine.

In addition to television, the Marines make careful use of magazine, billboard, and print ads, all in the hope of convincing young men and women to take the plunge and talk to someone like Gunny Hazzard. Other key tools of the Marine recruiter are school career day visits, booths at malls and military air shows and exhibitions, and even "cold calling" young people recommended by friends, parents, and school counselors.

It is tough and sometimes discouraging work. Right after Desert Storm, the U.S. armed forces almost had to turn away applicants, so many young people wanted to be part of a winning team. But times have changed. Just five years after the victory in the Persian Gulf, all of the services are scrambling to keep up the recruit pool required to sustain our forces. And to make matters tougher, the Marines have actually raised the enlistment standards for new recruits. Thus, right now, nine out of every ten applicants fail to qualify and cannot be accepted. The reasons range from problems with the law or drugs to failure to have a high school diploma. With all of the highly technical equipment required to run a modern fighting force, a high school dropout or even a student with a GED certificate simply will not do. This means that while the average Marine Corps recruiter used to have to meet 200 prospects to find one qualified recruit, now that number is over 250 and rising. Gunny Hazzard told me that the number is something between 300 and 400.

The process of qualifying a recruit involves lots of testing—medical, academic, and psychological. Then there is the candidate's personal situation. Life in the military may be hard, but to a potential recruit it may look like a way to escape an abusive family or a failed relationship. The recruiter must find out the potential recruit's motivation for joining the Corps, and whether the Corps really wants him or her. The Marines are surprisingly tolerant of past troubles with the law (as long as these do not exceed minor convictions, like traffic violations), or past casual use of drugs or alcohol. The recruiter becomes a coach and big brother of sorts, gathering background information to help the Corps waive any minor infractions. Some of the best Marine recruits come from such "problem situations," and thus are worth the extra effort.

Now, it should be said that not every person who walks into a recruiting station like that in Fairfax is a troubled kid with problems at home and school. One recruiter I spoke with was quite emphatic about this, and backed it with a recent success story. He was just finishing up a miserable month, without recruiting even one QMA. As he was walking out of the station to his car in the parking lot, on his way to get chewed out by a superior for not making his monthly mission, it happened. He saw a young man approaching the door. He looked like a recruiting poster Marine: hair "high and tight," with every button in place and a hard-body physique. The recruiter, thinking he was looking at a Marine, respectfully asked which unit he was assigned to. To his surprise, the young man told him he was walking in to join the Marines; he had wanted to do that since boyhood! The recruiter thanked God for his good fortune and took the young man inside, finding him to have an excellent school record, not so much as a speeding ticket, and near-perfect scores on the qualification tests. The young man was sworn in and on the bus to Recruit Training the very next day. As might be imagined, the recruiter's superiors forgave him for missing the meeting, and the Corps had another gold nugget to forge into a warrior.

Assume that a young person has decided to join the Marine Corps and has qualified. There is usually one more obstacle for the recruiter to overcome, and this frequently is the show-stopper. The parents. Despite the generally good image the Marines enjoy, many parents just cannot accept the idea that their son or daughter could join the Corps. Many parents from the generation of the 1960s and 1970s have a deep-seated anti-military bias rooted in the Vietnam War. Others resist the idea that their child is "giving up" on college and going into the military as an enlisted recruit. They see this as a "low class" career choice. Also in the back of every parent's mind is the fear their child may be killed or maimed in a far-away place. In a parent's mind, these are valid reasons to dissuade a child from enlisting. The recruiter thus finds himself in the role of family counselor, having to prove to a parent that the Marine Corps is not just a sump for the scum of American society. Recruiters frequently lose this round in the recruiting game.

Despite all these problems, Gunny Hazzard and his team do "win" their share. The week before our visit, they had enlisted three female QMAs, a real prize for any recruiting office. The following week, their office would swear in four more male recruits. Gunny Hazzard was quite candid when he told me that not every month went so well. Like salesmen, each month Marine recruiters start at zero and are judged on current, not past, performance.

When a candidate has been qualified, and all the paperwork is complete, the next step is to schedule a time to report for processing and transportation to one of the two Marine Corps Recruit Depots (MCRDs). MCRD San Diego, near Point Loma in the harbor district of San Diego, California, provides Recruit Training for all male recruits west of the Mississippi River, including Alaska, Hawaii, and the Pacific (Guam, Samoa, etc.). Folks in the Corps like to call the recruits trained there "Hollywood Marines" because of its proximity to that entertainment capital. The other MCRD, at Parris Island, South Carolina, handles Recruit Training for male recruits east of the Mississippi, as well as all of the Corps' female recruits.

The wait for a reservation at Recruit Training is short these days—unless you are a female recruit, as there is only one female recruit battalion at Parris Island, with a limited number of openings each year. When the time comes for the new recruit to report for training, he or she is transported to a Military Enlistment

Processing Station (MEPS), and then to the MCRD. In the mid-Atlantic region, the MEPS is located in Baltimore, and recruits are accompanied by the recruiter. After an entry physical, they are sworn in and driven to the airport for the flight to Charleston, South Carolina. From there, they are bused to their new home for the next three months or so, the MCRD at Parris Island. Let's visit this gateway to the Corps, and see what makes it such a special place in the hearts of Marines.

The Island: Parris Island and Recruit Training

Deep in the palmetto groves and scrub pines of tidewater South Carolina you can still find a land that looks little changed from the 1800s. When you arrive, you might swear that you have seen this place before, and you would be right. This is the home of the novels of Pat Conroy; and in fact *The Great Santini* and *The Big Chill* were filmed in the nearby town of Beaufort. The place is Port Royal Sound, the finest natural harbor between Virginia and Florida and home to several Marine Corps bases. Up the sound, Marine Corps Air Station Beaufort is home to Marine Air Group Thirty One (MAG-31), flying F/A-18 Hornet fighter bombers. Across from Hilton Head, with its beautiful golf courses and resorts, is our destination, Marine Corps Recruit Depot Parris Island.

Parris Island was fought over by French, Spanish, and English forces even before the Revolutionary War. Later, during our own Civil War, it was one of the first bits of Confederate territory taken by the Union, in 1861. Throughout the Civil War, the sound's superb natural harbor was a base for Union amphibious and blockade operations along the Southeastern coast. Later, during the Spanish-American War, the sound served as a naval base and staging area. The old stone dry dock near the commanding general's quarters is mute testimony to past naval activity. Parris Island became an MCRD during the run-up to World War II, when it supported the vast expansion of the Corps. Warm year-round weather makes it ideal for training, though it does get pretty steamy and tropical during the summer. One of the consequences of the climate is the profuse and voracious insect life, which must be seen (and felt!) to be believed. All the same, its close proximity to Charleston to the north and lack of encroachment by civilian development mean that it will probably be training Marines long after southern California real estate development has crowded MCRD San Diego out of existence.

Facilities at MCRD Parris Island are a mix of new and old, with modern mess halls and shooting-simulation galleries right beside old landing strips for World War II-era bombers. Even in these days of tight budgets, modernization and new construction of barracks continue. Parris Island is unique among East Coast Marine bases, having virtually no active Fleet Marine Force units. First, last, and always, Parris Island is dedicated to just one mission: taking raw, civilian recruits and making them into Marines. The core of this process is the Recruit Training Regiment (RTR), commanded in late 1995 by Colonel D.O. Hendricks. His senior non-commissioned officer (NCO) was Sergeant Major P.J. Holding, a veteran of over twenty years in the Corps. The RTR includes a support battalion and four training battalions—three for male recruits, with the fourth reserved for female recruits. At any one time, Parris Island is home to over seven thousand training and support personnel, and some 4,800 recruits. It is a busy place, and you can feel the energy as you enter the base.

The new recruit's first impression of Parris Island comes during the last stage of the bus ride down from Charleston. The MCRD is extremely isolated, connected to the rest of the world by a single two-lane causeway. Except for that, the entire depot is surrounded by salt marshes, swamps, and the sound. This makes security relatively simple, and "going UA" (Unauthorized Absence, the current term for AWOL) virtually impossible. Though the leadership of the Corps only smiles when you mention it, new recruits always seem to arrive in the middle of the night, around 2:00 AM. This intensifies the new recruits' sense of being cut off from their past and the outside world and focuses them on what is to come in the next few months. The buses roll up in front of the "receiving" building. There recruits are dumped onto a stretch of road marked by a line of yellow-painted footsteps. Each recruit stands on a set of the painted prints and takes part in his or her first formation on the way to becoming a Marine. It is a moving, memorable moment. Throughout the next few months of Recruit Training, the recruits will probably never again see this spot. Only afterwards do they always seem to find their way back to where their individual journeys into the Marine Corps began. From the yellow footsteps, they are marched inside the receiving building for a short orientation.

They spend the rest of the night with paperwork, haircuts, and gear issue, before they move on to a holding barracks for some rest. All personal belongings (civilian clothing, CD/cassette players, even combs) are taken from the recruits and placed into storage, to be returned upon completing or leaving Recruit Training. This has a further effect of cutting recruits off still further from their past lives, and makes any attempt by a rogue recruit to leave the island more difficult. Then there is "the moment of truth," where each new recruit is asked, for the last time, whether he or she really wants to be there, and if there is anything in their background which would keep them from serving as a Marine. This is important, for any lies detected after this point can result in immediate dismissal from the Marines. Admission of a past infraction means that if the problem can be worked out, the Corps will do so without damage to the recruit's career. The next few days are spent in further testing, physical examinations, an initial strength test, and appointments with various counselors. These activities are designed to alert the RTR training staff to any physical or psychological problems that might cause trouble with a new recruit. In the case of a physical injury or shortcoming, the RTR staff retains the recruit and tries to place him or her back into the training cycle later.

The other examinations can take a darker turn. Many young people in our society come from abusive families or destructive situations; and such people may choose the military as a way out of these situations. Although the Corps views its role as "making Marines and winning wars," as it accomplishes that it tries to provide a safe, positive place where qualified young men and women can get a clean start on life. Thus, when the RTR personnel find a young recruit with a problem, they work to help the person overcome it, rather than throwing that person back into society's reject bin. Throughout Recruit Training, you find examples of such interventions by RTR staff members. At times they have to physically place themselves between the recruits and dangerous situations. At other times they have to give a

Marine recruits on the Parris Island confidence course. This series of obstacles is designed to promote physical fitness and mental toughness in traversing assorted obstacles.

JOHN D. GRESHAM

young recruit an "assist" or "push" when they hit the "wall" that all recruits seem to hit somewhere in training. Like runners in a marathon, recruits often reach the point where suddenly the end goal seems unattainable; but with a little help and support it comes into focus and sight. Other interventions can be more hazardous, like having to drag a recruit clear of a mishandled grenade on the training range. A lost recruit hurts, and the RTR staff work hard to make sure as many as possible make it through. On another plane, the U.S. Navy Chaplain Corps looks after the spiritual well-being of the recruits, as well as that of the staff and their families. Through a program of lay readers, chaplains manage to cover virtually every religious tradition and denomination. They are a vital link back to the rest of the world for the recruits, also providing a liaison to the Red Cross in the event of a family emergency.

Following the orientation period, recruits are assigned to one of the four training battalions. They are assigned to platoons of about seventy to eighty personnel. Three or four platoons make up a "Series," which is the basic organizational unit within the battalion. Two Series make up a Company, with four companies per battalion. Each series is commanded by a 1st Lieutenant or Captain, with a Gunnery Sergeant as the senior NCO. Within each platoon, a team of four Drill Instructors is assigned to watch over the training and well-being of the recruits. The legendary Marine Drill Instructor (DI) is as revered by the Corps as he is misunderstood by the public. DIs come in two flavors, Senior DIs with their distinctive black patent-leather belts, and Junior DIs with green web belts. The Senior DIs are supervisors, taking charge of each platoon and the other sergeants.

Despite the popular notion that Recruit Training is a program of sadistic torture, and the DIs demented bullies, the truth is surprisingly different. Series commanders and DIs are selected volunteers whose mission is to get as many recruits as possible through Recruit Training successfully and safely. Now, this has not always been the practice of the Corps, and the 1956 incident at Ribbon Creek at Parris Island is always on the minds of DIs. They take care of their recruits the way a mother hawk watches over her young. This does not mean that Marine Recruit Training is easy or enjoyable. It is specifically designed to be neither. It is a tailored curriculum of physical, mental, academic, and skills training, designed to take recruits to their own personal limits, and keep them there for a long period of time.

During this training, the DIs work hard to keep the pressure on without losing any recruit who is capable of meeting the challenge of becoming a Marine. It is a very tough job. From the moment new recruits are assigned to a training platoon, there will be a DI overseeing every moment of their lives. This means that the Drill Instructors are running a twenty-four-hour-a-day watch schedule every day of Recruit Training. On average, the DI's day runs about eighteen hours long, with constant vigilance being the minimum requirement. Burnout is a common problem among the DIs and Series Commanders, and a program of rotation to non-training posts within the RTR helps keep them focused during their two year tours in the job.

The term "positive control" is used to describe the way DIs watch over each recruit. Designed to keep recruits safe and obedient, positive control is exercised through a combination of physical presence and what the Marines call the "command voice." The physical presence is a function of appearance, so DIs will wear an immaculate uniform topped by the famous Marine campaign cover (also known as a "Smokey Bear"). But the command voice is what really does it. Like the famous "rebel yell" of the Civil War, it is impossible to describe, but you know it when you hear it. Every DI and Series Commander has one; and some say it makes any order, comment, or statement presented to the new recruit sound like the voice of God himself. The DIs need the command voice, because the days of physical hazing and verbal abuse towards recruits is over. DIs use words the way a surgeon uses a scalpel to cut out a tumor. To an eighteen- or nineteen-year-old recruit, it is like being torn apart emotionally. One of my guides, public affairs officer Captain Whitney Mason, had just completed a tour as a series commander at Parris Island, and she confessed to having such a "voice" for those occasions when she needed it. Now, looking at this slender and slight lady, you might find it hard to believe, but she indeed does. The lessons delivered through the "command voice" last a lifetime. More than one Marine I've spoken with has told me that in the heat of combat, when he was so scared he was pissing in his pants, lessons delivered from a particular DI years before came through loud and clear through the terror, and saved his life.

Back to the new recruits. Much has been happening during their testing and transition period. During their early days at Parris Island, they rapidly lose the identity they knew in civilian life. Along with the mandatory "buzz cut" haircut for male recruits (female recruits enjoy a little latitude in permitted hairstyles), uniforms and other gear are issued. And just before they are assigned to their training platoon, recruits acquire the most important tool of the Marine rifleman, an M16A2 rifle. They will carry it throughout Recruit Training, and will learn to use it better than any other basically trained warrior in the world.

When the time comes to meet their DIs, this is done in a unique ceremony called a "Pickup Briefing." The new recruits march to their new platoon barracks (called a squad bay) and stow their gear. This done, the seventy or so recruits sit cross-legged on the floor and await the appearance of their DIs. Starting with the Company Commander, then the Senior DI with the distinctive black belt, the recruits are introduced to the Marines who will hold their lives in their hands for the next few months. The speeches are compelling, almost intimidating in their presentation. But if you look over the recruits then, you'll see that they are not just intimidated, they are actually frightened, as well they should be. The first time you experience a DI in full regalia and command voice is something you never forget. The DIs talk for a short time about what will be expected of the recruits and how things will run in the platoon, with special emphasis on safety and looking out for each other. Then things begin to happen. The recruits are lined up in front of high double bunks, and the DIs begin to drill them. At first they are ordered to dig into their duffel bags and footlockers and quickly find particular pieces of equipment or articles of clothing. Later, they begin to drill with their M16s. The idea is to get them to rapidly respond to the orders of the DIs and build the trust that will be required to make Recruit Training effective. In this way, more difficult and dangerous training tasks, particularly those involving firearms, can be safely accomplished.

Marine Recruit Training is accomplished in phases, spanning about three months (for female recruits, a few days more). It starts with the Forming Phase that we have been looking at. Lasting three to four days, it is designed to teach the recruits the basics of squad bay life and "getting green," as some of the recruits call it. During this time, the DIs take the time to interview each recruit, to get to know them better, and to establish what will be needed to lead a particular recruit through Recruit Training. It also is a final check to see if any personal problems need to be referred to professional counselors or medical personnel. The four DIs then split up the job of watching over the platoon; and one DI remains on duty in the squad bay at night during what is called the "fire watch." Recruits are enlisted to help with the fire watch, which further serves to indoctrinate them into the twenty-four-hour-a-day nature of military life. Though this experience is vital, for combat often requires going without sleep for extended periods, efforts are nevertheless made to ensure that recruits get adequate rest. It's usually lights out by 9:00 P.M., and up at 5:00 A.M. every day.

Following the Forming Phase is Phase I, which lasts approximately three weeks. This is mainly an orientation phase, where the recruits are given a daily regime of intensive physical training (PT), close-order drill, introductory classes in general military academic subjects, and their first experience with the obstacle course, which is a confidence-builder composed of assorted barriers to climb over, jump across, or crawl under. Recruits run the course repeatedly during Recruit Training, and by the time that they are done, they will know how to run it literally with their eyes closed. The daily PT is also vital, because the Corps requires a certain minimum level of physical fitness just to perform basic tasks. While many of the recruits are in good shape, PT makes them better, and it helps instill a desire for a daily regime of such exercise later. It's always easier to get in the exercise groove early in life. Take it from one who has discovered this too late, and is paying the price.

A Marine recruit armed with an M16A2 combat rifle maneuvers on the Parris Island combat assault course. The instructor in the background is monitoring the safety and perfomance of the recruit.

JOHN D. GRESHAM

Recruits also study various academic subjects. The public perception may be that the Corps is not a collection of intellectual heavyweights, but officers of other services who serve on joint staffs will tell you that some of the top military thinkers today are Marines. Subjects range from basic tactics to Marine Corps history. The Corps believes that smart recruits make good Marines. The stereotype of the ignorant "jarhead" is simply no longer accurate. The vast array of tasks and equipment required of even a basically trained Marine would make your head spin. As a nation, we ask a lot of our Marines, and they have to be properly trained if they are to deliver.

The attitude of recruits is constantly monitored as they adapt to life in the squad bay. Despite the best efforts of recruiters like Gunny Hazzard to prepare them mentally and physically, most recruits who "wash out" of Recruit Training are lost in the first three weeks of training. It is a tough thing to send a young person home from Parris Island, and the Corps does everything it can to minimize attrition. If recruits are injured in training, they are given time to heal and rehabilitate if possible. When recruits fall behind in academic or skills areas, they receive special help to make up so they can get back with the rest of the platoon. Through it all, the DIs watch over the recruits around the clock, making sure they stay safe.

Warfare training begins in Phase II, which lasts six weeks for male recruits, and seven weeks for female recruits. Here they practice marksmanship with the M16A2, to include their first experiences on the rifle range. The Corps takes marksmanship very seriously. If you cannot consistently hit targets on the range with a M16, you will *never* be a Marine. In Phase II, the general military knowledge taught in Phase I is tested, and there is the recruits' first experience with the Physical Fitness Test (PFT). Like proficiency with the M16, the successful completion of the PFT is mandatory to be certified as a Marine. Also in Phase II is the recruits' first experience with the new water-training facility recently completed at Parris Island. A surprising number of recruits have never seen a swimming pool, lake, or ocean, and they must learn to swim if they are to serve in an amphibious service. In the swimming facility the recruits learn how to float and move through the water, even when fully loaded with a rifle, uniform, boots, and pack. This training includes a series of drops from platform boards, which can be terribly unnerving to young people whose only experience with water may have been an open fire hydrant in an inner-city neighborhood.

Female Marine recruits await their turns at the Parris Island Grenade course. Female Marines have to qualify on all the same weapons and courses as their male counterparts.
JOHN D. GRESHAM

Finally, there is the tactical training that is so necessary to becoming a Marine warrior. This includes rudimentary small-unit and assault training, as well as traditional Marine training in "hitting skills" (fundamental to hand-to-hand combat), and training in the use of pugil sticks (large padded clubs). This part of basic training introduces the recruit to the unpleasant fact that life as a Marine can involve the very personal act of assaulting other people, and perhaps even killing them. Tactical training not only does this, it also teaches the new Marines how to measure and use force in combat.

Earlier, I pointed out that female recruits spend a week longer in Phase II than their male counterparts. Now is a good time to talk about why. Since World War I, women have augmented the strength of the Corps, freeing men for combat jobs. And like other branches of the U.S. military, the USMC has gradually expanded its range of opportunities for women. Today, something like 93% of all Marine MOSs (Military Occupational Specialty codes, which determine the jobs personnel are trained and certified to perform) are open to women Marines This even includes aviation jobs, such as flying fighter jets and attack helicopters. But for women Marines the official Defense Department definition of "combat" still restricts them from combat-related MOSs, the specialties senior leaders consider most necessary for promotion. This includes infantry, armor, and other ground combat positions. The stated reasons for this restriction are the same as those of the U.S. Army: Women are said to lack the strength and endurance necessary for the rigors of ground combat. But this situation is changing, as General Krulak is currently considering lifting the restriction from artillery and some other combat MOSs.

Now, despite the restrictions on women serving in front-line ground units, the Corps still has training and readiness standards for all Marines, and *every* Marine has to be prepared for combat, anywhere and at any time. This means that female recruits also train for combat. But the female recruits have a somewhat different training regime from their male counterparts. For starters, they are housed and trained in a separate training unit at Parris Island, the 4th Recruit Training Battalion. The 4th Battalion facilities make few concessions to alleged female requirements (personal privacy and such). Squad bays have roughly the same layout and equipment as the male ones. Some 4th Battalion executive officers and Sergeant Majors are male, but there are no male DIs or Series Commanders.

One significant difference between male and female training, however, reflects an ugly reality of our society: A high percentage of the women who enter the Corps report they have been physically or sexually abused, molested, or raped prior to their entry into Recruit Training. While Marine leadership is quite discreet in discussing this subject, its action on behalf of female recruits is specific and effective. The 4th Battalion has a psychiatrist on call to help deal with emotional problems, as well as a Licensed Clinical Social Worker at Beaufort Naval Hospital. Though the percentage of female recruits with previously confirmed histories of victimization is reported as 7%, something approaching 50% of these wind up telling of such experiences during initial Recruit Training interviews. You may think people scarred with such experiences should not be put in a position of responsibility (like becoming a Marine), but the Corps views this situation differently. Marine leadership sees any person who is mentally, morally, and physically qualified and who completes Recruit Training as someone worth having—a part of their family. Furthermore, it is the experience of the Marines that such women are survivors, exactly the kind of people who can succeed in the male-dominated culture of the Marine Corps. The payoff is that while the initial dropout rate among women has always been about 50% higher than that of men, the rate has been dropping rapidly over the last few years. As a bonus, the retention rate of women Marines who re-enlist for additional tours of duty is actually higher than that of their male counterparts.

Female recruits do everything at Parris Island that their male counterparts do. At the same time, the generally smaller build and lower body strength of women (compared to men) is taken into account. For example, on the obstacle course, a few (though not all) of the obstacles are scaled down slightly. It is just as difficult for women to get over them as it is for the men to get over the obstacles on their course. I should also say that the Corps is constantly reevaluating the curriculum of both the male and female recruits to see where improvements and/or additions should be made. For example, the Commandant recently merged the male and female requirements for distance running in the PFT, a change many leaders felt was long overdue.

When male Marines finish Recruit Training, they go to the School of Infantry at Camp Lejeune, North Carolina, where they learn ground infantry tactics and master heavy weapons. The Corps requires graduation from the School of Infantry before a Marine can join a ground combat unit. But because of the Congressionally-mandated DoD prohibition on women in ground combat units, female Marine recruits get an abbreviated course in heavy weapons and infantry tactics while they are still at Parris Island, adding one week to the training cycle. Thus, even before their male counterparts, the women recruits are firing machine guns and practicing rudimentary assault tactics!

For everyone, Phase III arrives as the most cherished part of Recruit Training. Once a recruit is in Phase III, he or she is over "the hump," and the DIs are working hard to ensure that every last recruit completes the course. Lasting two weeks, Phase III consists of final examinations and makeups. Final marks for PFTs, marksmanship, and other drills are scored; and records are updated. Included in all this are final inspection, drills, and rehearsal of the graduation ceremonies. It is a heady time for the young recruits. Frequently, new training series/platoons will be allowed to march to see the Phase III units, so they will know that it really is possible to become a Marine!

The official seal of the United States Marine Corps. The eagle, globe, and anchor in the center is the official emblem of the Corps.

Graduation week passes in a rush, with parents, friends, and loved ones coming down to visit, often for the first time since the recruits were taken in hand by their recruiters. Parents are usually amazed and proud to see what their son or daughter has accomplished. Their bodies will have become toned, their dress immaculate, and their manners impeccable. It is a wonderful thing when a parent sees a son or daughter leave as a child and reappear as a young man or woman. The day before graduation, there is a small ritual for each platoon—called an "emblem ceremony"—out on the parade ground. Standing in formation, the DIs award the recruits the eagle, globe, and anchor badge of the USMC for their dress hats. From that moment on, for the rest of their lives, no matter what they do or become, they will know the satisfaction that once in their lives, they were good enough to earn the title Marine.

The following morning, there is a large parade and ceremony on the parade ground for the graduating company. Awards are given for the top recruit and marksman in each platoon. And as their loved ones look on, there is a final parade. Then it is over, and you just have to watch what happens after to know that you have seen something special in the lives of several hundred young people. Hugs and kisses. Firm hand shakes and looks. Perhaps most impressive of all, new Marines rushing to introduce their families and friends to their DIs. "Thanks for getting me through Boot Camp" are words you hear often from former recruits to their DIs. Frequently, the parents also thank the training staff—for turning their child into something better, or different, or both. I defy you to watch this moment and not shed a tear or two. I did.

More School: Warrior Training and Beyond

Following graduation, the new Marines get a short leave, and then report for their next duty assignments. For male recruits, it's the School of Infantry at Camp Lejeune. There they are taught the use of heavy weapons and demolition and breaching gear, small-unit tactics, and other skills of ground combat. Every single male

Marine recruits graduate from basic training on the Parris Island parade ground. They are now basically trained Marines, ready to move onto their next school.
JOHN D. GRESHAM

Marine in the Corps completes this training, whether he is to become a crewman in a helicopter unit or a public relations specialist in the Pentagon. It is just as grueling as the Recruit Training course, and is a foundation of the combat ethos that makes every Marine a rifleman. From there they head out to their MOS schools, following their female counterparts, who received their warrior training during Basic School. Women recruits go directly to their MOS school, and from there on to their first unit assignment.

School is a common experience in a Marine's career, with some officers and enlisted personnel going through several dozen training courses by the time that they finish a twenty-year-plus stint in the Corps. Each school can last anywhere from two weeks to a year. The Intelligence Training School down at Dam Neck, Virginia, for example, lasts a full year and is considered to be among the best intelligence schools in the military. By the time they complete their first MOS school, most enlisted Marines will have made the rank of Private First Class (E-2) or Lance Corporal (E-3). Normally, this is the point where a Marine would start moving into combat assignments, such as a rifle platoon. Thirty months to four years after making Lance Corporal, Marines generally make Corporal (E-4), continuing to function in their chosen MOS, but with growing responsibility and more training.

There is also the option of transferring to other duties, which can give an enlisted Marine's career some balance and variety. While the concept of "career enhancing" or "joint" billets has yet to take hold in the enlisted ranks of the USMC, the Corps tries to provide Marines a chance to try different things and broaden their horizons. This might include serving as an embassy guard or on a General officer's staff. It may also mean going back to school, an activity that the Corps encourages all of its members to try. A surprising number of enlisted Marines even study for a college degree. The Marines have several ways to facilitate higher education for enlisted ranks: Some are paid to attend a university. And some few who choose to seek a commission as an officer are admitted to the Naval Academy at Annapolis, Maryland. In fact, the USMC recruits a larger percentage of its officer corps from the enlisted ranks than any other U.S. military service. Promoting from within (the "Grow Our Own" program) is a key feature of the Corps, and such avenues for advancement contribute a lot to the morale of enlisted Marines.

The Road to Gunny...and Beyond

There comes a point in the life of a Marine where he or she begins to think of the Corps as more than just a job and a paycheck; it becomes a career. This is when a Marine begins the drive to the magic rank of Gunnery Sergeant (E-7), or just "Gunny." It takes a Corporal about four to six years in grade to make it to Sergeant (E-5). When you make it, the level of responsibility rises quickly, and so does the workload. But the move up to Staff Sergeant (E-6), about four to six years later, is an even bigger step in a Marine's life, for it means that you have committed yourself to becoming part of the institutional "glue" that holds the Corps together. It also means a lot of hard work and patience, and a certain level of tolerance for the actions and views of those less experienced than yourself. As a Staff Sergeant, you will probably be assigned that most dreaded of duties, a new 2nd Lieutenant to watch over and hopefully make into a useful officer. You also will become a kind of parental figure to younger Marines assigned to your care. A Staff NCO never commands (that is the responsibility of officers), but a good Staff NCO is priceless as an advisor and partner to the officers who make up the leadership of the Corps. Good officers seek this kind of help as a matter of instinct.

At this point, a Marine is considered a middle manager and leader, with oversight over rifle squads, tanks and other vehicles, and aircraft. Finally, there is the drive to Gunny. Like making Staff Sergeant, it takes four to six years; and making it puts you in a different category within the Corps. Along with the almost mythic title that it carries, being a Gunny earns the respect of officers of any rank, and something like awe from younger Marines. You become one of the keepers of the "tribal knowledge" that keeps the Marine tradition alive from generation to generation. It also means more practically that you can look forward to a twenty-year-plus career, with retirement benefits and a pension. Trust me when I say that every Gunny I have known has earned the title. As an insight, consider that most of the Marine officers I have spoken with have told me more than once that Gunny is the *best* job in the Corps, with the widest ranges of responsibilities and duties.

When Marines make it to E-7, if they wish to continue their career in the USMC, they get to make a choice. The next step is Master Sergeant (E-8). And from there they can choose to take the route to Master Gunnery Sergeant (E-9), a career track which leads to greater opportunity and responsibility within technical fields. The other option is the command side of the NCO ranks, which leads to 1st Sergeant (also E-8). A First Sergeant is typically the senior NCO of a Company or similar unit. Beyond First Sergeant is the exalted rank of Sergeant Major (E-9). These extremely rare birds are the right hands of officers commanding MEUs, regiments, divisions, and the Corps itself. At the very summit of the Staff NCO pyramid stands the Sergeant Major of the Marine Corps, a post currently held by Sergeant Major Lewis Lee, a veteran with over thirty years of service. Sergeant Major Lee sits in an office near to that of General Krulak; and the Commandant would tell you that he is the voice of the enlisted personnel in the Corps. Lastly, there is also the possibility of a direct commission to Warrant Officer (and the slang title of "Gunner") for enlisted personnel with specialized skills, though this is extremely rare in the USMC.

The Mettle to Lead Marines: Officers

Though there are some subtle differences, the career paths of the small cadre of commissioned officers who constitute the leadership of this more-than-220-year-old institution are generally similar to those of the Army ranks described in *Armored Cav* and the Air Force ranks described in *Fighter Wing*. However, unlike the other services, the Marines don't get most of their officers from the service academy of their parent service. The USMC receives only a few of its new 2nd Lieutenants (O-1s) from the U.S. Naval Academy at Annapolis, Maryland. (A much larger percentage of Army and Air Force 2nd Lieutenants come from West Point and the Air Force Academy.) Every year, a portion of the Annapolis graduating class chooses a career in the Marines and is directly commissioned into the Corps. But this small group (no more than 175) fills only a fraction of the Corps' demand—it needs over 1,500 new officers per year. Most of the other officers the Corps develops are recruited from colleges around the country.

Whether they are Reserve Officer Training Corps (ROTC) graduates or join directly out of college, they all go to the institutional home of Marine officers, the USMC Officer Candidate School (OCS) at Quantico, Virginia. A few dozen miles south of Washington, D.C. along the lower Potomac River, Quantico is where the Corps makes the majority of its officers. Interestingly, some leaders at Quantico wish the Corps would require Academy graduates to go through OCS as well, so that all Marine officers would share a common initial training experience. The ten-week OCS is similar to the Recruit Training course at Parris Island. Though there is a greater emphasis on leadership training and basic command and control skills like radio procedures, land navigation, and calling in artillery and air strikes, the training is just as physically demanding, the hours just as long, and the tests just as challenging as those enlisted Marines must meet. To prove it, just watch the officer candidates (the officer equivalent of a recruit) go through a particularly demented Combat Obstacle Course nicknamed the "Quigly." It starts with a slime- and ooze-filled ditch that flows into a small creek. The course continues through dense woods, followed by climbs and descents on a steep hill. Other obstacles follow, ending with a crawl over ground under fire from a light machine gun (don't worry, the staff uses blanks!). The sight of a slime-covered group of officer candidates moving down a bone-chilling creek is bad enough. But when you see the instructors moving a few yards/meters ahead of them, to clear out poisonous water snakes that linger in the area, you get some idea of how much these young officer candidates want to lead Marines. They quickly come to understand that they are being entrusted with the most valuable asset the USMC possesses, its young men and women. Supervising them throughout OCS are the ubiquitous Gunnies.

Following OCS, officers go through another training course at Quantico called the Basic School. Here they learn the skills needed to run a rifle platoon. This training includes not only weapons and tactical instruction, but lessons in the inevitable supervisory and paperwork skills necessary to keep any bureaucracy running. Infantry officers must complete Camp Lejeune's twenty-six-week School of Infantry as well. From there, they head out into the Corps to their MOS schools and their first assignments. Like the enlisted Marines they will lead, there is one

A pair of Marine officer candidates transit the "Quigly" stream at Quantico, VA. This course is designed to train Marine officers how to transit water obstacles silently and still keep their weapons dry and ready to fire.

JOHN D. GRESHAM

common thread: Whatever their primary specialty (pilot, logistics officer, etc.), they are all riflemen first. They are all capable of fighting on the ground. This makes the USMC different from any other U.S. military service. It is also why the national leadership trusts Marines above any other military force to get a tough job done. You can trust Marines!

Taking young men and women and turning them into Marines is hard work, and General Krulak would tell you that the Corps only entrusts such work to its best members. From the recruiters like Gunny Hazzard at the Fairfax Station, to Series Commanders like Captain Whitney Mason at Parris Island, to the instructors at the Basic Warrior School, the process of building new Marines is the toughest job you can imagine. It goes on and on, and the process cannot be allowed to end, lest the very survival of the Corps be put into jeopardy. It remains in good hands.

When I was visiting the Fairfax recruiting office, Gunny Hazzard showed me a special corner. On a crowded bulletin board were dozens of letters, snapshots, and postcards from some of the young Marines he and the other recruiters had sent to Parris Island. Every letter I saw was a message of deep, personal gratitude from the new Marine, thanking the recruiter for showing the path to a new life. This is the payoff for a recruiter who has had too many rejections and not enough commitments. Or, as Gunny Hazzard likes to point out, this is what the Corps is all about—finding young people and showing them a path to a life of service and honor.

Small Arms

THIS IS MY RIFLE. There are many like it but this one is mine. My rifle is my best friend. It is my life. I must master it as I master my life.

My rifle, without me, is useless. Without my rifle, I am useless. I must fire my rifle true. I must shoot straighter than any enemy who is trying to kill me. I must shoot him before he shoots me. I will. . . .

My rifle and myself know that what counts in this war is not the rounds we fire, the noise of our burst, nor the smoke we make. We know that it is the hits that count. We will hit. . . .

My rifle is human, even as I, because it is my life. Thus, I will learn it as a brother. I will learn its weakness, its strength, its parts, its accessories, its sights, and its barrel. I will keep my rifle clean and ready, even as I am clean and ready. We will become part of each other. We will. . . .

Before God I swear this creed. My rifle and myself are the defenders of my country. We are the masters of our enemy. We are the saviors of my life.

So be it, until victory is America's and there is no enemy, but Peace.

—My Rifle: The Creed of a United States Marine,
by Major General William H. Rupertus, USMC

The ethos of the Marine Corps is not found in the technology of its weapons, but in the character and morale of the individual Marine with a rifle in the presence of an enemy. Back in the 1970s, when the Marines were still short on the new anti-tank guided missiles (ATGMs), there was a Marine officer training a class on anti-armor tactics. When the instructor was asked what weapon was best against heavy enemy armored vehicles, he showed a slide of the Marine Corps emblem, saying, "Gentlemen, this is your best weapon." Just being Marines was their best weapon. Themselves.

While better equipped than a quarter century ago, today's Marine Corps is still taking young men and women and making each one into a lethal fighter. Marines are also taught that they are likely to find themselves thinking and acting on their own in situations requiring great responsibility—operating alone, making decisions, and taking actions that represent American policy. A recent recruiting poster showed a Marine sniper and his rifle in full camouflage, with the words "Smart Weapon."

The Corps ideal. A Marine on exercise at Camp Lejeune, holding a position with his M16A2 combat rifle. The USMC still values the individual Marine with their personal weapons as their basic building block.

JOHN D. GRESHAM

Tribal Elders:
The USMC Weapons Training Battalion

We're going to spend some time telling you about the weapons that Marines carry into battle. We'll visit an outfit dedicated to the idea that, even in a world full of laser-guided bombs and missiles, there is still a need for one well-aimed shot from a weapon held by human hands. The place is the Marine Corps Base at Quantico, Virginia, and the unit is the Marine Corps Weapons Training Battalion. On the Quantico reservation, inland from Interstate 95, stands a small cluster of buildings, mostly of World War II vintage. This is the home of the Weapons Training Battalion, the U.S. Marine Corps' premier shooting unit. Established in 1952 after the nightmare of the Korean War proved how much the Marine Corps needed to hone its shooting skills, the battalion operates sixteen different shooting ranges, classroom facilities, an ammunition loading and packing plant, and a complete gunsmithing and machine shop. Here the Corps trains the best shooters in the U.S. military, while maintaining a capability to build and maintain customized firearms. If you are a gun enthusiast like me, this is Firearms Heaven.

Colonel Mick Nance commands the Weapons Battalion. He will tell you that he has one of the best jobs in the Corps. Backing him up is Sergeant Major F.W. Fenwick, command NCO for the battalion. The unit is the Corps' repository of corporate knowledge on the subject of shooting all kinds of portable weapons and using explosives as breaching tools. Preserving and improving the shooting skills of the Corps is no small job, and Colonel Nance's Marines work hard. Some of their missions include the following. They:

- Write and maintain all the training courses for marksmanship and small-arms training in the USMC.
- Run the Marine Marksmanship Training Program and supervise the Common Skills Qualification Data Base across the Corps.

- Train every Marine officer candidate from the officer's school (on the other side of the Quantico base) in marksmanship. Both men and women come to the Weapons Training Battalion in mixed companies to master weapons skills.
- Train and qualify Marine personnel in several Military Occupational Specialty (MOS) codes related to marksmanship and small arms.
- Participate in operational testing and evaluation of all new small arms, ammunition, and breaching and demolition systems fielded by the Corps.
- Assist in training and arming Marine Corps rifle and pistol competition shooting teams.
- Run an ammunition load and pack facility. Every year, this facility loads over 100,000 rounds of ammunition for the Marine shooting teams.
- Develop specialized weapons, demolition, and breaching tools for unique Marine applications.
- Manufacture, modify, issue, and maintain a variety of Marine Corps firearms, including the M1911 .45-caliber MEU (SOC) pistol.
- Conduct the "High Risk Personnel" anti-terrorism course for diplomats and other personnel assigned to overseas posts.
- Maintain weapons and ammunition storage for the FBI, CIA, DEA, and other agencies that utilize the Quantico range complex.

The Weapons Training Battalion has an impressive cadre of trained and experienced personnel. Like the elders of a tribe, the men and women of the battalion have a broad and deep base of practical knowledge, whether acquired in the classroom, at the workbench, or on the battlefield.

Consider the training of rifle marksmanship for new officers at Quantico. The course looks like this:

Marine instructors on the "High Risk" training course at Quantico, VA. This course is designed to teach diplomats and other high-risk personnel defensive field techniques.

JOHN D. GRESHAM

Phase I– Familiarization: Officer candidates are introduced to the M16A2 combat rifle, with particular emphasis on cleaning, maintenance, and aligning the sights (called "zeroing" the sights). The classroom basics of shooting are taught, as well as some practice in shooting house simulators, which use modified weapons firing compressed gas.

Phase II– Known Range Firing: This is actual range training and qualification at known ranges in a variety of postures, with fixed (stationary) targets. During this phase, the proper grips on the weapon, use of the sights, and compensation for crosswind, elevation, and weather are taught and certified.

Phase III– Unknown Range Firing ("Ironman") Training: This is the really hard part of the training, with firing against moving targets at unknown ranges. The officer candidate must rapidly assess the range and crossing rate of a pop-up target. Each candidate is given two magazines, with a total of thirty-five rounds, and twenty-nine targets to hit. A score of twenty-five out of thirty-five is considered good; sixteen is poor.

By teaching basic concepts, mixing in a dash of simulated skills training (Phase I), building upon these with actual dynamic training (Phase II), and then testing in a real-world context (Phase III), the Marines produce a rifle combatant who can take and hold a position, and make an enemy think twice about trying to take it back.

The Marines of the battalion pass along the hard-earned knowledge that goes with their trade to the new generations on the way up in the Corps. Some of the courses (designated by MOS numbers) that they run include:

A Marine with an instructor on the Quantico rifle range. This new range uses computer-controlled targets to teach combat shooting skills.

JOHN D. GRESHAM

- **MOS 8531—Rifle Range Coach/Instructor:** This course qualifies an enlisted Marine to safely run a firearms range and to teach the current doctrine and skills to recruits or officer candidates.

- **MOS 8532—Small Arms Weapons Instructor:** An advanced version of the 8531 course, it emphasizes additional skills and concepts over a wider range of weapons and environments, particularly follow-up and proficiency training. Each MEU (SOC) would likely have one or more of these instructors.

- **MOS 9925—Range Officer:** Assigned to supervise and manage the official training and shooting ranges of the Corps. Only thirty-two Marines can hold this designation at one time.

- **MOS 0306—Infantry Weapons Officer:** The officer version of the 8532 course. A MEU (SOC) or regiment would likely have one such officer assigned.

- **MOS 8541—Scout/Sniper:** This is the famous eight-week course that turns a Marine into the most deadly shooter in the U.S. arsenal, an 8541 Scout/Sniper. With a 40% dropout rate, it is one of the toughest courses in the U.S. military. Once a Marine completes this course, he is qualified to be assigned to a Scout/Sniper platoon in a MEU (SOC) or other unit.

- **MOS 8542—Advanced Scout/Sniper:** This five week follow-up to the 8541 course teaches more advanced leadership, tracking, navigation, shooting, and weapons skills.

- **MOS 2112—Gunsmith:** This is, perhaps, the most traditional course in the Weapons Training Battalion curriculum. It is designed to make a Marine into a completely qualified machinist and gunsmith. You would likely find a 2112 in every MEU (SOC), regiment, or major training base in the Corps. More than a course, it is a virtual apprenticeship. The first six months are spent teaching trainees to build their own tools and jigs. After that, they learn everything from welding broken parts to turning blanks into rifle barrels.

Marines are not limited to taking and qualifying on just one of the MOS courses listed above. During an enlisted Marine's career he may qualify for many MOS codes, not unlike the way a Boy Scout collects merit badges on the way to Eagle Scout rank. The Corps values weapons skills, and encourages Marines to master them, ensuring that individual marksmanship will continue to be a living part of the Marine ethos.

The Weapons Training Battalion is both an armory and a schoolhouse. Yet the battalion is not just sitting on its laurels. Innovations during the past year included moving targets on the qualification courses at Quantico, firing from inside nuclear/chemical/biological (NCB) suits, and a new night combat syllabus. Colonel Nance and his Marines are looking forward to the 21st century. In the next ten years, they expect to specify, test, field, and train a new combat rifle, a new combat shotgun, ammunition, and other systems.

Firearms

A Roman centurion evaluated his legionaires by their proficiency with sword and javelin. Genghis Khan judged his Mongol warriors by their skill at archery from horseback. Air Force pilots judge one another according to the quality of their "hands" on the stick. Among Navy aviators flying skills are judged by how well a pilot can "trap" during carrier landings. Every Marine is a rifleman, and the measure of a rifleman is marksmanship—the ability to cause a weapon to project a metal pellet across a volume of space so that it strikes a target with precision. I happen to like this way of sizing people up, because it is a skill that no one is born with. Shooting skills have to be learned. Unlike baseball or other sports which use the same innate reflexes as throwing rocks or swinging branches, there is no natural equivalent to shooting a firearm. Doing it well requires speed and precision—as well as stress and risk—greater than nature could ever evolve. Shooting skills are also gender-independent. The upper body strength required to shoot well is minimal. Despite the cultural traditions and legal barriers that restrict them from combat, women can learn to shoot just as well as men. Some of the top-scoring Russian snipers of World War II were women, and women compete equally with men in a number of Olympic shooting events.

Within the Marine Corps, the ability to put metal onto a target is taught as a common skill. Every officer and enlisted Marine who graduates from the OCS or the Basic School learns to fire and qualify on a variety of firearms. Without an acceptable level of marksmanship, they cannot graduate, or for that matter, stay in the Corps. This emphasis on shooting benefits the Corps in many ways, both obvious and hidden. Most evident is the reluctance of our enemies to face Marines in combat. Before the first shots of the 1991 Gulf War were even fired, many Iraqi soldiers expected to be annihilated by the Marines facing them, so they surrendered when the ground war began. More practically, Marines who can accurately deliver aimed fire will use less ammunition, reducing the load on hard-pressed combat logistical systems.

What follows is a look at Marine small arms today and tomorrow. We'll explore the heavier stuff later, but first we will learn about the weapons that define "Marine."

M16A2 Combat Rifle

The M16A2 rifle is the standard weapon in Marine combat units. Basic marksmanship skills are established and evaluated with this rifle; and every Marine in the Corps, from the newest Private to the Commandant, can fire the M16A2 with precision. The M16 had its origin in German assault rifles, like the MP44, developed during World War II. The MP44 combined the precision of a semi-automatic bolt-action rifle with the firepower of a fully automatic submachine gun or machine pistol. The assault rifle allowed troops to lay down a heavy volume of fire with good accuracy and still have the mobility of light infantry.

Following the war, many armies developed their own assault rifles (today called combat rifles), but with mixed results. The Russian AK-47, designed by Mikhail Kalashnikov, set the pattern for the modern combat rifle. Designed for cheap mass production, the AK-47 could fire semi-automatic (single-shot) or full-automatic

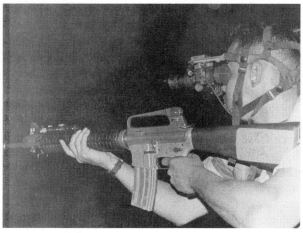

A 26th MEU (SOC) Marine test-fires an M16A2 combat rifle in the hangar bay of the USS *Wasp* (LHD-1). He is wearing the new AN/PVS-7B night-vision-goggle system, and the PAC-4C night-spotting system is attached to the top of the rifle's barrel.

(pull the trigger and get a stream of bullets). Because it was simple and rugged and easy to obtain, it became the symbol of Third World "popular liberation" movements during the Cold War. Western armies lagged behind Russia in combat rifle design during the 1950s, but began to catch up in the 1960s. Belgium's Fabrique Nationale (FN) and Germany's Heckler & Koch (H&K) produced 7.62mm combat rifles on the AK-47 model, but the United States still lagged. Because the U.S. Army had sunk a huge amount of money into a new semi-automatic rifle, the 7.62mm M14, the Army rejected an experimental FN-type weapon, the T-48. The M14 could be readily assembled by the same plants that built the Garand M-1 during World War II, while the T-48 would have required massive industrial retooling.

In the late 1960s, the North Atlantic Treaty Organization (NATO) standardized upon a smaller lightweight cartridge for future small arms, allowing more rounds to be carried by an infantrymen. Though this high-velocity 5.56mm/.223-inch round provided lethal hitting power (engineers use the gruesome term "wound ballistics"), there was strong resistance in the U.S. military to switching over to a new weapon firing it. What convinced the U.S. military to accept the new caliber was the Armalite AR-15, an automatic rifle designed by the brilliant Eugene Stoner in the late 1950s. Lighter and easier to fire accurately than the M14, the AR-15 was a revolutionary weapon. It caused such a stir that Colt Manufacturing Company of Hartford, Connecticut, arranged a license to produce it as the CAR-15. Military and government agencies including the Air Force Security Police, Secret Service, and FBI bought CAR-15s commercially. The CAR-15's popularity put pressure on the Army and Marines to adopt it as well. By 1966, Colt produced an Army version, the M16, which was quickly issued to Army and Marine Corps units. It was a mistake.

The first troops to receive the new weapons were already embroiled in the jungle war of Southeast Asia. But the M16 had a troubled start there, being both loved and despised by the troops. On the plus side, the M16 was 1.2 lb/.55 kg lighter than the M14, and soldiers could carry more ammunition. Troops also liked having "personal machine guns," and developed the habit of using full-automatic suppressive fire in the close confines of the Vietnamese jungles. This was gratifying—when it worked. But then there was the down side: Almost as soon as the troops switched to the new weapon, they found that the M16 was prone to jamming and fouling, particularly in the muddy lowlands of South Vietnam. This was not just a minor annoyance. In combat, a jammed weapon will get you killed. Rumors spread among the troops that this was a common occurrence. It was the start of one of the worst ordnance scandals in U.S. military history.

Congressional investigators later found that the reliability problems resulted both from the way the Army redesigned the CAR-15 into the M16, and from the way the troops had been trained to maintain it. Against Stoner's advice and Colt's specifications, the Army had substituted a lower-than-recommended grade of propellant in the 5.56mm cartridge used by the M16. This led to fouling and internal corrosion of the weapon. There were also reliability problems with the cartridge primers (the tiny explosive charge struck by the firing pin). The Army had accepted lower-quality standards in machining weapons parts, and it showed. Finally, due to shortages of cleaning kits and lubricants, at least some troops in the field were told, incorrectly, that the M16 was a "self cleaning" weapon. In fact, the M16 is a precision machine, requiring regular inspection and cleaning. As a result of Army mismanagement and inept fielding, the reputation of the M16 was seriously tainted. For a time, Marines in Southeast Asia were reissued their old M14s, until the Army could fix the M16.

Meanwhile, a clean-burning powder was substituted for the inferior propellant, and more reliable primers were produced. In addition, the Army had Colt modify the basic M16 to the M16A1 configuration with a chrome-plated chamber (to avoid fouling) and a stiffer buffer spring to decrease and stabilize the automatic firing rate. The extractor mechanism was also modified, to keep fired cartridges from jamming. And a program of intense training taught troops deployed in the field to properly clean, lubricate, and maintain their M16s. In consequence, the reliability of the M16 improved dramatically, along with the attitude of the soldiers and Marines using it. Eventually, the M16A1 became the standard combat rifle for the U.S. Army and Marine Corps and many allied countries during the late 1960s and 1970s. After its early problems were resolved, the M16A1 developed a solid reputation for performance and reliability. It may not have had the glamor of combat rifles like the H&K-91 or the Israeli Galil, but the M16A1 did the job during the lean years after Vietnam.

In the late 1970s the Army began a major update on the M16. Topping the wish list were a better forward grip, more accurate sights, and an automatic burst limiter to conserve ammunition. Introduced in 1983, the M16A2 is in use by the U.S. armed forces today. The features added to the A2 were:

- A heavier and stiffer barrel, for improved accuracy and reduced wear. In addition, the rifling on the barrel has been optimized for new NATO standard M855/SS 109-type 5.56mm (.223-in.) ammunition used by the M249 Squad Automatic Weapon (SAW). It can also fire earlier M193 5.56mm/.223-in. ammunition without modification to the weapon.
- A three-round burst limiter, which restricts "automatic" firing to only three shots per trigger pull.
- A muzzle compensator designed to reduce barrel rise and displacement during automatic firing.

- A plastic handgrip with a round contour which is tougher and easier to grip.
- A plastic buttstock that is lighter and tougher than that on the A1 model.
- An improved rear sight deck, with adjustments for range and windage.
- A modified upper receiver assembly which can be easily adapted to deflect ejected cartridges away from the face of left-handed shooters.
- Fittings for the new combat bayonet.

For $624.00 per unit, the M16A2 is quite a bargain for the American taxpayer, as results from Desert Storm proved.

The first time you pick up an M16A2, you are struck with the feeling that you are holding a serious piece of machinery. Weighing 8.8 lb/4 kg, the M16A2 feels good in your hands—well balanced and deadly. It is 39.6 in./100.7 cm long, and consists of four major assemblies:

- Lower receiver and buttstock.
- 5.56mm/.223-in. bolt carrier.
- Upper receiver and sight.
- Barrel and forward grip.

The four assemblies break down quickly for cleaning and maintenance. This is easy to learn, even in the dark with your eyes closed. Keeping the M16A2 clean is vitally important, because the components fit very tightly, and any grit or dirt can easily jam or foul the weapon. The Marine Corps is lavish in supplying cleaning kits, pads, and CLP lubricant/cleaner. You can always tell a seasoned combat Marine, because he will be the one in the group who cleans and lubricates his weapon, even before he eats or sleeps.

The 5.56mm/.223-in. ammunition feeds from a reusable spring-loaded magazine which is loaded from the bottom of the lower receiver/buttstock assembly. Today, thirty-round units are the standard, but twenty-round magazines are also used. The usual load for a Marine might vary from ten to sixteen of these, though the combat vest only has room for six ready thirty-round magazines. To reload empty magazines you take a supply of 5.56mm/.223-in. ammunition (called "ball rounds"), and methodically insert them one after another into the magazine, being careful not to scratch the cartridges or bend the springs. Snap the magazine into the bottom of the M16A2, and you are ready to go.

Firing the M16A2 is very simple. When you're ready to fire, you pull back the T-shaped cocking handle to load the first round into the chamber. Once this is done, you move the firing selector from the Safety position to either the Semi or Auto positions. At this point, you have a live weapon with a round in the chamber. Take aim on the target and pull the trigger. In the Semi setting, you fire one round for every pull of the trigger. If you are using the Auto setting, the M16A2 will fire a three-round burst every time you squeeze the trigger. The burst limiter was developed after Army researchers found that accuracy fell off rapidly when more than three shots were fired. Also, the tendency for troops to hold down the trigger in "rock and roll" bursts was wasteful of ammunition. Once a magazine is empty, you push the release button to eject the expended magazine, snap another into its place, and are ready to fire again.

Firing is one thing, but hitting the target is another. The Marine Corps has always prided itself on a tradition of marksmanship, and that tradition continues today with the M16A2. Two new features of the weapon improve its accuracy. The first is a ribbed tubular foregrip (replacing the "Mattel Toy" grips of earlier models). The second is a new sight deck and sight, which makes it easier to put rounds onto a target. You simply turn a dial to the required range setting, align the forward bead with the rear sights, and fire. If you have properly compensated for wind or temperature variations (which they teach you), the rounds should be hitting the targets with regularity. The Corps requires that Marines be able to hit targets with accuracy (50% or more of the rounds fired for hits) at 200, 300, and 500 yards/182.9, 274.3, and 457.25 meters, from a variety of firing position and postures. By comparison, the U.S. Army qualifies basic recruits at 100 yards only. Take it from me. Hitting targets at 100 yards/91.4 meters is easy. Although Marine recruits are taught to fire automatic, three-round bursts, single-shot firing is emphasized. Economy of ammunition is a key factor. When you fire in the burst mode, the muzzle tends to climb up, due to recoil, so only your first and second rounds will usually be on target. One way to avoid this is to steady the weapon against a tree or rock.

The M16A2 is probably the most accurate combat rifle in general service today. In fact, the Army competition shooting team recently moved from the M14 to a modified M16. One variant being procured today is the M-4 short-barreled carbine, with a folding stock. This weapon, identical to the M16A2 in performance, but smaller and lighter, is issued to vehicle and helicopter crews and support and service units, where space and weight are at a premium. The shorter barrel creates louder noise and a slightly different balance. The Marines are procuring over ten thousand of these handy little weapons from Colt. New kinds of ammunition being considered include a tungsten-cored armor piercing 5.56mm/.223-in. round from Sweden. In 1996, the M16 entered its third decade as the primary combat rifle of the U.S. armed forces. Continuous improvements and variants will keep this classic weapon lethal into the 21st century.

Another major development is night sights, to make the M16 more capable in darkness or bad weather. The Marines already have the AN/PVS-4 light-intensification sight for the M16, but they are rapidly developing and fielding newer systems. For instance, a new night fighting/spotting system, the PAC-4C, utilizes a special shoulder sling and red laser dot. But what Marines (especially reconnaissance and scouting units) really want is a thermal-imaging sight. The Marines have already adapted the thermal-imaging sight from the man-portable Stinger surface-to-air missile (SAM), though it is a bulky, expensive device which drains batteries rapidly. Both the Marine Corps and the Army are evaluating the NiteSight, a miniature thermal sight from Texas Instruments (TI). Small and lightweight, it draws much less power than earlier thermal sights. The key is a TI-designed imaging system. Unlike most thermal sights, it does not have to be chilled far below zero. Because it functions at 70° F/21° C, size and cost are greatly reduced. TI has plans to adapt NiteSight for motor vehicles and commercial aircraft.

MP-5N Submachine Gun

Okay, I'll admit it. When we visited the Weapons Battalion, my mouth really started to water when I saw *it* at the firing range, with as many loaded 9mm magazines as I wanted to blow off. *It* is the Heckler & Koch (H&K) Machine Pistol-5 Navy (MP-5N), the world's finest submachine gun. If you enjoy shooting, then the MP-5 is your dream weapon. Considering that a submachine gun is designed to spray an area with bullets, it's lightweight, deadly, and surprisingly accurate. The MP-5N derives from the German machine pistols feared and respected by opponents during the Second World War. These early machine pistols, called "burp guns" by Allied soldiers, were lightweight, simple, and deadly, particularly in street fighting or inside buildings. Since the end of World War II, many nations and companies have tried to produce their own machine pistols, with varying success. The U.S. M-9 "Grease Gun" was wildly inaccurate and only marginally reliable. The little Israeli Uzi is a worldwide best-seller, favored by VIP bodyguards, because it can easily be concealed under a jacket. But H&K has produced the world's finest submachine gun: the MP-5N.

The Marine Corps bought the MP-5N for what it calls close-quarters battle (CQB). This includes actions by units in MEU (SOC)s, Force Recon, and Base Security, as well as the various USMC Special Weapons and Tactics (SWAT) teams that they maintain. The need is simple: to get in close, then rapidly and accurately put a 9mm round through a target before the other guy can return the favor. The MP-5N has been adopted by law enforcement and special operations units around the world. Elite military hostage rescue units (like the SEALs, Delta Force, GSG-9, SAS) and police SWAT teams (FBI Hostage Rescue, German Police, New Scotland Yard Special Branch, etc.) make the MP-5 *their* close-combat weapon. The MP-5N is just that good. Let's fire one and see why.

When you pick up an MP-5N, you can feel German quality and engineering (you get the same kind of feeling when you drive a Mercedes Benz sedan). As you would expect from some of the best firearms engineers in the world, everything

A Quantico instructor holds an MP-5N submachine gun. This weapon is used by the Marines for close-quarters combat.

John D. Gresham

about the MP-5N has a function, yet there is a comfort and elegance to the whole thing. The basic weapon is 19.3 in./49 cm long with the stock folded (26 in./66 cm with it extended), and weighs about 7.4 lb/3.4 kg with a thirty-round magazine loaded. In addition, there are fittings for a flashlight (for use in night fighting) and a flash/noise suppressor (this adds about a foot to the overall length of the weapon).

The MP-5N uses the same NATO Standard 9mm ammunition as the M9 Beretta handgun and many other automatic pistols. This ammunition has excellent stopping power at short ranges (less than two hundred yards/meters), and is readily available anywhere in the world. You load MP-5N in much the same way as the M16A2. You insert a thirty-round magazine into the lower receiver until you feel (and hear) a satisfying "click." You then pull back the cocking handle, switch the firing selector from Safety to Single Shot or Automatic, aim, and fire away. With typical German efficiency, H&K stamps symbols on the side of the weapon for each mode, which makes it almost "idiot proof"!

Single-shot firing is even easier than with the M16A2, and there is almost no barrel displacement when you fire. Out to about two hundred yards/meters, you just put the sights on the target, and then you hit it. Automatic fire is even better. The barrel rise, so common on automatic weapons, is almost non-existent on the MP-5N, and keeping the weapon on the target is easy. In fact, other than a heavy machine gun, nothing I've ever fired compares to the experience of automatic shooting on the MP-5N. While I was reloading the weapon with a fresh magazine (just press the release button and push in a new one), Colonel Nance came up behind me and said, "Go ahead, I'd do it too!" This said, I let loose with a thirty-round burst, emptying the magazine in less than 2.3 seconds. Astonishingly, about half of the rounds actually hit the target, about one hundred yards/meters downrange. While I was shooting, I could hear the sound of the bolt and slide cycling, but almost nothing from the actual firing of the 9mm rounds. It was a bizarre sensation until I realized that this was a result of the superb flash/noise suppressor screwed onto the muzzle of the MP-5N. It was amazing to pump out almost eight hundred rounds per minute and scarcely hear it!

There are no current plans to replace the MP-5N. It is an almost perfect weapon for the CQB role, and will likely stay that way for years to come. If you want to know perfection in firearms, find a way to get some "trigger time" on an MP-5N. You will not be disappointed.

M40A1 Sniper Rifle

For decades, the Marine Corps has been famous for its sniper program. Sniping—to kill or disable enemy leaders—is an integral part of infantry combat. Since the first Marines climbed into the rigging of sailing ships to sweep the decks of enemy frigates with musket fire, the Corps has valued accurate shooting. But Marine historians tell us that systematic emphasis on marksmanship only began in the early 1900s under the influence of Commandant Heywood and under the direction of Captain William Harllee.

The core of this capability today is the M40A1 sniper rifle. First fielded in the 1970s, this bolt-action heavy-barreled rifle fires a 7.62mm Match Grade round out to

A Marine sniper shows off an M40A1 sniper rifle. This weapon is used for long-range shooting by specially trained Marine personnel.

1,000 yards/914 meters, with enough accuracy to hit a man-sized target in the head. The M40A1 is built from stock parts by the armorers of the Weapons Training Battalion at Quantico. Based on the Remington Model 700 rifle, it is "accurized" to an almost unbelievable degree by adding:

- A commercial competition-grade heavy barrel.
- McMillan fiberglass stock and buttpad. Each stock is blasted in a glass bead machine at Quantico to improve accuracy.
- A modified floorplate and trigger guard, as well as a lightened trigger.
- A 10-power Unertl sniperscope.
- A five-round magazine.

With these features, the M40A1 can fire with an accuracy of less than one minute of arc. That's less than 1/60 of a degree. At 1,000 yards/914 meters, this means an error of less than 10 in./25.4 cm! With a little work, Colonel Nance's gunsmiths and armorers at Quantico usually get the error down to a third of that. Much of the technology that makes the M40A1 so accurate derives from the efforts of the USMC competition rifle team, which uses similar rifles and heavily modified M14s in contests with the shooting teams of the Army, Navy, Coast Guard, Secret Service, DEA, and FBI.

Sniping is an art of extremes, and just shooting well will not get you through the Scout/Sniper course. Land navigation, spotting, and concealment are just as important, but unfortunately beyond the scope of this book. I did get to fire the M40A1 to gain an appreciation of this arcane shooting science. As a rule, sniping is done from the prone position, with pairs of snipers working together. One sniper is "on the scope" as a shooter, with the other using an M49 spotter scope to pick out targets and monitor the tactical situation. About every half hour, the two trade off, to avoid fatigue on the shooter.

The first trick to hitting a target with the M40A1 at long distance is holding the weapon properly. To do this, you jam the rifle butt (with its special buttpad) hard into your right armpit. You then wrap the sling tightly around your other arm, using your left hand to cradle the rifle along the forward part of the stock. When the sling is about to cut off the circulation to your left hand and the buttstock is hurting your armpit, you've got the M40A1 about rigid enough to start sighting. You then look through the 10-power Unertl sniperscope, and begin to work the crosshairs. At 600 yards/548.6 meters, a target with an 18-in./45.7-cm kill zone is just a dark dot which appears to dance around the scope. You quickly realize that this is caused by your own breathing and heartbeat; experienced snipers learn to regulate these when shooting. Once you have the dot of the target reasonably lined up, you gently pull the trigger, and then the world explodes in your face. The kick of the M40A1 is like a shotgun, and the sound is like a bullwhip cracking in your head. In less than a second, the round flies out to the target, and then the adjustment process begins. You look at the grass and dust clouds to evaluate the wind and heat shimmer to help adjust the scope to compensate for crosswinds and heat updrafts that "loft" the round. This done, you pull back the bolt, eject the spent casing, and push the bolt forward to drive the new round home. The fascinating thing, though, was that with only a few rounds of practice and some skilled help from one of Colonel Nance's sniper instructors, I was putting rounds regularly through the target some 600 yards distant! This is over a quarter mile/half kilometer away, and the effect of hitting an object that far away with a hand-held weapon has to be felt to be believed.

Now before you get too impressed with my performance, consider that Marine Scout/Snipers are required to do the same thing at almost twice the distance, with only one shot (that is all a sniper will usually get!) and no chance to make adjustments. All this on a mission that may last days, in any weather, against an enemy trying to kill you like a pesky varmint. It is a bizarre way to make a living, and the men who do it are strange birds. But to an enemy, the M40A1 in the hands of a skilled Marine Scout/Sniper is a hellish weapon, more feared than even a bomber loaded with napalm! It is scary to know that *you* might be hunted by another man; and this makes it tough to do *your* job. The overwhelming psychological impact of the sniper helps to explain why the Corps invests so much in maintaining this capability.

Barrett M82A1A .50-Caliber Special-Purpose Sniper Rifle

When you see it the first time, it just looks evil, like a preying mantis ready to strike an aphid. It could be the star of its own action/adventure movie with Stallone and Schwarzenegger as supporting players. It is the Barrett M82A1A .50-caliber special-purpose sniper rifle, the most unusual weapon in the Marine small-arms inventory. The M82A1A is designed to augment the M40A1 sniper rifle when longer ranges and greater hitting power are required. The Barrett fires the same ammunition as the M-2 Browning .50-caliber machine gun. If you have ever fired the M-2, you know that it kicks like a mule, and requires a very firm mount or a heavy tripod dug into the ground. The whole M-2 machine-gun/tripod combination weighs several hundred pounds, hardly convenient as a sniper weapon. Nevertheless, Marine snipers like the legendary Staff Sergeant Carlos Hathcock (ninety-three con-

The amazing Barrett M82A1A sniper rifle. This weapon fires the same round as the M2 .50-caliber machine gun, and is used for extreme range shooting by the Marines.

JOHN D. GRESHAM

firmed kills in Vietnam) mounted special sniper scopes on standard M-2s and scored hits at ranges over a mile/1.6 km.

The M82A1A traces its roots to the rebel war in Afghanistan in the 1980s. The United States, mostly through the CIA, aided the Mujhadeen rebels fighting Soviet forces in Afghanistan. A well-publicized part of this aid included *Stinger* shoulder-launched SAMs to knock down attack helicopters and strike aircraft. The Mujhadeen also asked for a man-portable long-range armor-penetrating sniper weapon. (Sniping is a traditional art of the Afghan mountain tribesmen.) The answer was a weapon designed by Ronnie Barrett from Murfreesboro, Tennessee. Barrett, a builder of homemade weapons for many years, designed a system of springs to buffer the recoil of a .50-caliber machine gun. By spreading the recoil energy over a longer duration, the springs reduce the peak load on the weapon and the gunner. Barrett built a weapon that could be broken down and carried as several man-sized loads. The CIA bought a number of these heavy sniper rifles for the Afghan Mujhadeen, who used them to terrorize Soviet troops. The Barrett performed so well in Afghanistan that the Marine Corps evaluated and eventually adopted it as the M82A1A sniper rifle. Today, the M82A1A (produced by Barrett Firearms Manufacturing) is deployed by Marine Force Reconnaissance units in three-man fire teams. Each team member carries one part of the weapon (upper receiver, lower receiver, or scope and ammunition). The team alternates the jobs of shooter and spotters.

The semi-automatic M82A1A is 57 in./128.25 cm long, and weighs 32.5 lb/14.8 kg unloaded. It fires a .50-caliber bullet (Raufoss Grade A, DoDIC A606) against targets defined as "equipment-sized" (like a jeep or tent), at ranges of up to 1,800 meters/1,968.5 yards. A sniper team with a Barrett can reach out and hit useful targets at ranges of over a mile/1.6 km. During Operation Desert Storm, M82A1A teams were knocking out things like artillery-spotting radars and communications equipment, raising hell with Iraqi command and control. The M82A1A is basically a .50-caliber machine gun spring-mounted inside an aluminum housing. This gun-inside-a-box design allows a sniper to safely and comfortably fire the weapon with accuracy. A folding bipod and a special buttpad help to absorb the recoil. In fact, the peak recoil load is actually lower than the M40 because of the buffering system, the bipod, and a high-efficiency muzzle brake (which gives

The Beretta M9/92F 9mm pistol. This is the standard-issue personal side arm of the Marines.

JOHN D. GRESHAM

the Barrett its insect-like appearance). Mounted on top of the M82A1A is a 10-power Unertl sniperscope matched to the Raufoss .50-caliber ammunition. The M82A1A is chambered to accept any NATO-standard .50-caliber/12.7mm ammunition, though currently, only the Raufoss round is issued. The Barrett has a ten-round box magazine, which feeds through the lower receiver housing. Like the M40, it only fires single shots, after which the team rapidly breaks down the weapon, slides the various parts into specially designed backpacks, and exits the engagement area.

Firing the Barrett is almost as easy as the MP-5N. You load a magazine into the bottom of the weapon, pull back the cocking handle, sight the weapon (adjusting for windage and other factors), and pull the trigger. The weapon fires with a distinctive "crack," and then pushes back gently into your shoulder. It is surprisingly comfortable. Like the M40A1, the key to accurate firing is steadiness and patience. The Barrett M82A1A is a unique, specialized weapon. It enables Marine snipers to disrupt enemy units and make life miserable for them in their own rear areas. This degrades enemy morale and paralyzes their leadership. Unlike the dramatic gun camera video of laser-guided bombs hitting targets, you won't see this weapon on CNN, but the effect can be just as devastating.

Beretta M9/Model 92F Combat Pistol

No single piece of combat equipment is more personal to a combat soldier than a handgun. Not all personnel require one, but to those who do, the Marine Corps issues a combat side arm, the Beretta M9/Model 92F Combat Pistol. Selected to replace the classic M1911A1 Colt .45-caliber pistol which served for more than half a century, the Beretta has been a lightning rod for critics. These include advocates of the .45 and Congressional supporters of competing handgun manufacturers that lost out to the M9/92F. Nevertheless, the M9/92F is a fine handgun with excellent design features. Let's take a closer look.

For nearly five hundred years, the Beretta family has been making firearms for soldiers and sportsmen (one customer was Napoleon's Grande Armee). Today,

Beretta manufactures shotguns and automatic pistols that are among the best available. In 1985, the Italian firm was selected to supply the U.S. military with a common, non-developmental ("off-the-shelf") handgun compatible with NATO Standard 9mm ammunition. With a multi-year production contract for over 500,000 pistols, the losers in the competition came out with fangs bared, taking shots at any perceived problem.

One complaint was that the U.S. military was buying foreign weapons, depriving Americans of jobs. In fact, the contract required production in a U.S.-based factory (Beretta operates a plant in Maryland). But the design did have its share of real problems, for, like any design, this one had its share of teething pains. During endurance testing, for instance, some slides on the test weapons began to crack. This resulted from an extremely hard mounting fixture which put too much strain on the weapons (strengthening the slides was relatively easy). Now, with over a decade of production and operational service behind it, the M9/92F is in its prime, filling most of the combat handgun requirements for the U.S. military. Let me show you how to fire one.

The M9/92F is a large-frame semi-automatic 9mm pistol with a fifteen-round magazine. It is an ambidextrous weapon, equally handy for right- or left-handed shooters. The M9/92F is lighter than the old Colt M1911 .45-caliber that it replaced, weighing 2.55 lb/1.16 kg with a loaded magazine. It fits nicely in the hand; my rather large palm and digits make it easy to grip. The M9/92F has exceptional safety features to minimize the risk of accidental firing. These include:

- An open slide with an ambidextrous magazine-release button to speed up and simplify reloading.
- A double-action trigger. When you start to pull the trigger, you feel a resistance; the weapon only fires when your finger provides additional pressure.
- A visible firing pin block to show the user that a round is chambered.

You have to want to shoot this gun to make it fire. A fumble or mistake is very unlikely to result in accidental discharge. This is critical when you are in a CQB situation.

To show us how to properly handle the M9/92F and several other firearms, Colonel Nance graciously loaned me the expertise of Sergeant Kenneth Becket, an instructor from the High Risk Personnel training course at Quantico. Stepping up to the firing line, he handed me an empty M9/92F with the slide open and the chamber empty. The first thing you are expected to do is look up into the chamber to make sure it is empty. This done, you slide a magazine up into the grip until it clicks home. Now you firmly grasp the slide and cock it to the rear. This chambers the first round, and you are ready to fire.

The key to hitting targets with a semi-automatic pistol like the M9/92F is correctly holding, or gripping, the weapon. The subject of proper pistol grip provokes endless debate among shooters, and there is probably no best way to hold a pistol, but the grip currently favored and taught by the Corps works well. Sergeant Becket had me firmly grip the pistol in my right hand, and then grip over the holding hand with the fingers of the left hand, making sure that the palm of the grip hand is on the surface of the pistol grip. The idea is to create a rigid mount for the weapon, as well as to maximize the surface area of your hands in contact with the weapon. Once you have the proper grip, you thumb the safely to the Off position, and are ready to shoot.

As with shooting combat and sniper rifles, the Marine Corps teaches pistol shooters to use the sights to get aimed fire. This is not just to save ammunition. In a pistol shootout, the first shooter to score a hit almost always wins. The USMC theory of pistol shooting requires that every shot be aimed from the sights, even if it takes a bit more time. Even with trained shooters like policemen, pistol shooting is, in a word, hideous. Forget what you see on television and in the movies. Accurate pistol fire from beyond about five yards/meters is almost unheard of. For example, in the last twenty years there are painfully few recorded instances of New York City policemen hitting anything beyond twenty-five feet/eight meters with a pistol. For this reason, the Marines teach pistol shooters to carefully get the proper grip, calmly line up the target through the sights, and then squeeze off one round. Repeat the procedure until the target drops. This procedure will almost guarantee victory and survival in a showdown at close quarters.

With the target in the sights, you gently squeeze the trigger until the weapon fires. This can be a little disconcerting to new users of the M9/92F, because of the double action for safety on the first shot (you have to pull the trigger over a cam to fire). There is a feeling of pulling the trigger forever before the first round fires. But when the M9/92F fires, it is smooth and clean, with the round hitting a white "witness plate" target about 6 in./15.25 cm square placed about 16 feet/5 meters away. Once the M9/92F fires its first round, the trigger becomes single-action (short pull) and the shooting much easier. After each shot, Sergeant Becket coached me to line up and check my grip. And soon I was consistently hitting with round after round. After the fifteenth shot, he had me thumb the magazine release, and rapidly slide in a fresh one. At this point, the weapon is still cocked, so all you have to do is check that the safety is Off and fire the first round of the magazine as before. By the time we were done, the white paint of the target witness plate was scarred and worn, testament to the sergeant's coaching skill!

Though there are equally good weapons from manufacturers like Glock, FN, and Colt, I like the Beretta. While I personally favor a single-action weapon like the 9mm Browning Hi-Power myself, the safety and reliability of the M9/92F make it an excellent weapon for military use. With minimal training, a shooter can expect to hit a target within killing range. And the 9mm NATO standard ammunition makes it fit well into the logistical chain of almost any nation.

Colt .45-Caliber M1911 MEU (SOC) Pistol

The USMC has almost a reverence for the old M1911 Colt .45-caliber semi-automatic pistol. Its stopping power is legendary: It was designed to take down charging machete-wielding Filipino insurgents. The original M1911 was replaced by the M1911A1 in 1925, and nearly all existing weapons in Federal armories were upgraded to the new configuration. After that, the Colt became such a fixture that when the Department of Defense decided to issue the M9/92F in 1985, many Colt .45 users considered it just short of treason. Despite the .45's reputation for kicking

The specially built MEU (SOC) pistol, constructed from an M1911 Colt .45. These unique handguns are issued to Marines for close-quarters battle.

JOHN D. GRESHAM

like a mule and having the accuracy of a blunderbuss, it was loved by generations of American fighting men, particularly Marines. Thus, in 1986 there was general delight when the USMC decided to bring back a special version of the Colt, the MEU (SOC) pistol. The MEU (SOC) is a rebuilt and modified M1911A1 Colt .45, issued as a backup weapon to reconnaissance units equipped with the MP-5N. It was selected over other pistols for its inherent reliability and the greater lethality of the .45-caliber projectile, which weighs about twice as much as a 9mm bullet. Despite the limited inventory of five hundred units maintained by the Corps, the almost spiritual attachment of Marines to the M1911A1 guarantees support for this weapon.

The MEU (SOC) pistols are manufactured from existing Colt M1911A1 .45-caliber pistols (there are thousands in storage). They are rebuilt at Quantico by the armorers of Colonel Nance's Weapons Training Battalion. After each M1911A1 frame is stripped and checked for structural soundness, the following modifications are made:

- A commercial competition-grade ambidextrous safety.
- A precision barrel and trigger assembly.
- Extra wide, rubber-coated safety grips.
- Rounded hammer spur.
- High-profile combat sights.
- Stainless-steel seven-round competition-grade magazines with a rounded plastic follower and an extended floor plate.

These improvements make the MEU (SOC) pistol more "user friendly." They also make the MEU (SOC) pistol one of the most comfortable and accurate handguns I have ever fired.

I was given the chance to fire one of the MEU (SOC) pistols at the same distance and target as the Beretta. I've fired my share of .45-caliber pistols before, and the M1911A1 has always been a beast. Even with my size and weight, the M1911A1 always left me bruised and battered, with little damage to the targets. The MEU (SOC) pistol is different. Using the same grip and sighting technique as I used on the Beretta, I got a string of hits on my first magazine. A single-action trigger makes it smoother to fire than the Beretta, and the reduced recoil is easy on even

An M249 Squad Automatic Weapon (SAW). The SAW is a 5.56mm fully automatic machine gun, and one is assigned to each four-man Marine fire team.

JOHN D. GRESHAM

small-handed shooters. Seeing the damage .45-caliber rounds were doing to the target witness plates, I could only imagine what they would do to a human target. This weapon is more than accurate and deadly; it is fun to fire, much like the MP-5N. I could have spent the whole day firing it under Sergeant Becket's coaching. Eventually I had to regretfully give it back. The MEU (SOC) is the finest large frame pistol you will *never* be able to buy. And I want one!

M249 Squad Automatic Weapon (SAW)

When the first machine guns appeared in the late 1800s, they revolutionized warfare. Until the introduction of the tank, machine guns ruled the battlefield. For many years infantry leaders longed for a machine gun that a man could carry, to set up a base of fire to support squad-level operations. As early as 1916 Marines used the French M1909 Benet-Mercie, license-built by Colt, in the Dominican Campaign; and by 1917 they had some British Lewis guns. During World War I, the U.S. Army resisted the idea of a light machine gun, fearing that it would lead to excessive ammunition waste. Instead it adopted the famous M1918 Browning Automatic Rifle (BAR), which entered service in the last two months of the war. This 22-lb/10 kg weapon fired standard .30-06 ammunition from a twenty-round clip. Even though the twenty-round clip limited the rate of sustained fire to about sixty rounds per minute—half the firepower of a typical belt-fed bipod-mounted light machine gun—and the effective range was also shorter, the BAR was robust and reliable. The Marines liked it so much they made it the centerpiece of the fire team. Unfortunately, the BAR stayed in service too long, a problem most often noted by those who had to lug the damned thing around the battlefield.

In 1957 the BAR was replaced by the M60, a close copy of the World War II German MG42 light machine gun. The Army "improved" that design, which led to frequent stoppages and jams, poor durability, and barrels prone to overheating. It fired 7.62mm ammunition instead of the 5.56mm/.223-in. round used by the M16. Thus, a platoon with both weapons had to manage two separate ammunition supplies, complicating logistics. Also, the M60 was still very heavy (at 18.75 lb/8.5 kg) to be lugging around with 10 to 20 lb/4.5 to 9 kg of ammunition. Thus, M60 gunners dreamed of a lighter weapon which would be easier to carry and operate, use the same 5.56mm/.223-in. ammunition as the M16, and carry more rounds for the weight.

By the late 1970s, the Army and Marines agreed to procure a non-developmental (i.e., "off-the-shelf") replacement for the M60 in rifle squads. After many models were evaluated, the winner was a weapon from Fabrique Nationale (FN) of Belgium. This became the M249 Squad Automatic Weapon (SAW), first issued to Army and Marine units in the mid-1980s. Since then a "Product Improvement Program" kit has modified the barrel, grips, stock, buffer, and sights. The M249 is an attractive little weapon, not much larger than an M16A2. With a folding bipod and tool kit, it weighs only 15.2 lb/6.9 kg and is some 40.9 in./103.8 cm long. A sling allows the gunner to fire it from over the shoulder when on the move. It can accept either the thirty-round 5.56mm/.223-in. magazines of the M16A2, or a two-hundred-round belt (which is preferred). The belted ammunition comes in a plastic box, which weighs only 6.9 lb/3.1 kg. This is a vast improvement over the M60, in terms of the weight a fire team has to lug around the battlefield. Marines issue one M249 to each four-man fire team. The other three team members have M16A2s, and one of these comes with a M203 40mm grenade launcher, so that each fire team has a machine gun, three combat rifles, and a grenade launcher. Quite a lot of firepower for just four men.

For my demonstration, Colonel Nance's instructors had flipped down the folding bipod legs at the front of the M249 so that I could fire from a prone position. This is the most comfortable and accurate way to fire the M249, because it tends to spread the recoil over three points (the two bipod legs and your shoulder), limiting the movement of the weapon. As I mentioned earlier, you can load the weapon either from the bottom with a 30-round M16 magazine or a 200-round belt which feeds across the top of the SAW. To load, you attach a plastic belt box to the left side of the SAW. This done, you raise the receiver cover and pull the belt over and across the receiver feed tray, align the first round over the feed tray, and close the cover. Then you pull back the cocking handle to load the first round, release the safety, and pull the trigger.

The SAW fires at a satisfying 725 rounds a minute. While you are putting a lot of rounds onto the target, the weapon is not cycling so quickly that you cannot control it. You can fire single shots or short bursts easily, or empty a whole box of two hundred rounds in just over 16.5 seconds. Accuracy of the M249 is quite good. The sights are more complex than those on the M16A2 (with adjustment knobs for elevation and windage), but when properly adjusted, they help you to consistently put rounds on target out to an effective range of about 1,000 meters/3,281 feet. I was able to put a stream of bullets right into the chest of a man-sized target at 200 yards/183 meters without difficulty. When you fire the M249, there is a solid feel with very little kick or travel. Firing the SAW is so nice that before long, you begin to feel invulnerable and omnipotent. As an SAW gunner, you have to deny yourself this feeling, because you are no better protected than any other infantryman, just better armed. If the SAW has a vice, it is the one common to all machine guns, a tendency to jam during long bursts. This is one reason why short bursts are encouraged (the obvious desire to conserve ammunition is another reason). The SAW is easily cleared in the event of a jam, simply by lifting the cover plate and pulling the jammed round clear. The M249 SAW is an excellent light machine gun. Its standard M988 5.56mm/.223-in. ammunition means that every Marine in a four-man fire team now fires the same ammunition, simplifying logistics and maximizing the utility of a team's load. I like it!

M240G Light Machine Gun

When the Army and Marines replaced the M60 at the squad/fire team level, they also had to replace it in other medium-machine-gun roles. In its final version, the M60E3, it had been used as a pintle-mounted weapon on vehicles and aircraft (M-1 tanks, trucks, helicopters, etc.), as well as in heavy weapons platoons. In these roles, the 5.56mm/.223-in. round really does not have the hitting power and range required, so the M60E3 with its 7.62mm round was retained well past its prime.

The Army and Marine Corps finally found the ultimate replacement for the M60E3 in the M240G. Gunners like it for its reliability and reduced maintenance requirements. The M240G is basically a scaled-up M249 SAW, firing 7.62mm ammunition. Designed and built by FN of Belgium, the M240G is a lightened version of the original M240. The 240G is functionally identical to the M249 SAW, except for the following features:

- It is longer (47.5 in./120.6 cm) and heavier (24.2 lb/11 kg) than the M249, or for that matter, than the M60E3. This is the main "down" side to the M240G.
- The M240G fires the NATO-Standard 7.62mm ammunition instead of the 5.56mm/.223-in. rounds. This makes for better hitting power and greater effective range (out to 1.1 miles/1.8 km).
- It has three selectable rates of fire, between 650 and 950 round per minute.

Aside from these differences, the M240G is almost identical to the SAW. Now every medium machine gun in the U.S. military inventory will come from the same basic family. Like its little brother, the M249 SAW, the M240G is popular with the troops, though the Marine recruiters joke that they are looking for bigger recruits to lug it around the battlefield!

Combat Shotguns

In really close combat, there is nothing better than a shotgun (except maybe a flamethrower!) for hitting power. Marines use three different though similar commercial shotguns for CQB missions. The Remington 870, Winchester 1200, and Mossberg 590 have all been adapted for combat by adding a bayonet attachment, sling, and a phenolic buttplate to soften the recoil. Shotguns are not carried as a primary weapon (like the M16A2 or MP-5N), but as special secondary weapons for use at close quarters. In addition to the obvious anti-personnel role, they can also be used to blow open a door (by blasting the lock or demolishing the hinges); and they make a fine "non-lethal" riot-control device. A new family of shotgun shells from MK Ballistic Systems, called Flexible Baton-12, fires projectiles that look like small rubber beanbags. These deliver enough force to knock down a human being, without the blunt trauma often associated with so-called "rubber" bullets.

Colonel Nance and his staff are now preparing to evaluate a more capable combat shotgun. Though the actual weapon has yet to be selected, it will certainly have a large magazine (thirty rounds or more), and provide a semi-/fully automatic firing capability. When the Marine Corps puts its stamp of approval on this new shotgun, it is likely to be procured by law enforcement agencies all over the world.

Foreign Weapons

Quietly and discreetly, Colonel Nance's Weapons Training Battalion introduces new Marines to some of the weapons that they may face or capture on future battlefields. The first reason is obvious: Marines in the field should recognize the sound of an enemy weapon being fired, and know to get down out of the line of fire. Many weapons, like the ubiquitous AK-47 combat rifle, have a highly distinctive sound signature, and knowing this can help you locate its firing position. In addition, knowing an enemy weapon allows you to identify its weaknesses, possibly giving you an edge in combat. Finally, Marines have to be ready to fight with what they can get if they are lost, cut off, or even abandoned (remember Wake Island and Guadalcanal). To this end, new Marines are indoctrinated in the characteristics of weapons used by other nations. Many of the foreign weapons that Marines learn about at Quantico are of crude but effective design like the AK-47. Thus, knowing how to use them will continue to be an important battle skill for Marines.

Grenades, Mines, Explosives, and Breaching Tools

While firearms are the primary tools of an infantryman, there are times when a gun will not do. Ordnance engineers like to say that there is no condition in the human experience that cannot be solved by an appropriately shaped, sized, timed, and detonated charge of high explosive. Explosive weapons have had an important place in close combat since the invention of the grenade several hundred years ago. Today's Marines can carry a variety of grenades, mines, and other devices in their rucksacks, and we're going to take a look at them here.

Hand Grenades

Shortly after gunpowder reached the West in the Middle Ages, some creative warrior took a handful of the new explosive, packed it into a container, lit a fuse, and threw it at his enemy. This was a good idea when it worked. The problem was that it didn't work all that often. Early grenades were frequently more dangerous to their users than their intended victims. Because of the unreliability of the explosive and fuses, you could never really be sure they were going to go off, or how big an explosion ("lethal blast radius") you would get.

Modern grenades commonly used by Marines include:

- **M67 Fragmentation Grenade**—Weighing 14 oz/.4 kg, it carries an explosive charge of 6.5 oz/184.6g of Composition B. When you pull the pin and release the safety handle (called a "spoon"), there is roughly a four-to-five-second delay prior to detonation. When it goes off, it spews fragments out to a lethal distance of around fifteen meters/forty-nine feet. The user must either be under cover when it explodes, or throw it far enough to be safe from the blast.

- **M7A3 CS Riot Hand Grenade**—This is a "non-lethal" device, designed to deter or incapacitate a rioting crowd. Weighing only 15.5 oz/.44 kg, it is loaded with a mixture of pelletized CS (tear gas) and a burning agent, which helps atomize and disperse the gas. When inhaled, ingested, or exposed to mucous membranes (eyes, mouth, etc.), it incapacitates the victim within fifteen to thirty seconds, with an effect duration of less than ten minutes following exposure to fresh air and, if necessary, water to flush the eyes and mouth. Troops using the M7A3 normally wear a gas mask to avoid exposure themselves.

- **M18 Colored Smoke Hand Grenade**—The M18 is not designed to kill or wound anyone. It simply marks areas for helicopter landing zones and no-shoot areas during strikes by aircraft and helicopters. Weighing some 19 oz/.54 kg, these grenades come in four varieties: red, green, yellow, or violet smoke. Each M18 will generate smoke for approximately fifty to ninety seconds, and the volume of smoke is sufficient for screening squad movements, if the wind is not too strong.

There is little research on improved hand grenades, since these do exactly what is required. The Marines maintain a stock of more than 1,138,000 grenades of all types, showing how important they are to the firepower of the Corps.

M203 40mm Grenade Launcher

One of the problems with hand grenades is that a human being (even Dan Marino) can only throw one so far. In World War I grenade-throwing attachments were developed for bolt-action rifles to provide more standoff range for the infantry. These were not direct-fire weapons, and they were not terribly accurate; the grenades had to be lofted, like a mortar round. During the Vietnam War the U.S. Army introduced the M79 grenade launcher (nicknamed the "thump gun"). This stubby weapon, resembling an oversized, sawed-off shotgun, fired a 40mm shell, called a grenade, to a range of about 150 meters/492 feet. At this range, a good thump gunner could put a round through a door or window. Each 40mm projectile has about the same lethality as a hand grenade, but with considerably more accuracy and range. There are several different types (smoke, fragmentation, gas, flechette, etc.), with various effects.

The M79 was used extensively in Vietnam, and is still favored by law enforcement agencies for riot control and SWAT teams, but it is an extra weapon the soldier has to lug around that is not useful for anything else. Thus, the M203 grenade launcher was created. The M203 is a "clip-on" device, which attaches to the bottom

An M203 40mm grenade launcher attached to an M4 5.56mm carbine. The M4 is the shortened version of the M16A2 combat rifle.

JOHN D. GRESHAM

of the forward receiver of an M16A2 combat rifle. A Marine with the M203 still has full use of his M16A2, but he can also launch 40mm grenades. You load it by pushing the barrel of the M203 forward, and then sliding a round into the breech. By pulling the barrel tube backwards, you lock the weapon shut and are ready to fire. All you have to do is release the safety, aim the weapon, and pull the M203's trigger, located just forward of the magazine loading chute of the M16A2. Surprisingly, the M203 is quite accurate, and gunners can put rounds through a door or window at quite a good range. Each fire team includes one M203 gunner. It is a deadly little weapon, well liked by Marines.

Mines

Mines are weapons that wait, sometimes for decades. Combat soldiers both love and hate land mines. They love to sit behind a minefield and watch the enemy blunder into it. But they hate the feeling of helplessness and terror that comes from being caught in a minefield, seeing their friends suddenly and horribly maimed. And then once the war is over, the winners get to pick all the damned things up and disarm them. Unfortunately, this doesn't always happen, and large areas of luckless countries like Cambodia, Angola, and Afghanistan have been rendered uninhabitable by millions of land mines. Although some European countries that profited hugely from the sale of land mines are beginning to ban their export for humanitarian reasons, mines are so effective and cheap that there is little hope of a workable international law prohibiting their manufacture and use.

The Marines deploy a variety of different mines, including the following, which are man-portable:

- **M16A1 "Bouncing Betty"**—This is a "bounding" anti-personnel mine. When someone steps on one of the firing prongs (which are left exposed when the M16A1 is buried), a small propelling charge fires it about 6f/1.8m into the air, at which point it detonates. The M16A1 contains a 1-lb/.45-kg explosive charge, which produces a lethal fragmentation range of around 88f/27m.

- **M18A1 "Claymore"**—This is a flat curved plate filled with steel balls embedded in plastic explosive. It has folding metal prongs that stick into the ground and a chilling label embossed on the housing: "This side toward enemy." It functions like a huge shotgun shell. Once the M18A1 is emplaced, it can either be fired by a trip wire or command-detonated from a distance. When detonated, the 1.5-lb/.68-kg C-4 charge fires a 60° fan-shaped pattern of fragments, each the size of a ball bearing. The fragments are lethal out to a range of around 328f/100m. The Claymore is primarily used for ambushes, but it can also function as a "silent sentry," covering ground that cannot be brought under direct observation and fire.

Anti-personnel mines are effective against opposing infantry, and Marines can carry them in sufficient numbers to make them a real threat. Though there are larger mines like the M15 and M18 used against tanks, they are too heavy to be man-portable.

Explosives and Breaching Tools

In addition to grenades and mines, Marines frequently carry supplies of plastic explosives and detonation gear for demolitions. They may use explosives to breach doors and other obstacles. In most cases, these are improvised devices, tailored to a particular situation. Today, C-4 is the most common explosive used by U.S. forces. With the consistency of modeling clay, it is extremely powerful, clean, and quick-burning. Another common explosive tool is detonator cord, which burns so hot and fast it can cut through metal. These explosives are usually detonated electrically, with a positive control whenever possible. Explosives experts hate time fuses, because they are just one more thing to fail, or to be disarmed later.

The growing menace of domestic terrorism raises legitimate concerns about showing people how to build homemade explosive devices. For that reason I will not give you specifics. That said, explosives have some positive uses; they are not always used to kill or injure people. Consider a door. Any cop will tell you that going though a door with a perpetrator on the other side is a good time to have your insurance policy paid up and your sins confessed. Door-busting quickly and safely is vital, especially in the tricky business of hostage rescue. So consider this little improvised device.

Cut a large coffee tin or other institutional food container in half, down the long axis. You now have a concave container, into which you loop a length of detonator cord and a detonator. On top of the detonator cord, you pack the remaining space with soft plastic packages of saline solution from the medical supplies carried by your Navy corpsman. Once this is done, the open side is sealed with duct tape. Now apply double-sided sticky foam tape over the duct tape. Then slap the sticky side onto the door you want to go through and step back. When the detonator is fired, it drives the saline liquid forward with such force that the door is knocked off its hinges. Since the explosion is quick and clean, and the area is drenched with the saline solution (its just salt water, remember), there is virtually no danger of fire.

Marines learn dozens of such tricks for taking down different kinds of structures. To a properly trained Marine, explosives are another tool, like a saw or bulldozer, to get a job done. In the arts of combat, Marines are world-class masters of creative improvisation.

Tools of the Trade

"We must not be lulled into complacency because we have always been ready, relevant, and capable. What might be ready, relevant, and capable today may be less so the day after tomorrow. We must anticipate change, adapt to it, and foster it. We shall remain relevant only if we are willing to meet future challenges and adapt to new needs."

—General Charles C. Krulak,
Commandant of the Marine Corps

Even though the Marines focus on building better personnel and giving them superb personal combat skills, the Corps still lugs around a fair amount of stuff. Perhaps not as much per capita as an armored unit or an Air Force wing, but even a small Marine Expeditionary Unit–Special Operations Capable–MEU (SOC)–must operate in many environments and roles. On one day, you might see an MEU (SOC) staging an embassy evacuation or rescue. On another day, the mission might be disaster relief or peacekeeping. Meanwhile, a MAGTF still has to be able to perform traditional combat missions, such as amphibious and helicopter assaults. A battalion landing team (BLT) like that in an MEU (SOC) might operate about two dozen armored vehicles, while an equivalent Army unit like a cavalry squadron would own three times that many. The difference is like the one between a draft horse and a thoroughbred. You can ride both, but the draft horse can also pull a cart or plow. The MAGTF is a shock unit (a thoroughbred), which requires reinforcement to conduct really long-term operations.

The money that buys Marine weapons and equipment comes from three sources. First is "Blue" (Navy) money, which buys landing craft and amphibious ships, operated and maintained by sailors. Second comes "Blue" Navy dollars which buy "Green" equipment for Marines, like aircraft, helicopters, and communications and electronic equipment. Finally, there are "Green" Marine Corps funds, to purchase tanks, uniforms, missiles, etc. Marines only control the last category; they have to request the other two from the Navy. The Marines are technically part of the Department of the Navy, after all.

In Fiscal Year 1995 (FY-1995), the Marines only received about $554 million in "Green" dollars. Even with the other "Blue" dollars from the Navy, the total Marine Corps procurement budget is under a billion dollars a year. This level of funding will have to increase if procurement of new systems like the tilt-rotor MV-22B Osprey transport helicopter and Advanced Amphibious Assault Vehicle

(AAAV) is to begin in the next few years. The Corps still enjoys a strong base of public and legislative support, and it is lobbying hard for what it needs.

Most Marine equipment is not designed specifically for the Marines. The Corps must depend on technologies and systems developed by other services, such as the Army and Air Force. The Air Force might define its key technologies as stealth, airframe structures, jet power plants, avionics, and precision guided weapons. The Army knows all about armor systems, ordnance, vehicle power trains, and command and control networks. By comparison, the Marines have only a few limited areas of technical specialization. These include tilt-rotor aircraft propulsion technology, high-speed water-planing hull designs, and lightweight man-portable anti-armor systems. By taking advantage of other services' technology investments, plus a few key investments of their own, the Marines have become the world's most capable sea soldiers. Remember, though, that the Marines have been on the cutting edge of technology since before World War II. Precision weapons delivery (dive bombing) and vertical envelopment (helicopter warfare) are Marine Corps inventions. Systems like AAAV, the MV-22B Osprey, and the Predator anti-tank rocket may represent the shape of the future for the other services as well.

In reviewing the heavy equipment of the Corps, we'll look closely at only those that are Marine specific. For others, like the M1A1 Abrams tank and the TOW anti-tank missile, you can refer to *Armored Cav* and *Fighter Wing*. Unlike the other services, the Marines are not defined by the equipment they use. They are defined by how they use the tools that they have, and the missions they perform.

Personal Equipment and Sustenance

The best personal weapons are of little value to the soldier without food, clothing, navigation equipment, and the like. Many such items used by the Marines are developed in Army laboratories and centers. For this reason, many Marines sometimes feel their requirements are held captive by their "big brother" the Army. Let's take a look.

Clothing and Sleeping Gear

The dress uniforms of the Marine Corps may be the smartest and best-looking of all the services, but the basic Battle Dress Uniform (BDU), or "Utilities" as they are known, is nearly identical to what the Army wears. BDUs come in a variety of camouflage patterns, including Woodlands (greens and browns), Desert (beige, brown, and gray), and Urban/Arctic (white, black, and gray), which doubles as a good winter/mountain uniform. BDUs come in various weights, from light knit (a fifty-fifty cotton/nylon rip-stop mix) to quilted high-technology fabrics (Gore-Tex, Supplex, Thermex, and FiberFill) for cold weather. They can be also treated with a waxy substance so that they do not absorb or pass chemical agents onto the skin of the wearer.

Boots are a *big* problem. Though this situation is changing, the Corps has traditionally had inferior boots for the all-important feet of its Marines. New boots are finally being evaluated and fielded for the Marines. These include the Dannon

The Trimble Navigation Miniature Underwater GPS Receiver (MUGR), which utilizes a floating antenna to allow swimmers and divers to obtain highly accurate surveys and tactical positions.

JOHN D. GRESHAM

desert boot, popular in the Persian Gulf in 1990 and 1991, as well as a new winter/wet boot system designed to keep feet dry in the worst conditions. The helmet is still the Kevlar "Fritz" design used by the Army, though the first new lightweight Kevlar-29 units are beginning to arrive.

The biggest current challenge for outfitting Marines is clothing for cold- and wet-weather conditions. Historically we associate Marine operations with tropical weather, or more recently, with Middle Eastern deserts, but the Corps has faced arctic missions for over half a century. Since the U.S. occupation of Iceland in 1941, Marines have operated in high latitudes and altitudes. Even today, a Marine brigade's set of equipment is prepositioned in caves around Oslo, Norway, for operations on NATO's northern flank. The Corps is upgrading its mountain and cold-weather equipment, with new pants, parkas, mittens, socks, underwear, and balaclavas (hoods). There is a new four-part sleeping bag system, with inner and outer bags, liner and bivy sack (outer cover), certified for temperatures as low as -40° F. Along with special cold-weather rations, these make combat operations in alpine regions and cold weather both possible and livable for Marines.

Navigation

In the last few years, navigation has been revolutionized by the NAVISTAR Global Positioning System (GPS). A constellation of twenty-four satellites in medium Earth orbit (about 11,000 mi/17,700 km in altitude) transmits calibrated signals that generate accurate three-dimensional positions. GPS receivers are increasingly portable, rugged, and cheap. Those receivers saw their first military use in the 1991 Persian Gulf War, where more than five thousand such systems in aircraft, ships, vehicles, and even handheld units contributed to victory over Iraq. Marines used GPS receivers in aircraft like the F/A-18 Hornet fighter bomber and landing craft like the LCAC, and handheld units in the air-ground liaison control (ANGLICO) teams that controlled artillery fire and airstrikes. GPS gave U.S. forces a major advantage on the battlefield, where knowing the exact time (from the satellite's onboard atomic clocks) and your own position is critical. GPS has emerged as a new kind of public utility, with ever-increasing military and civil applications. While the baseline civilian version is limited to 3-D accuracy of about 100 ft/30.5 m, military GPS signals are accurate to about 9.8 to 16.4 ft/3 to 5 m. Utilizing a code which must be punched into the receiver each day (called P(Y)-code), the military signals have proven so accurate and reliable that guided missiles and bombs can use them for guidance.

Marines have embraced GPS with excitement and anticipation, as systems with embedded GPS arrive in greater numbers each year. Because the Corps is always interested in what technology can do for individual Marines, to make them more dangerous to enemies and safer to themselves, the Marines have worked hard to deliver P(Y)-code man-portable GPS receivers down to the squad level. This is a tough objective, because it requires procuring and fielding tens of thousands of such receivers. There are two current models: the Small, Lightweight GPS Receiver (SLGR, built by Tremble Navigation) and the Portable, Lightweight GPS Receiver (PLGR, from Rockwell International). The "Slugger" and "Plugger" are about the size of portable stereos. Combined with a radio, they enable every Marine (theoretically) to call in artillery and air strikes with accuracy. By the dawn of the new century, every USMC aircraft and vehicle will have a GPS receiver, many of them embedded in navigation and fire-control systems. The eventual goal is to give every Marine an individual GPS navigation capability. General Krulak likes to talk about building a GPS receiver into the butt of every M16, and he is serious about it.

One top priority is a new rescue radio for combat aviators. Current rescue radios assigned to U.S. combat flyers frankly stink. During Desert Storm, by simply direction-finding on their radios, Iraqi forces captured downed pilots before rescue forces could reach them. In the short term, there is a modification of the basic PRC-112 radio, called the Hook-112. The Hook-112 involves the addition of a GPS receiver and a burst transmitter to the basic PRC-112, beaming coordinates to rescue forces without betraying the position of the downed flyer. Further on, there is a system known as the Combat Survival/Evader Locator (CSEL), which will combine a GPS receiver with an almost undetectable satellite terminal into a small, handheld package.

In addition, the Marines will soon deploy a mobile survey system based around a GPS receiver to assist expeditionary units in emplacing artillery sites and other position-critical units. Designed and produced by Trimble Navigation, 40 of these systems have already been bought, with an additional 203 planned for future buys. Trimble is also supplying the Marines with a new generation of super-rugged, P(Y)-code GPS units for use by reconnaissance forces. Called the Miniature Underwater GPS Receiver (MUGR), it is about the size of a Walkman radio. MUGR is fully waterproof, and can actually operate underwater! By using a floating antenna attached by a wire tether, the MUGR allows a reconnaissance force to survey a beach or harbor covertly. These systems represent only the tip of the GPS iceberg. In the near future, expect to see the "Fritz" Kevlar helmets of American troops sporting flat satellite antennas with the ability to send and receive signals.

Communications

By the fall of 1996, the Marines will finally begin their long-awaited move to the Army's Single Channel Ground and Airborne Radio System (SINCGARS). SINCGARS utilizes "frequency hopping" to make its signals difficult to intercept or jam. The 2nd MEF will get the entire suite of SINCGARS radio systems for aircraft, vehicles, and personnel in FY-1996 and FY-97. SINCGARS will be taken to the field by 26th MEU (SOC) during their 1996/97 Mediterranean cruise. The current SINCGARS variants are shown in the table below:

Version	Designation	Description
V1	AN/PRC-119	Portable Manpack
V2	AN/VRC-87	Short Range Vehicular
V3	AN/VRC-88	Short Range Vehicular/Dismountable
V4	AN/VRC-89	Long Range Vehicular/Dismountable
V5	AN/VRC-90	Long Range Vehicular
V6	AN/VRC-91	Long Range Vehicular/ Dismountable Short Range
V7	AN/VRC-92	Vehicular Dual Long Range Retransmission
V8	AN/ARC-201	Airborne Transceiver
N/A	AN/PRC-6725E	Short Range Handheld

Marines deploy a number of satellite communications systems, ranging from large fixed systems for command posts to backpack models for on-the-scene commanders. The key to military satellite communications is access to the proper frequency channels, which are usually overbooked and the subject of intense competition by users, all of whom need to communicate *right now*. The Department of Defense maintains a number of satellite communications systems to support military operations. But the high tempo of U.S. military deployments has saturated existing military systems. Every communications satellite has a number of transponders, which provide television or radio channels. Each transponder is assigned according to priorities determined by theater commanders, or even by the Joint Staff at the Pentagon. There are simply not enough to go around. As a result, the Defense Department is also a major customer for commercial satellite communications air time from commercial suppliers like INMARSAT and Hughes. The Marines have equipment that operates on most standard satellite frequencies, though the most common is the man-portable UHF TACSAT system. This version, known as the PRC-117D, is carried by a communications specialist, with a backpack battery and transceiver and an attached antenna. Able to transmit voice or data, it works well in the field, though it is a battery hog.

While the Marine Corps has a robust and effective communication architecture today, things are going to be changing fast. Already on the horizon are direct-broadcast/receive commercial satellite phone systems, and military communicators are drooling to get some. Global handheld satellite phones will create a telecommunications revolution that makes current-generation cellular phones look like soup cans connected by a string. For example, Texas Instruments has already developed a two-way satellite antenna that is just a flat square a few inches/centimeters on each side. Requiring only minuscule power to operate, it can be fitted to the roof of an HMMWV, or possibly even the top of a Kevlar "Fritz" helmet. The dream of tying every Marine into a global communications net is now within sight.

Food and Water

Marines might be able to hold a position without fuel and with just the ammunition they are carrying, but without food or water, they will have to surrender or die within a few days. Water is usually no problem; Marines have a ready supply of pure water from the ships that bring them ashore. The Corps has also made a significant investment in portable reverse-osmosis water-purification systems that can be delivered via transport aircraft or prepositioned ships. As a result, other services and coalition allies frequently depend upon Marine units to supply their water needs until follow-on logistics forces arrive.

Food is a different matter. The Corps is a virtual hostage to the meal systems produced by the U.S. Army; it must order food items from the Army logistics system. Options are limited. To begin with, there are Meals Ready to Eat (MREs), heavy, bland, but nourishing rations. Since Desert Storm, MREs have actually gotten heavier, for the Army has chosen to pack more stuff into the brown plastic packages, rather than make what was already inside more appetizing. The result is that field troops tend to throw much of the MRE away, and thus fail to take in the nutrients and calories they need. Though MRE manufacturers like Star Foods already have better products on hand, the Army is not willing to buy them at this time. It is working to issue better MREs, though, and expects to field several new kinds in FY-2000. Because MREs are so unappetizing, American peacekeepers in Bosnia have been using their own money to buy nutritional snacks or freeze-dried camping food; and if they're lucky, they can get some French or British rations. The French version of the MRE, for example, contains fresh bread and pate!

The Marine food service system falls into three levels. The first or "A"-type rations are boxes with three trays of prepackaged food (meats, vegetables, and starches), which are heated in tray boilers and served cafeteria-style to troops. The "B"-type rations are actual meals that are cooked in field kitchens made from locally purchased ingredients as well as dehydrated/freeze-dried ingredients shipped from the U.S. Finally, there are the field rations, normally composed of MREs. I say normally, because when troops enter cold-weather and high-altitude areas, they begin to burn calories at an incredible rate. While a typical Marine might burn about three thousand calories per day under normal environmental conditions, cold weather can double this rate. Since Marines routinely throw out much of the stuff inside the four MREs issued each day, something else is clearly required for cold-weather operations.

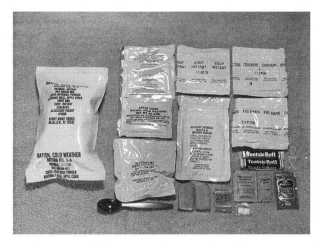

A cold-weather ration, with its contents displayed on the right. This is a two meal, freeze-dried ration, containing a breakfast and lunch. It includes oatmeal, soup, granola, crackers, cocoa, coffee, candy, and a powdered drink. The breakfast rations are designed to provide maximum sugar/carbohydrate content on cold mornings.

JOHN D. GRESHAM

That is the cold-weather ration. Produced by Oregon Freeze Dry, Inc. (they also produce Mountain House brand camping food) and packaged by Right Away Foods, these rations take up only half as much space and volume as a comparable diet of MREs, and deliver more calories. The ration itself is composed mainly of freeze-dried foods which are contained in a sealed plastic bag. These only require rehydration to make them ready to eat. Given a supply of snow for melting and a heat source, the cold-weather ration can provide an excellent source of hot food for field units. As an added benefit, it is very high in calories (about three thousand per issued ration), and quite light in weight. Compared to MREs, cold-weather rations are quite tasty, and this means that the troops eat everything in the packs issued each day.

As the Marine Corps moves towards the 21st century, it is looking forward to the new varieties of MREs due to be fielded by the Army. But don't be surprised if the USMC finally begins to produce rations to its own design and specification. The Commandant's Battle Lab at Quantico, Virginia, is studying the problem from a purely "Marine" point of view, and may yet produce field rations with an "expeditionary" flavor.

Fire Support

Marine units are primarily infantry-based formations, which depend upon the fire of supporting units to achieve their objectives. Supporting fire must be both accurate and lethal to allow lightly laden Marines to stand up to everything they might have to face, from irregular forces (as encountered in Somalia and Liberia) to conventional military units like those in the Persian Gulf. Without firepower, Marines have to trade their lives to take objectives; and the American people simply will not accept excessive casualties. Thus, Marines have a great professional interest in fire support. Almost every Marine can read a map, use a radio, and call in fire from ships, aircraft, or artillery. A single rifle platoon might receive air support from AV-8B Harrier IIs or AH-1W Super Cobras, and artillery support from a battery of 155mm guns, or an offshore destroyer or cruiser. The Corps is currently suffering a severe shortfall of fire support. In the five years following Desert Storm, the Marines

A Marine mans
an M2 Browning
.50-caliber machine gun
mounted on an HMMWV.
OFFICIAL U.S. MARINE CORPS PHOTO

and the Navy lost over half of their total fire-support resources with the decommissioning of the *Iowa*-class (BB-61) battleships and retirement of many support aircraft and artillery units. This is a source of severe concern to Marine and Navy leaders.

Browning M2 .50-Caliber Machine Gun

To listen to an old Marine "Gunny," you would think it was the most beautiful of women. The M2 .50-caliber machine gun is a favorite heavy weapon of Marines and ground troops everywhere. This heavy machine gun provides a base of fire for the rifle platoon and company. It forces the enemy to keep his head down and confronts him with a threat he must neutralize. While he is trying to knock out the damned machine gun, Marines can maneuver onto his flanks or close with his position. A heavy machine gun can shred dry-wall or wooden buildings, or unarmored vehicles. At short ranges and favorable angles it can even penetrate the side or rear plating of armored vehicles. This makes it a *very* dangerous piece of equipment to have in your pocket.

The "50 cal" first entered service with the U.S. Army in 1919, too late for service in World War I. During the Second World War it was standard armament on American fighter and bomber aircraft, and was widely employed as an anti-aircraft weapon on every kind of ship and ground vehicle. The M2 is an automatic recoil-operated, air-cooled machine gun that weighs 84 lb/38 kg. Recoil-operated means that it uses an ingenious arrangement of levers, cams, and springs to capture part of the recoil energy to extract and eject the spent cartridge case, feed the next round, load it, and fire it. This cycle repeats as long as the gunner holds down the V-shaped trigger located between two hand grips at the rear of the gun. Release the trigger and a latch secures the mechanism in the "open bolt" position, ready to fire again.

The .50-cal can be found in the turret of the AAV-7/LVTP-7 amphibious tractor, on the simple pintle mount on the HMMWV, and on the high-tech coaxial mount on the Avenger air-defense vehicle. The weapons platoon of a Marine rifle company fires it from a hefty 44-lb/20-kg tripod. It takes at least two Marines to carry the weapon, plus men to carry cans of ammunition. The ammunition is assembled into belts with reusable spring clips called "disintegrating links," which are stripped off by the gun's

An HMMWV on patrol with an Mk19 40mm grenade launcher mounted on top. This weapon can fire all of the same rounds as the M203 grenade launcher.

feeder mechanism. The rate of fire is 550 rounds per minute, and gunners are trained to fire short bursts to conserve ammunition. The theoretical maximum range is 4.22 mi/6.8 km, and the M2 has even been used for "indirect fire" at high angles of elevation to create a "fire-beaten zone" on the other side of a hill. In typical battlefield conditions the practical range is about 1.1 mi/1.8 km. The legendary lethality of the M2 derives from the heavy charge of propellant in the cartridge and the superb ballistic shape of the projectile, which has a distinctive "boat tail." There are several ammunition types. These include target-practice (TP), armor-piercing solid-shot, armor-piercing incendiary (API), and high-explosive (HE).

Over the years, many firms have produced the M2 on license from the holders of John M. Browning's original patent. The current contractor producing the M2 for the U.S. Department of Defense is Saco Defense, Inc., and the FY-1994 unit cost was $8,118.00. Its unique combination of range, lethality, durability, and simplicity guarantees that the M2 will soldier on well into the next century. In fact, the last Marine M2 gunner has probably not yet been born.

Mk 19 40mm Machine Gun, Mod 3

Back in the 1960s, deep in the swamps of the Mekong Delta where a well-concealed and heavily armed Viet Cong ambush might lurk around the next bend in the river, crews of U.S. Navy patrol craft discovered that .50-cal machine gun fire was often insufficient to break up an attack. They needed a weapon that could spew out a stream of explosive grenades to suppress enemy forces. To meet this need, the Navy developed the Mk 19, officially classed as a "machine gun," but actually an automatic grenade launcher. The Mk 19 had a long and troubled development cycle, earning the nickname "Dover Dog," after the Delaware arsenal where it was designed. After a series of modifications, it has proven itself in service with the Army, Navy, and Marine Corps. The Mk 19 is an extremely simple weapon using the "blowback" principle. The barrel and receiver assembly recoil against a heavy spring, and as they rebound, the next round is loaded and fired. The weapon fires the same family of 40mm grenades as the M203 launcher attached to the M16 rifle.

An M198 towed howitzer
assigned to BLT 2/6,
buttoned up and ready
for deployment.

JOHN D. GRESHAM

By itself the weapon weighs 72.5 lb/33 kg. It was designed to use the same tripod as the M2 .50-cal. machine gun, but is also found in the turret of the AAV-7/LVTP-7 amphibious tractor. The cyclic rate of fire is from 325 to 375 rounds per minute, but the practical rate of fire is about 40 rounds per minute in short bursts. To achieve the maximum range of 2.2 km/1.37 mi, you have to elevate the weapon to loft the grenades and forget about real accuracy. Practical range for flat-trajectory fire is about 1,500 m/4920 ft. There are several types of ammunition, assembled into disintegrating link belts and transported in metal canisters. The HEDP (high-explosive, dual-purpose) grenade will pierce 2 in./51mm of armor, and spray metal fragments that can kill within 5 m/16.4 ft and wound within 15 m/49.2 ft. Other types of ammunition include incendiary, smoke, and tear gas rounds. The Mk 19 is usually found in the weapons platoon of a rifle company and the weapons company of a rifle battalion. One Marine can load and fire the weapon, but it requires a team of three to four to carry it, along with a supply of grenades. It is manufactured by Saco Defense, and the 1994 unit cost was $13,758.00.

Mortars

Mortars are the company or battalion commander's own personal artillery. The mortar is a portable, cheap, and simple weapon: just a metal tube with a bipod elevating bracket and a heavy base plate. You assemble the weapon, aim the mortar at the target, and drop the mortar round down the barrel. The round strikes a firing pin at the bottom of the tube, and off it goes. Limitations of the mortar are its relatively short range and inaccuracy. But this old weapon is now gaining new respect, thanks to the development of precision guided ammunition.

Marines employ two different kinds of mortars. The M224, used in the heavy weapons platoon of the rifle company, is a 60mm weapon weighing only 46.5 lb/21 kg. Maximum range is 2.2 mi/3.5 km. A good crew can sustain a rate of fire of around twenty rounds per minute. The other model, the M252, is used in the heavy weapons company of the infantry battalion. An 81mm weapon, it is based on a 1970s

British design, weighs 89 lb/40 kg, and has a maximum range of 3.5 mi/5.6 km. The sustained rate of fire is sixteen rounds per minute. There is a wide variety of ammunition types in each caliber, including high-explosive, smoke, and incendiary rounds. High-explosive rounds can be fitted either with an impact fuse or a proximity fuse that detonates at a preset altitude, showering the target with fragments.

M198 155mm Towed Howitzer

This big gun is one of the more controversial weapons in the Marine arsenal. While it is the Marines' primary field artillery piece, the Corps leadership feels that the M198 is simply too big and too heavy. Also, it takes up too much space on amphibious lift ships, and in firing position it is too vulnerable, especially when a quantity of ammunition is stacked near the gun. In addition, the M198 has a high center of gravity, which makes it prone to tipping over and being difficult to handle. On the other hand, it uses standard, widely available 155mm ammunition with terrific lethality. Weighing 15,758 lb/7,154 kg, it requires a heavy (5-ton) truck to tow it, along with its eleven-man crew and a supply of ammunition. It can be lifted as a sling load by the CH-53E helicopter. The M198 can hurl a projectile up to 14 mi/22.4 km, and a special rocket-assisted projectile extends this range to 18.6 mi/30 km. The 566 guns in the Marine inventory will serve for at least another decade, until the introduction of a new lightweight howitzer which is under development.

Mk 45 5-in./54 Naval Gun Mount

With the retirement of the *Iowa*-class (BB-61) battleships, the Navy's gunfire support capability is reduced to one or two of these rifled 5-in./127mm weapons on each major surface combatant (cruiser, destroyer, and a few amphibious ships). Built by United Defense's Great Northern Division, the Mk 45 5-in./54 turret has a high degree of automation, sustaining a rate of seventeen rounds per minute. The turret normally operates unmanned, with the six-man Navy crew working below decks. The Mk 45 can throw a 70-lb/31.75-kg projectile to a maximum range of 14 mi/23.6 km, though extended-range ammunition is under development. The main ammunition types are high-explosive and incendiary (white phosphorus). A ship generally carries several hundred rounds per gun in its magazines, and major task forces are accompanied by ammunition ships, which can rapidly replenish the supply, using a UH-46 helicopter.

The Future: The Lightweight Howitzer and Arsenal Ship

Solving the problem of replacing the fire-support assets lost since Desert Storm is a joint Navy/Marine Corps challenge. The most urgent fire-support upgrade is replacement of the M198 155mm howitzer. Six different industrial teams have produced competing designs for a new lightweight howitzer. These include United Defense, Lockheed Martin, Royal Ordnance, and VSEL. In addition to lighter weight, the Marines want a weapon with much longer range (which means a longer barrel) and smaller crew requirements, and a higher rate of fire (which means power-assisted ramming and loading.) Expect to see deliveries in the early years of the next century.

An artist's concept of the proposed "Arsenal Ship." The vessel would be packed with vertical launch cells for missiles that would provide bombardment and fire support for Marines ashore.

A bigger problem is offshore fire support. Marines really miss those old *Iowa*-class (BB-61) battleships. Nothing will ever match the spectacular effect of 16-in./406mm shells falling on a target within 25 mi/40 km of a coastline. Over a hundred ships with 5-in./127mm guns have left U.S. Navy service, gutting naval gunfire capability. To make up for this drawdown, the Chief of Navel Operations and former Deputy Chief of the Joint Chiefs of Staff Admiral Bill Owens conceived the idea of the Arsenal Ship. The Arsenal Ship would replace the lost firepower of the retired *Iowa*-class (BB-61) battleships by constructing a simple, relatively inexpensive ship packed with missile launch cells—as many as 732 tactical missiles, including Tomahawk and perhaps a version of the Army TACMS. In effect, the arsenal ship would win the war in one salvo, and then reload for the next war. The ship would rely entirely on off-board sensors for targeting. Covered with radar-absorbing coatings, an Arsenal Ship would have virtually no superstructure. Some design studies envision ballast tanks that could be flooded to give the ship extremely low freeboard, making it a very difficult target for enemy anti-ship missiles. Unfortunately, all this thinking hasn't gone very far; and there are practical problems. Not the least of these: The Navy has done virtually nothing to integrate and procure the TACMS missile for naval service, perhaps because it's reluctant to use an Army missile aboard Navy ships (the "not-invented-here" syndrome). Only nuclear submariners have done substantive work on TACMS, since they are desperately looking for new missions for their subs in the post-Cold War period. Whatever happens, supporting fires will be the make-or-break item for continued forced-entry capabilities into the 21st century.

Anti-Armor/Aircraft Systems

Cambrai, Northern France. 0620 hours on November 20th, 1917. In the misty dawn, the soldiers of the Kaiser's 2nd Army looked out over "No-Man's-Land" and saw over two hundred primitive British tanks lumbering toward them. The Germans opened fire with the Mauser rifles and Maxim machine guns that had made them nearly invincible during three long years in the trenches, and watched in horror as the bullets bounced off of the armor plate. Then, surprisingly, and most uncharacteristically for German infantry, they ran away.

Almost thirty-five years later, near Osan, Korea, on July 5th, 1950, soldiers of the 24th Infantry Division's Task Force Smith had held their roadblock stubbornly for almost five hours against a superior force of invading North Koreans. They were mostly young draftees, but their sergeants were tough World War II combat veterans who knew their business. Then they heard a low rumble that grew to a roar as thirty Russian T-34/85 tanks came down the road. The bazooka teams fired, and watched in horror as the 2.75-in./70mm armor-piercing rockets bounced off the tanks' sharply angled armor plates. Then they did something surprising and uncharacteristic of American infantry. They ran away.

There is a common lesson in these two stories. Tanks scare the crap out of infantrymen who have no way to fight back effectively. To stand up against tanks, foot soldiers need two things: courage and an anti-tank weapon they trust. Good leadership and training will supply the courage. Good ordnance engineers and technicians can supply the weapons. Early tanks were practically blind on the battlefield, and even the best modern tank designs (like the M1A1 Abrams) are visually handicapped. Men on foot can exploit this weakness with great effect. During the Hungarian Revolution in Budapest (1956), Russian T-34s were knocked out by Hungarian freedom fighters, who immobilized the tanks by jamming steel pipes between the tracks and the road wheels, then bombarded them with firebombs made from bottles and gasoline.

Modern portable anti-tank weapons fall into two categories: those light enough for one soldier to carry, and specialist weapons that require a crew and possibly a motor vehicle to haul them around. The Marine Corps has usually followed U.S. Army doctrine, equipment, and tactics for anti-armor combat, but has a few ideas of its own. Let's take a quick look at the portable anti-armor systems used by the Corps.

AT-4

The Marines have always been willing to acquire foreign-made weapons when they are the best of their breed. The AT-4 was acquired to replace the very light and inexpensive 70mm M72 LAW (Light Anti-tank Weapon), which is increasingly becoming ineffective against modern battle tanks. The AT-4 is a lightweight, single-shot, disposable version of the "Karl Gustav" 84 mm anti-tank launcher manufactured by FFV in Sweden. The AT-4 can be carried and shoulder-fired by one Marine, but is typically employed in the heavy weapons platoon of a rifle company with a two-man fire team. The second Marine serves as a spotter and carries additional AT-4s for the team. Weighing 14.75 lb/6.7 kg, the 40-in./1.01-m.-long rocket launcher has a nasty back-blast. Maximum effective range is 300 m/984 ft, and the shaped-charge projectile can penetrate 400mm/15.75 in. of armor plate. The FY-96 unit cost is about $1,100 per AT-4 rocket.

Marines of the
26th MEU (SOC)
prepare to fire an
SMAW rocket launcher.
This Israeli-made weapon
is used for bunker-busting
and demolitions.
OFFICIAL U.S. MARINE CORPS PHOTO

An HMMWV of BLT 2/6
mounting a TOW anti-tank
missile launcher on maneuvers
in Israel in 1995.
OFFICIAL U.S. MARINE CORPS PHOTO

SMAW

The Shoulder Launched Multipurpose Assault Weapon (SMAW) is a high-tech descendant of the World War II bazooka—a portable rocket launcher that can disable a tank or knock out a bunker. It was introduced in 1984 as a unique Marine Corps item, because the Army's M72 LAW lacked the accuracy and punch the Marines wanted, and other anti-tank rockets were too heavy. The SMAW is based on an Israeli weapon called the B-300. The 16.6-lb/7.54-kg fiberglass launch tube is 30 in./76 cm long when carried. For firing, you snap a rocket in its disposable sealed canister into the breech end, which extends the total length of the weapon to 54 in./137 cm. The Marines carry 1,364 of these unusual weapons in inventory, and they cost about $14,000.00 each. The SMAW fires two kinds of 83mm rockets—HEDP for use against lightly armored vehicles or buildings, and High-Explosive Anti-tank (HEAT) for use against heavily armored vehicles. Maximum range against a tank is 500 m/1,640 ft, but the SMAW is intended for use at close ranges. Accuracy

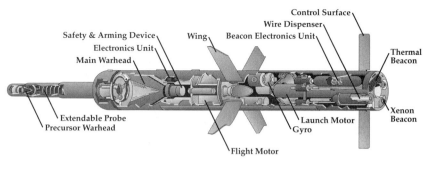

Safety & Arming Device
Electronics Unit
Main Warhead

Wing

Control Surface
Wire Dispenser
Beacon Electronics Unit

Thermal Beacon

Extendable Probe
Precursor Warhead

Flight Motor

Launch Motor
Gyro

Xenon Beacon

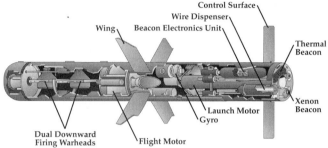

Wing

Control Surface
Wire Dispenser
Beacon Electronics Unit

Thermal Beacon

Dual Downward Firing Warheads

Flight Motor

Launch Motor
Gyro

Xenon Beacon

ABOVE: The Hughes Missile Systems TOW-2A anti-tank missile. The precursor warhead on the extensible probe helps defeat reactive armor. *JACK RYAN ENTERPRISES, LTD., BY LAURA ALPHER*

LEFT: TOW-2B Anti-Tank Missile. *JACK RYAN ENTERPRISES, LTD., BY LAURA ALPHER*

is ensured by a "spotting rifle" attached to the side of the launcher. This is a British-made 9mm semi-automatic weapon that fires a special tracer round that is ballistically matched to the flight characteristics of the rocket. You hoist the weapon to your shoulder, look through the sight, and fire a spotting round. When you see the spotting round impact on the target, you fire the rocket, with a very high probability of a hit. SMAW works so well that during Desert Storm the Army "borrowed" 150 launchers and five thousand rockets from the Marines.

Hughes MGM-71 TOW-2 Anti-Tank Missile

"TOW" stands for "Tube-launched, Optically-tracked, Wire-guided." This famous family of missiles originally entered service in 1970, and has been continuously improved and upgraded through a series of modifications. TOW first saw combat in 1972 in Vietnam, where it was successfully fired by U.S. Army helicopters against North Vietnamese tanks. In the Marine Corps, TOW is mainly used by specialist anti-tank platoons of heavy weapons companies, mounted on HMMWVs (which carry six missiles), or by anti-tank variants of the eight-wheeled Light Armored Vehicle (LAV-AT, carrying two missiles ready to fire with ten stowed).

The TOW-2 missile is 3.8 ft/1.2m long, about 6 in./150 mm in diameter, and weighs 65 lb/29.5 kg. There are four spring-loaded, pop-out guidance fins at the tail and four wings at mid-body. Like most anti-tank missiles, TOW has two rocket motors, a small kick motor that ejects the missile from the launch tube, and a sustainer that ignites at a safe distance. An unusual feature on TOW is that the rocket exhaust nozzles are on either side of the missile body, to avoid interference with the

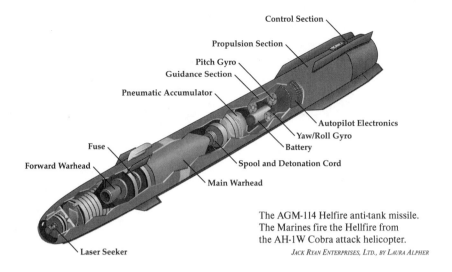

Control Section

Propulsion Section

Pitch Gyro

Guidance Section

Pneumatic Accumulator

Autopilot Electronics

Yaw/Roll Gyro

Fuse

Battery

Forward Warhead

Spool and Detonation Cord

Main Warhead

Laser Seeker

The AGM-114 Helfire anti-tank missile.
The Marines fire the Hellfire from
the AH-1W Cobra attack helicopter.

JACK RYAN ENTERPRISES, LTD., BY LAURA ALPHER

fine steel guidance wires that stream out from the tail. TOW launchers can interface with a variety of different sighting and control units, and the Marines are currently acquiring an Improved Target Acquisition System (ITAS) from Texas Instruments, which combines a laser range finder, FLIR (Forward Looking Infrared), modular software, and a rechargeable ten-hour battery. TOW-2A uses a tandem warhead for direct attack, and TOW-2B uses a pair of explosively forged projectile warheads from a top-attack flight profile. Otherwise the two versions are identical. Maximum effective range is 3,75 km/2.3 mi.

Rockwell International AGM-114 Hellfire

Hellfire is a long-range high-speed laser-guided missile and it is used exclusively by Marine Cobra attack helicopters, although the U.S. Army and Navy have experimented with firing it from ground vehicles and ships, and Sweden has acquired a coast-defense version fired from a portable tripod mount. Hellfire is primarily an anti-tank missile, with a 20-lb/9-kg dual shaped-charge warhead that can essentially defeat any imaginable tank from any angle. It can also be used successfully against other targets. For example, the opening shots of the 1991 Persian Gulf War were Hellfire missiles fired by Army AH-64 Apache helicopters against Iraqi air defense radar sites.

Hellfire is a big brute of a missile, measuring over 5 ft/1.625 m long, 7 in./178 mm in diameter, and weighing almost 100 lb/45.3 kg. Maximum range depends on the speed and altitude of the firing aircraft, but 5 mi/8 km is claimed. The solid-propellant rocket motor rapidly accelerates the missile to supersonic speed. The seeker in Hellfire's nose is similar to the seeker of a laser-guided bomb. It is programmed to home on a spot of laser light, pulsing with a particular pre-set code. As far as the missile is concerned, it does not matter who or what is lasing the target. The missile can be programmed to "lock on after launch," enabling the designator to remain hidden until the last few seconds of missile flight. The missile can fly a straight-line (direct-

A Marine Stinger SAM team of the 26th MEU (SOC) stands alert on the USS *Wasp* (LHD-1). Such teams frequently stand watch on U.S. Navy ships to help catch any "leakers" through the ship's air defenses.

JOHN D. GRESHAM

attack), or a "lofted" flight path, which provides extended range and an advantageous "top down" impact angle against an armored target.

The Army's Apaches can "self-designate," but Marine AH-1W Cobras do not presently carry a laser designator. In 1996, though, a Night Targeting System will start entering service with the Cobras. But until these system are installed, the Cobras face a tricky tactical coordination problem. They have to rely on "buddy-lasing," which can be performed by a ground-based forward observer, or a Marine UH-1N helicopter equipped with one of the three surviving Nite Eagle laser-designator packages salvaged from the Army's failed Aquila RPV program. During Desert Storm, Marine Cobras, teamed in tank-killing units with these few UH-1Ns, successfully fired 159 Hellfires. Each Cobra can carry up to eight Hellfires on launch rails attached to its stub wings. In FY-1994, Hellfire had a unit cost of abut $35,000.00

Hughes MIM-92 Stinger Surface-to-Air Missile (SAM)

The last time American fighting men had to face an enemy who held air superiority was in 1942 in Tunisia against the German Nazi Luftwaffe and the Fascist Italian Regia Aeronautica. Indeed, the main "air threat" to our ground troops in Vietnam and the 1991 Gulf War were mistaken attacks by "friendly" pilots. Yet, even the most obsolete Third World air force could inflict serious damage on a Marine landing force during the first few critical hours of an operation. While the ground-pounding Marines have great confidence that their brother and sister Marines who fly will be there to help in a pinch, they have always taken the problem of short-range anti-aircraft defense seriously. Each expeditionary Marine unit will normally have an assigned air defense platoon, equipped with the MIM-92 Stinger SAM, which began to replace the much less effective 1960s-vintage Redeye missile in 1982. The platoon includes three HMMWVs, each carrying three-man Stinger teams. The Stinger is sealed in its disposable launch tube at the factory and has a long shelf life. The launch tube clips onto a reusable gripstock assembly, an IFF antenna (this is

An Avenger SAM vehicle assigned to the 26th MEU (SOC) in Tunisia during 1995. Based on an HMMWV chassis, it is armed with eight Stinger SAMS and a .50-caliber machine gun.

OFFICIAL U.S. MARINE CORPS PHOTO

optional) is attached to the front of the assembly, and the gunner hoists the entire 34-lb/15.4-kg assembly to his shoulder. The gripstock incorporates an audio cueing system, to tell the gunner when the missile seeker is "locked" onto a target. Normally the team will be alerted to the approach of hostile aircraft via radio from a ground-, air-, or ship-based surveillance radar.

Stinger is 5 ft/1.5 m long, 2.75 in./7 cm in diameter, and weighs 12.5 lb/5.7 kg at launch. Range is highly dependent on the speed and direction of the enemy aircraft, but the official specs are 1 km/.6 mi minimum to 8 km/5 mi maximum. Stinger's seeker has an "all-aspect" engagement capability. This means that it does not need a direct line of sight to the hot metal of the engine exhaust; it is sensitive enough to sense that the aircraft is warmer than the sky behind it. Developed by Hughes Missile Systems, the seeker also incorporate a reprogrammable microprocessor, so that software changes can be rapidly implemented to cope with ever-changing enemy countermeasures.

In FY-94, the unit cost of a Stinger missile was $38,000.00, and there were 13,431 in the U.S. Marine inventory. Stinger's first taste of combat was with the British Special Air Service Regiment in the 1982 British-Argentine war. A large number of Stingers were also supplied to Afghan freedom fighters during their long war against Soviet occupation; and they proved incredibly effective in the hands of uneducated but highly motivated gunners. Stinger has an impact fuse for direct hits and a proximity fuse that can turn a near miss into a kill by showering the target with fragments. There is also a timed self-destruct, so that live missiles do not come down on the heads of friendly troops.

The most exciting new Stinger development for the Marines is the Avenger air-defense vehicle. This is integrated by Boeing using the chassis of an HMMWV with a rotating turret that incorporates a FLIR, a laser range finder, an M2 .50-cal.

machine gun, and reloadable canisters for eight missiles. A pair of Avengers will be normally be assigned to the Stinger platoon of a MEU (SOC). Combined with the three man-pack teams, it gives the MEU (SOC) a rudimentary air-defense capability. When combined with an offshore SAM umbrella from escorting surface ships, and perhaps the air-to-air capabilities of the MEU (SOC)'s embarked Harrier detachment, it gives the Marines a fighting chance against air attack until follow-on forces arrive to take over the job.

The Future: Texas Instruments (TI)/Martin Javelin

Javelin represents a new generation of precision-guided fire-and-forget antitank weapons. The joint Army/Marine Corps program, now in production, began in 1989 under the acronym AAWS-M (Advanced Anti-tank Weapon System—Medium). The Marines will receive a small initial batch (140 missiles) in 1997, and expect to field a full operational capability in the heavy weapons platoon of the rifle company and the heavy weapons company of the battalion by 1999. The joint Army/Marine requirement is 31,269 missiles and 3,541 Command Launch Units through the year 2004, but in the absence of a war, procurement targets rarely survive successive rounds of budget cuts.

At first glance, what Javelin does seems impossible. "Precision guidance" usually requires a human being in the loop to control the flight of the weapon up to the moment of impact. A good example is the Marines' current portable anti-tank missile, the hated McDonnell Douglas M-47 Dragon, which entered service in the early 1970s. The Dragon gunner, crouched in an awkward and uncomfortable position, must keep the target centered in his telescopic sight during the missile's entire time of flight, as long as twelve seconds out to 1,000 m/1,094 yd. Steering commands travel down twin steel wires that uncoil from bobbins on the missile and the launch tube. If the enemy detects the smoke and flash of the missile launch, he will quickly fire back in the general direction with everything he's got. If the Dragon gunner ducks, or even flinches, the missile will probably fly into the ground or pass harmlessly over the target.

Javelin does things differently. Because it uses an intelligent imaging-infrared seeker, the new missile combines precision guidance with fire-and-forget operation. In effect, the missile software "remembers" the thermal signature of the target it locked onto when it was launched. It also "knows" how to follow a moving target, and how to perform tricky maneuvers during its last few milliseconds of "life." The missile performs a climb and dive to strike the top of the target, where the armor is thinnest. If the target is inside a building, or under some kind of top-cover, the gunner can select a direct flight path.

The Javelin system has two components: the missile round in a disposable launch tube, and the reusable 14-lb/6.4-kg Command Launch Unit (CLU), which looks rather like a big box camera with trigger-grip handles. The CLU snaps into a connector on the launch tube, and the gunner hoists the entire 49-lb/22.4-kg weapon up onto either shoulder, activates the replaceable battery (which powers the system for up to four hours), and looks through the eyepiece. In daylight, this functions as a four-power telescopic sight; and at night, or in blowing sand, smoke, fog, or other obscured conditions, it functions as a Forward Looking Infrared (FLIR) viewer,

A pair of infantry men launch a "fire-and-forget" Texas Instruments/Lockheed Martin Javelin anti-tank missile. This man-portable system will come into service with the Marines in several years.

TEXAS INSTRUMENTS

presenting a green-and-black thermal image of the battlefield, with a 4-power wide field of view or a 9-power narrow field.

Javelin can be fired safely inside an enclosure, since there is no back-blast per se. A small kick motor, which burns for only a 1/10 of a second, ejects the missile from the launch tube to a safe distance before the main rocket motor ignites. Maximum range is over 2,000 m/1.25 mi. Javelin uses a "tandem warhead" to defeat spaced armor or explosive-reactive protection systems. A small shaped charge detonates first to strip away any outer layers; then, microseconds later, the main shaped charge detonates to penetrate and destroy the target. It is effective and deadly, as well as being the first of a new generation of "brilliant" guided weapons to enter U.S. service. So excited is the Marine Corps about this system that even before it is fielded, the Corps is thinking about using it as the primary anti-armor system on the new AAAV amphibious tractor. Keep your eye on this one, folks. It's going to be a winner!

The Future: Lockheed Marine Loral Aeronutronic Predator

For all of its shortcomings, the Marines generally miss the old M72 LAW. Light and compact, it gave them the ability to hit and destroy, albeit at short ranges, almost anything short of a heavy tank. In addition, it could be (and was) carried by every Marine in a rifle squad, meaning that a unit had a bunch of them to use in combat. Unfortunately, by the late 1970s the LAW was going out of service and was being replaced by heavier and more specialized systems like the AT-4. Nevertheless, the Marines have always wanted another "wooden round" heavy weapon like LAW, and they began a program to give them a 21st century version. Originally known as

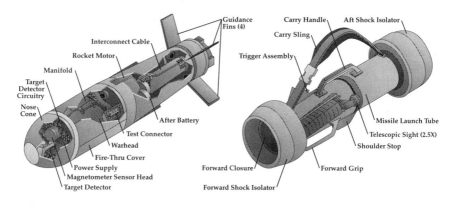

A cutaway view of the new Predator anti-tank missile being developed for the Marine Corps by Lockheed Martin Loral Missile Systems. The launcher is shown to the right.

JACK RYAN ENTERPRISES, LTD., BY LAURA ALPHER

SRAW (Short-Range Assault Weapon), Predator has been under development since the 1980s, and will enter service around the year 2000. Weighing only 19 lb/8.6 kg, and measuring 35 in./89 cm in length, the missile and its disposable launch tube will be issued like a round of ammunition that any rifleman can carry and fire. Like Javelin, Predator has a "soft launch" motor that allows it to be fired safely from inside an enclosure.

System costs are kept low (about $5,000.00 per unit in FY-96) by dispensing with costly precision guidance and thermal-imaging components. For the required maximum range of 600 m/1,970 ft, it is sufficient to have a few microchips and mechanical components that function as an "inertial autopilot." Against a stationary target, this automatically compensates for crosswinds, uneven terrain, and variations in thrust as the rocket motor burns out. Against a moving target (up to speeds of 22 mph/35.4 kph), the missile's autopilot senses the slew (crossing) rate as the gunner tracks the target for about a second before launch, and then automatically computes the correct lead angle for target intercept. All the gunner has to do is keep the crosshairs of the 2.5 power telescopic sight on the center of the target and pull the trigger. The Predator does the rest.

In its nose Predator carries a highly sensitive "target detection device" that combines a tiny range-finding laser, angled downward and forward to sense the edge of the target, and a magnetometer that senses the mass of the target. When the software concludes that the missile is directly over the target, it detonates the 5-lb/2.25-kg warhead, which projects an explosively formed heavy metal penetrator (like that of the TOW-2B) at almost Mach 5 down through the thin roof of the target. In tests on old M-48 tanks, the projectile even continued downward to blow a hole through the hull floor! Loral has also proposed a "direct attack" version for the Army, with a simple, massive high-explosive or incendiary warhead. Minimum range, determined mainly by the safe arming distance for the warhead, is only 56 ft/17 m, making this an ideal weapon for ambushes in urban or wooded terrain. Maximum velocity of the missile is 984 fps/300 m/s, and the time of flight to 500 m/1,640 ft is only 2.25 seconds. While its size and weight will probably mean that only one Predator per Marine will be carried, it will give a rifle squad back its lethal-

An M1A1 Abrams main battle tank assigned to BLT 2/6 in the well deck of the USS *Whidbey Island* (LSD-41). Note the openings on the left rear and aft deck for the air inlet and exhaust stacks.

JOHN D. GRESHAM

ity against armor and other heavy targets. In addition, the growth potential of Predator, as well as the Javelin system, means that these systems will be in service well into the 21st century.

Armored Fighting Vehicles

The Marine Corps today has a small but vital force of armor, which is designed to provide support to the rifle units that are at the core of its being. It is a force focused on supporting Marines in the field and helping them accomplish their missions. Amphibious tractors are used to deliver troops to the shore under armor. The wheeled force of Light Armored Vehicles (LAVs) is used to provide screening and reconnaissance, as well as an under-armor anti-tank system. And the small force of main battle tanks (MBTs) provides a hard edge to the rest of the force, both in offensive and defensive operations. All of these vehicles are part of the TO&E of the Corps because they are needed on a modern battlefield, not because they are easy to support and move around. That perhaps is why the Corps is asking the question about whether or not MBTs and other armored vehicles will actually be needed in the future. This question is part of the ongoing Sea Dragon project at the Commandant's Warfighting Laboratory at Quantico, Virginia, and will be under study for some time to come. Meanwhile, armored vehicles will remain part of the Corps.

General Dynamics M1A1 Abrams Main Battle Tank

The Marines acquired their first tanks during World War II as hand-me-downs from the U.S. Army. Though tanks have seen action with the Corps in virtually all of their combat actions since that time, they never have been the center of the Marine combat force. Always used to support rifle units, they have mostly been deployed in small units like platoons or companies. From the 1960s to the Gulf Crisis in 1990, the armored fist of the Marine Corps was based around the M48 and M60-series Patton tanks. These were the last U.S. MBTs that utilized cast-hull-and-turret construction,

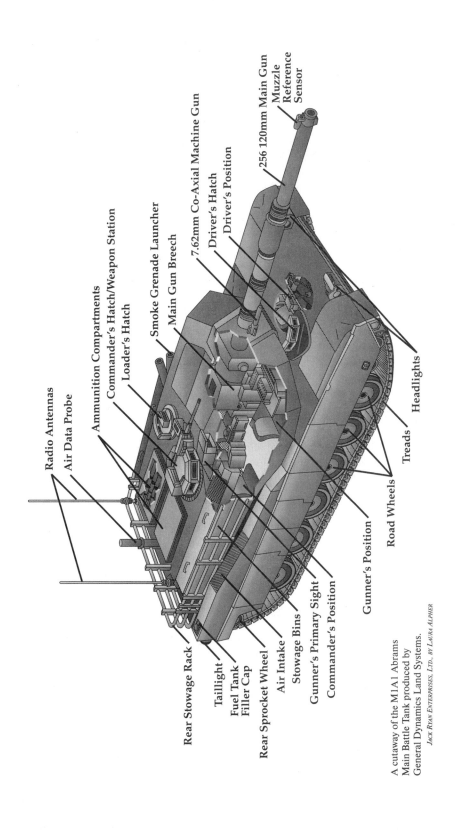

Radio Antennas

Air Data Probe

Ammunition Compartments

Commander's Hatch/Weapon Station

Loader's Hatch

Smoke Grenade Launcher

Main Gun Breech

7.62mm Co-Axial Machine Gun

Driver's Hatch

Driver's Position

256 120mm Main Gun

Muzzle Reference Sensor

Headlights

Treads

Road Wheels

Gunner's Position

Commander's Position

Gunner's Primary Sight

Stowage Bins

Air Intake

Rear Sprocket Wheel

Fuel Tank Filler Cap

Taillight

Rear Stowage Rack

A cutaway of the M1A1 Abrams
Main Battle Tank produced by
General Dynamics Land Systems.

Jack Ryan Enterprises, Ltd., by Laura Alpher

and served with honor for almost three decades. But by 1990, they were badly dated in terms of mobility, firepower, and protection. This is not to say that they were not a welcome addition to the forces that served in the Persian Gulf. On the contrary, when the M60 tanks of the First Marine Expeditionary Force's (I MEF) 3rd Tank Battalion rolled off of the ships of Maritime Preposition Squadron Three (MPSRON 3), they represented the first heavy armor to arrive in support of Operation Desert Shield (in August 1990). Equipped with reactive armor, they held the line until the M1A1 Abrams MBTs of then-Major General Barry McCaffrey's 24th Mechanized Infantry Division arrived in September.

While other Army armored units arrived in the fall of 1990, the Marines continued to use their elderly M60s. Still, the limitations of the old Pattons were not lost on the leadership at Central Command (CENTCOM) headquarters. For this reason, the British 7th Armored Brigade (the "Desert Rats"), and later the 2nd Armored Division's "Tiger" Brigade, augmented I MEF with their more modern tanks and armored fighting vehicles. As the run-up to Desert Storm started, the leadership of the Marine Corps decided to do something about the shortcomings of the MBT force, and decided to request an early introduction of the M1A1 Abrams into service.

The story of the M1A1 coming into service with the Marines started in the late 1980s, when they ran compatibility trials with the Abrams. Marine Corps requirements had not really been considered when the M1 was being designed and developed by the Tank and Automotive Command (TACOM) in Warren, Michigan. In fact, the Marines have usually had very little to say when it came to the design of MBTs, and the M1 was no exception. This is not to say that the M1 was in no way compatible with Marine requirements. It was. But the Abrams was developed to be transported in the C-5 Galaxy and C-17 Globemaster heavy transport aircraft, without any particular eye to future use in the Corps. By the late 1980s, though, the obsolescence of the M60 was obvious to the Corps leadership, and moves were begun to bring the Abrams into Marine service.

The major additions and changes to accommodate the Marine mission involved the addition of a fording kit, which provided the M1's gas turbine engine with a steady supply of water-free air. This involved the addition of several tall stacks that are installed whenever the Abrams is involved in crossing streams or other water hazards, or emerging in the surf-line from a landing craft. Plans went forth to begin procurement of a small force (about four hundred) of the M1s to upgrade the Marine MBT force in the early 1990s. The 1990 crisis in the Persian Gulf short-circuited these plans. When it became clear in November that an offensive to evict Iraq from occupied Kuwait would be required, and not wanting his Marines to fight in obsolete MBTs, General Al Gray (the Commandant at the time) requested that TACOM send the Marines in the Gulf an allotment of M1A1 MBTs to flesh out one tank battalion (the 2nd) of I MEF. The 2nd Tank Battalion fought their way through the flaming hell of the Kuwaiti oilfields in February of 1991. Since that time, every tank battalion in the Marine Corps has received the M1. Meanwhile, the Marines procured enough extra tanks to flesh out the embarked tank battalions aboard the ships of the three MPSRONs stationed around the world. The last of these did not come easily, since they were diverted from U.S. Army stocks of the tank. The Army's position was that they needed all of the big iron beasts that they could get, though the diversion of several hundred to meet the needs of the Marines seems a small inconvenience for the Army. In any case, the money for the Marine Abrams program went right back into producing new state-of-the-art M1A2s, which are much more advanced than the A1 models handed over to the Corps.

The M1A1 model lacks the advanced digital data links and electronics of the later M1A2s, but it has the same heavy depleted uranium armor, special M829 "silver bullet" ammunition, and engines as its more modern brethren in Army service. For the Marines, this is hardly a problem, since they tend to use their tanks in four-tank platoons, and are not in need of the extra command-and-control systems designed into the M1A2. This is not to say that they may not desire to have some of the more modern versions later on. They might. The new AAAV is planned to have the same kinds of interconnects into the so-called "digital battlefield" planned for the 21st century, so don't be surprised if the Marines don't have General Dynamics Land Systems remanufacture their M1A1s into A2s sometime down the road.

One of the more interesting M1A1 developments has been the first deployment of M1s with the 26th MEU (SOC) in August 1995. This is the first tank deployment with an afloat amphibious unit in almost five years, and represents a new acceptance of the MBT by those who practice amphibious operations. The unit's commander, Colonel Jim Battaglini (whom we will meet later), wanted the edge that a platoon of four M1A1s might give his unit, especially if they were required to operate in the Balkans. This request was based on a careful evaluation of the Abrams's different assets and detriments. On the plus side was the incredible armor, firepower, and mobility that four such vehicles would give him. With its highly accurate and powerful 120mm smoothbore gun, the four tanks would have more gun firepower than a pair of Aegis cruisers with their twin 5-in./127mm guns. After the incredible reduction in supporting firepower that has occurred over the last five years, this is an important reason for taking the 67-ton steel monsters along. The downside of this has to do with the weight issue. That is, each one of the M1A1s weighs so much that a Landing Craft, Air Cushioned (LCAC) can carry only one M1, while a conventional Landing Craft, Utility (LCU) can carry two. Furthermore, both types of landing craft are limited to delivering them in fairly calm seas and surfs. Finally, the M1A1 has a big logistics tail, requiring regular refueling (it gets about 1 mi/1.6 km for every two gallons/7.6 liters of diesel fuel/JP-8 burned), lots of spare parts, and an M88 recovery vehicle. All this is a significant addition to the load carried by an amphibious ready group. Despite the problems, Colonel Battaglini felt the gains were worth the price, and the first deployment with the tanks has been completed successfully. There will be more to follow. For now, though, plan on seeing the M1A1 in Marine service well into the 21st century.

Light Armored Vehicle (LAV)

Back in the late 1970s, the Marine Corps began to be concerned about its lack of a good, general-purpose armored reconnaissance and personnel carrier. What was required was something smaller, faster, and more agile than an MBT like the M60 or a large personnel carrier like the LVTP-7/AAV-7. Traditionally, the Marines have lacked the kind of armored cavalry units that the Army considers essential to its operations, and the coming of large Warsaw Pact armored forces in the late 1970s worried the Corps leadership. They feared that without an armored reconnaissance and screening force, MAGTFs might be overrun before they could be made ready to repel an armored assault. It was in this context that the Marines began a program to build a family of light armored vehicles to support their operations. The requirement was rigorous, because it specified that the winning design would have to be both armored and capable of dishing out enough firepower to kill an enemy armored personnel or reconnaissance vehicle. In addition, it had to be capable of being lifted by transport aircraft as small as a C-130 Hercules, or carried as a swing load by the new CH-53E Super Stallion helicopter. This meant that the new LAV could weigh no more than sixteen tons, and this almost guaranteed that it would have to be wheeled instead of tracked. Thus, the new vehicle would have to be an unusual kind of armored fighting vehicle these days, an armored car. What sets armored cars apart is that they carry fair armor and weapons, but on a chassis only half the weight of a tracked vehicle. In addition, they are very fast on roads and good terrain, though somewhat less so in poor terrain and driving conditions (snow, mud, etc.). Dating back to World War I, they have been used by reconnaissance and screening forces with great success.

A total of eight contractors submitted bids on the LAV contract, with the winner being declared in 1982. The winning team was composed of Detroit Diesel, General Motors (DDGM) of Canada—which supplied the chassis, and Delco Electronics (part of Hughes/GM)—which built and integrated the weapons turrets. The vehicle itself was based on the Swiss Piranha (designed by MOWAG), a diesel-powered, eight-wheeled vehicle which would carry an M242 25mm Bushmaster cannon and an M240G 7.62mm machine gun in the turret. Fast and agile, it would also be capable of carrying six Marines in the rear compartment, thus allowing it to act as a small armored personnel carrier. While it would not be as capable or as sophisticated as the new M2/3 Bradley Infantry Fighting Vehicle (IFV) that was also just coming into service, it would do its job for about half the cost ($900,000.00 at the time). In addition, it would be far more deployable and mobile across a variety of conditions than the Bradley. Because the LAV was based upon an off-the-shelf design, procurement was fast and the first units were in service by the mid-1980s.

So successful was the initial version that a number of variants were procured. All of them were based upon the same basic DDGM chassis, and generally have a driver and commander, as well as gunners and other crew as required by their respective roles. The driver is located in the left front of the vehicle, where he (USMC armor personnel are currently male) steers with a conventional steering wheel. Other controls (accelerator, brakes, etc.) are also fairly conventional, and the LAV family drives very well. All versions of the LAV are armed with a single M240G 7.62mm machine gun (with two hundred ready rounds and eight hundred additional stowed) on a pintle mount and eight smoke grenade launchers (with eight ready grenades and eight stowed), and are fully amphibious (with only three minutes preparation) for crossing rivers, lakes, and other water obstacles. The LAV family is driven by a 275-hp General Motors diesel engine with all eight wheels being powered (8X8). Thus, even across broken or steep terrain, the LAV is a very quick vehicle.

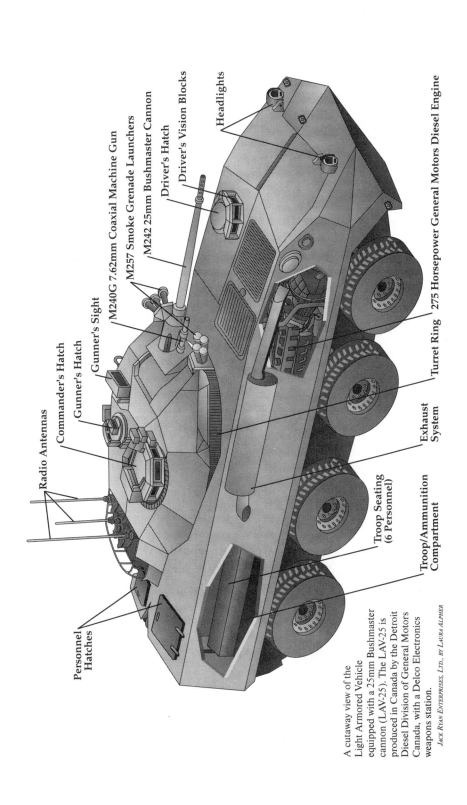

Radio Antennas

Commander's Hatch

Gunner's Hatch

Gunner's Sight

M240G 7.62mm Coaxial Machine Gun

M257 Smoke Grenade Launchers

M242 25mm Bushmaster Cannon

Driver's Hatch

Driver's Vision Blocks

Headlights

Personnel Hatches

Exhaust System

Turret Ring

275 Horsepower General Motors Diesel Engine

Troop Seating (6 Personnel)

Troop/Ammunition Compartment

A cutaway view of the Light Armored Vehicle equipped with a 25mm Bushmaster cannon (LAV-25). The LAV-25 is produced in Canada by the Detroit Diesel Division of General Motors Canada, with a Delco Electronics weapons station.

JACK RYAN ENTERPRISES, LTD., BY LAURA ALPHER

Speeds of up to 62 mph/99.8 kph on hard-surface roads are possible, while the LAV can swim a calm body of water at 6 mph/9.6 kph. Armor protection might be described as "basic," which means that while it can stop shell fragments and fire from heavy machine guns and light cannons, it will probably not survive a hit from an anti-tank missile or an MBT gun. On the other hand, the LAV's high mobility and maneuverability make it capable of running away from everything but an attack helicopter or aircraft.

There are many versions of the LAV; they include the following:

- **LAV-25**—This is the baseline version of the LAV, which is equipped with the M242 25mm Bushmaster cannon and a M240G 7.62mm machine gun. An additional light machine gun can be mounted on a pintle mount. The two-man (commander and gunner) turret has 210 ready rounds—150 high-explosive (HE), 60 armor-piercing (AP)— of 25mm ammunition, as well as stowage for 420 more in the rear compartment if troops are not carried. There are 400 and 1,200 rounds of 25mm and 7.62mm ammunition respectively. The weapons are sighted through an optical sight with a light-image intensifier for night operations, though no FLIR system is yet carried. The turret is powered by an electrically pumped hydraulic system, which is fully stabilized so that it can fire on the move. A total of 401 LAV-25s are in service with the Marine Corps.

- **LAV-AT**—The LAV-AT (for Anti-Tank) uses the same chassis as the LAV-25, and is equipped with a two-man "hammerhead" mount for a twin Hughes Tube-launched, Optically sighted, Wire-guided (TOW) missile launcher in place of the 25mm cannon turret. In addition, a M240G pintle-mounted machine gun with four hundred 7.62mm rounds is carried. Thanks to the erectable "hammerhead" design, the LAV-AT can hide behind a hill or rise and still sight and fire its missiles. A pair of missiles are stored as ready rounds in the launcher, with room for fourteen more in the ammo compartment. A total of ninety-five LAV-ATs are in service with the Marines.

- **LAV-AD**—The newest version of the LAV is the LAV-AD (for Air Defense). The weapons station is armed with two four-round packs of Stinger SAMs as well as a 25mm GAU-12 three-barreled Gatling gun. Equipped with a FLIR targeting sensor and a digital data link for queuing, it is a significant improvement over the existing Avenger system which is based on an HMMWV. Currently, seventeen are being procured by the Marines, with additional procurement likely.

- **LAV-C2**—Every unit needs secure positions where commanders can receive reports and issue orders. Unfortunately, fixed command posts rarely last long in combat, because either they fall too far behind an advancing force, or they are quickly destroyed by enemy artillery or air strikes when their positions are determined by radio-direction-finding equipment. Thus, the armored mobile command post. To give this capability to LAV units, the Marines have purchased a force of fifty command variants. In the LAV-C2, the weapons turrets are deleted, the crew and ammo compartments are made into a single space and equipped with a shelter tent extension for the rear of the vehicle, and there is a battery of radio gear. This includes four VHF sets, a combined UHF/VHF unit, a UHF position-location reporting set, one HF radio, and a single portable VHF set.

- **LAV-L**—Armored units need a lot of supplies in order to accomplish their crucial jobs. Since logistics vehicles of LAV units come under the same kinds of fire as the combat vehicles, they need to be armored as well. For this reason, 94 LAV-L logistics versions were purchased. Based upon the LAV-C, the LAV-L is basically an open compartment for carrying supplies; and it is equipped with a 1,100-lb/500-kg manually powered crane for lifting heavy items like pallets and engines.

- **LAV-M**—One of the shortcomings of Marine armored units is that they have no organic armored artillery units like the Army's M109A6 Paladin 155mm self-propelled howitzer. However, the Marines have developed and deployed fifty armored mortar carriers, based on the LAV. Called the LAV-M, it is equipped with an M252 81mm mortar and carries ninety-nine (five ready, ninety-four stowed) 81mm projectiles. Using the same open-compartment chassis as the LAV-L and C variants, it has a hatch over the rear compartment for the mortar to fire through. The LAV-M also carries a baseplate and bipod for operating the M252 dismounted.

- **LAV-R**—Nearly every family of armored vehicles breeds a recovery version, which can be used to haul broken or damaged vehicles to the rear for repair, and the LAV is no exception. The Marines have acquired forty-five of this type, designated LAV-R. Each LAV-R is equipped with a 9,000-lb/4,086-kg boom crane, a 30,000-lb/13,620-kg

A Marine LAV-C2 (command and control) of BLT2/6 disembarks from an LCAC in Tunisia in 1995.

OFFICIAL U.S. MARINE CORPS PHOTO

winch, a battery of floodlights, an electric welder, a 120/230-volt generator, and a 10-kw hydraulic generator. The crew consists of a driver, commander, and rigger who is cross-trained in welding and other maintenance/repair skills.

Other versions are currently in development, including an electronic-warfare (EW) version that has an array of direction-finding, intercept, and jamming equipment packed onboard. Watch for this LAV-EW version to appear before the turn of the century in USMC service. Other countries using versions of the LAV include Australia, Canada, and Saudi Arabia.

In combat, the LAV has acquired a reputation for reliability and effectiveness, in spite of its light armor and lack of a FLIR thermal sight system. During Desert Storm, LAVs acted as the armored cavalry for the units of I MEF, fixing and finding Iraqi units from the Battle of Al Kafji to the final liberation of Kuwait City. Tragically, the bulk of the LAV losses occurred from friendly fire: One LAV-25 was mistakenly destroyed by a TOW missile from an LAV-AT; and an errant AGM-65 Maverick missile from an Air Force A-10A killed another.

United Defense LVTP-7/AAV-7A1 (Landing Vehicle, Tracked, Personnel)

There is no more traditional Marine mission than to land on a beach and then storm inland to an objective. Doing this mission right calls for an extremely specialized kind of vehicle—the amphibious tractor. The amphibious tractor is a strange hybrid mixture of landing craft and armored personnel carrier, a seemingly impossible mix if you think about it. The first requirement for an amphibious landing craft is that it be a seaworthy boat. It needs to handle well in rough seas, and to be able to come ashore in plunging ocean surf—up to 10 ft/3 m high—without swamping or getting stuck. On top of that, the armored personnel carrier must have good cross-country mobility, all-around firepower, and protection for the crew, at least from small-arms fire and shell fragments. All of those requirements make for a design problem with daunting contradictions. Consider the following. You need to design a machine that can deliver a platoon of twenty-five Marines from a landing ship some miles offshore to a hostile beach, making at least 8 mph/13.5 kph. Then, the machine has to be able to crawl inland at 40 mph/64 kph. And it has to have both protection and firepower. The resulting design was neither subtle nor pretty. But it was a great improvement over previous Marine amphibious tracked vehicles.

The Marines call it an "amtrac" (amphibious tractor), and it's the product of an evolution that began way back in the 1930s in Clearwater, Florida. Donald Roebling was an eccentric millionaire, the grandson of Washington Roebling, the visionary engineer who designed and built the Brooklyn Bridge. One of Roebling's pet projects was the "Alligator," an amphibious crawler designed to rescue hurricane survivors or downed aviators in the cypress swamps of the Everglades. Engineers at the nearby Food Machinery Company (FMC, which built orange juice canning equipment) helped him fabricate parts for the contraption in their spare time. In 1938, the Marines sent an officer to request a demonstration, but Roebling wasn't

A pair of AAV-7A1s
moves to contact
during an exercise.
UNITED DEFENSE

interested. Then came Pearl Harbor. And Roebling changed his mind. Even so, he maintained his quirky integrity: He refused to accept any royalties from the Government for his design patent, and when he discovered that the cost of building the first military prototype, the LVT-1, was $4,000 less than the Navy Department had allocated, he insisted on submitting a refund!

By the end of the war FMC (now the managing partners of United Defense) had built over eleven thousand LVT "Water Buffaloes" in dozens of different types and modifications. They first saw action with the Marines at Guadalcanal in 1942 as cargo carriers, but their moment of glory came in the invasion of Tarawa in November 1943. Planners had miscalculated the tides and underestimated the difficulty of crossing the jagged coral reefs that encircled the tiny atoll. But the amtracs waddled ashore while the normal landing craft were stranded and shot to pieces, thus saving the day and the invasion. The Marines eventually organized a dozen amtrac battalions in the Pacific, and the U.S. Army even formed a few in Europe (these spearheaded the assault crossing of the flooded Rhine in the spring of 1945). Later, in the Korean War, amtracs played a key role in the Inchon landing.

When Marines were deployed in force to Vietnam in 1964, the standard amtrac was the LVTP-5, a forty-ton steel monster that carried thirty-seven men, with a ramp door at the bow and a gasoline engine in the rear. It was a good landing craft, but impractical for the jungles and rice paddies of Southeast Asia. The fuel tanks were located under the floor, which made the vehicle a death trap if it struck a mine. As a result, Marines generally preferred to ride on top, and contemporary photographs often show LVTP-5s decorated with improvised forts on their roofs made of sandbags and chain-link fence.

Even before our direct involvement in Vietnam, the shortcomings of the LVTP-5 were well known, and plans were afoot to make good its shortcomings. In 1963, the Marine Corps asked industry to develop a smaller, less costly amtrac with better cross-country performance. FMC's first LVTPX-12 prototype was finished in 1967; and with minor modifications, it entered production in 1971 as the LVTP-7.

Production eventually ended in 1983 when an improved version, designated the LVTP-7A1 (also known as the Amphibious Assault Vehicle Seven—AAV-7A1), came into service. A total of 995 of the original vehicles have been rebuilt to the AAV-7A1 standard, joining 403 new production units. LVTP-7s also serve with the naval infantry of Argentina, Brazil, Italy, South Korea, Spain, Thailand, and Venezuela.

The AAV-7 is a huge box of welded aluminum alloy, slightly pointy at one end, 26 ft/7.9 m long, 10 ft, 9 in./3.3 m wide, and 10 ft, 3 in./3.1 m high at the deck. Its EAAK armored version weighs 46,314 lb/21,052 kg empty. There is a lumpy weapons station/turret to starboard and smaller lumps for the platoon sergeant's and driver's hatches to port. Marines enter or exit through an enormous hydraulic-powered ramp at the stern, or through a small hinged door in the ramp itself. The accommodations inside can only be described as "austere," with a row of seats along each wall and a removable bench in the middle. The 400-hp Cummins diesel engine is mounted in the right front, where the massive engine block provides some measure of protection to the crew compartment behind it. The Marines who ride it like to complain about the ventilation system, which seems to suck the exhaust fumes directly into the crew compartment. Diesel fumes may smell awful, but diesel fuel is much less explosive than gasoline if your vehicle is hit. The same basic engine is used in the Army's M2/3 Bradley armored infantry fighting vehicle (IFV). The tracks on each side run over six road wheels, each with a torsion-bar suspension system. In the water, the vehicle is propelled by twin jet-pumps which draw water from above each track and spew it out at a rate of 14,000 gallons/53,000 liters per minute. Steering deflectors on the jet pumps allow the vehicle to turn completely around in its own length.

It was originally intended that the powered one-man turret would carry a German-designed 20mm automatic cannon and a 7.62mm coaxial machine gun. But this proved impractical and the production turret carried the classic and reliable M2 .50-caliber heavy machine gun, with a simple reticle gunsight. A thousand rounds of belted ammunition are stowed in two-hundred-round cans which can be reloaded internally. The improved turret of the LVTP-7A1 is powered by an electric motor rather than a hydraulic drive, and supplements the M2 machine gun with a stubby 40mm Mk. 19 automatic grenade launcher with ninety-six belted rounds as the basic load. Eight externally mounted smoke-grenade launchers can deploy a dense white obscuring cloud over a wide arc in a matter of seconds. A switch on the driver's panel can also be activated to dump raw fuel into the engine exhaust manifold, which generates dark obscuring smoke, at the cost of very high fuel consumption.

On land, the AAV-7 can reach a maximum speed of 45 mph/72 kph. It can climb a 3-ft/.9-m vertical obstacle or cross a trench 8 ft/2.4 m wide. In addition, it can climb an astonishing 60% grade and negotiate a 40% slide-slope without tipping over. At 25 mph/40 kph, the maximum endurance is 300 mi/483 km. The driver has an AN/VVS-2 night-vision device, an electro-optical image intensifier, or "starlight scope," that amplifies even the weakest light. On water, the maximum speed is rated as 8 mph/13 kph, but this assumes a calm sea. The AAV-7 is not, however, limited to placid waters; it can operate in higher sea and surf conditions than any other landing craft used in the world's amphibious forces. Theoretically, the AAV-7 can cruise for up to seven hours at 6 mph/9.6 kph, but Marine doctrine, based on considerations of fatigue, control, and navigation, prescribes a run-in to the beach of no more than an hour. An efficient bilge pump serves to keep the crew compartment dry, even in rough seas. Standard navigation equipment is limited to a crude magnetic compass, but today many vehicles have a GPS receiver added to their equipment fit.

In an assault from the sea, Marine platoons will typically embark on their AAV-7s inside the docking well of amphibious transport such as an LPD, LHD, or LSD. The transport will then flood its ballast tanks and open the stern gate, creating a gentle incline for the amtracs to crawl down into the water. They then turn on their water jets and head for the beach. The goal of the amtrac unit is to deliver its passengers safely as close as possible to the objective, where they will dismount and secure the area. The amtracs might then return to their mother ship to pick up a second wave of Marines, or a load of supplies—up to five tons of food, ammunition, and equipment. Note that the amtrac with its driver and gunner "belong" to an amphibious tractor battalion, while the passengers will generally be an embarked Marine infantry platoon belonging to a rifle company. During Desert Storm, most Marine platoons stayed with the same amtrac for the four days of the ground war, using them just like conventional armored personnel carriers.

On dry land, the AAV-7 has severe tactical shortcomings, mostly because of its large size and limited armor protection (in Kuwait, the amtracs were supported by LAVs and main battle tanks). By itself, the AAV-7 is terribly vulnerable to enemy anti-armor weapons, since its thin aluminum armor was designed only to keep out small-arms fire and shell fragments. A bolt-on Enhanced Appliqué Armor Kit (EAAK) has been developed, which adds several thousand pounds of weight, but defeats the Soviet KPV 14.5mm armor-piercing machine gun, which is carried by many threat helicopters, light armored vehicles, and heavy weapon teams. One of the greatest threats to the crew of an armored vehicle is fire resulting from the penetration of a rocket-propelled grenade or anti-tank guided weapon. The amtrac now carries an automatic fire-suppression system, which combines super-fast-acting infrared sensors with quick-discharge bottles of Halon, an inert gas that snuffs out the fire before the fire can snuff out the crew. In practice, combat vehicles usually spend much of their time in battle with the engine idling, to keep the batteries charged and the radio operating while they wait for orders. The standard vehicle has three secure voice radios; but a special command version has six VHF, one UHF, and one HF set, plus a ten-station intercom system. Very soon, it will be fitted with the new SINCGARS-series radios, which will greatly improve the range and quality of communications for the big craft. As currently planned, the fleet of AAV-7s will have to serve about another fifteen years until the arrival of the new Advanced Amphibious Assault Vehicle, which is currently under development. Until roughly 2006, they will have to hold the line, and continue doing their uncomfortable, dangerous job.

The Future: The Advanced Amphibious Assault Vehicle (AAAV)

A quiet little program run out of an office building in Arlington, Virginia, will provide a replacement for the long-serving AAV-7s. The Advanced Amphibious Assault Vehicle (AAAV) is going to be the world's most advanced armored fighting vehicle, with capabilities previously undreamed of by Marines, or by soldiers of any

An Advanced Amphibious Assault Vehicle (AAAV) prototype during high-speed water trials. Production AAAVs will be able to transit over 25 nm/48 km fully loaded through heavy seas in less than an hour.

nation. Our story begins back in the late 1970s when the Marines began to reevaluate their doctrine for forced entry amphibious operations. Ever since "Brute" Krulak took the first unit of amphibious tractors out on their evaluation trials, higher speed through the water has been a desired goal. There even was a stillborn program, the Landing Vehicle Assault (LVA) back in the 1970s, that was designed to achieve that goal. Unfortunately, the technology to achieve the lofty requirements of the LVA specification simply was not there, and the program was terminated in 1979.

Now, you did not need to have a Ph.D. in Systems Engineering back then to figure out that the nature of naval warfare was changing. These changes, though, did not invalidate the high-speed amphibious tractor requirement. On the contrary, it was being rapidly confirmed by current events. One look at any one of a dozen trade publications would have shown you the variety of weapons and systems being developed to attack surface vessels from ships, subs, planes, and shore bases. In short, the closer an amphibious task force approached an enemy shore, the more dangerous it was getting. Take, for example, the British experience in the Falklands War of 1982: In less than a month of amphibious and support operations, the Royal Navy lost two destroyers, two frigates, a pair of landing ships, and a container ship to Argentine air and missile attacks. Several times this number were damaged. The lessons were made clear for the whole world to see: Put yourself within visual range of a hostile shore, and you'll get shot with weapons that will likely hit you and hurt you.

If the Falklands experience was bad, everyone who dealt with such matters knew that the future was going to be worse. They knew that within a few years, you would need to stand away from an enemy shore and deliver your forces from a long distance if your large amphibious forces were to survive. Thus, the Marines and Navy began to develop new ships and delivery systems that would allow a greater standoff from the shoreline during amphibious operations. The Marines' part in this revolution in amphibious warfare doctrine is centered around three systems. The first of these was the LCAC, which allowed the amphibs to stand over 50 nm/91 km from the shoreline. Following the LCAC will be the MV-22B Osprey tilt-rotor transport aircraft, which is designed to replace the CH-46E Sea Knight. With greater speed, range, and payload (by roughly 300%) than the Sea Knight, it allows a ship like the *Wasp* (LHD-1) to stand over 200 nm/366 km offshore and still put its cargo ashore in about an hour. The final system designed to exploit standoff from the beach will be the AAAV.

The AAAV is designed to move at speeds over 25 kn/45 kph, so that the ship that launches it can stand over the visual horizon from the beach. And that's very good. But more important, the AAAV is going to be the finest armored IFV ever built, better even than the Army's M2/3 Bradley fighting vehicle. This is a tall claim for a system that has just had its prime contractor (General Dynamics, Land Systems) selected, but you have to understand the Marine Corps' approach to a design problem like this one to appreciate why. To repeat something I've said before: The technology base of the Marines is very narrow and specifically tailored to the missions of the Corps. Well, the technology elements of the AAAV fall into just that category, which means that the Corps has invested much of its hard-fought research and development (R&D) budget in the AAAV effort. Now, you might ask what it takes to give a high-performance IFV the characteristics of a high-speed powerboat. Well, the following is a list of some of the systems that had to be developed to make the AAAV possible:

- **High-Speed Hull—**Over a series of fifteen years, a series of high-speed-planing-hull designs has been developed to test the feasibility of the AAAV concept. Through the use of three subscale test models (built by AAI Corporation), a basic design utilizing a retractable bow flap, which acts like a surfboard, has been settled upon as the basis for the AAAV design. Called "Skimming Bricks," they are providing a solid database of experience with which to develop the AAAV hull.

- **Dual Mode Propulsion System—**The AAAV will be equipped with an incredible 2,600-hp MTU/Detroit Diesel turbocharged diesel engine. Sealed as a self-contained power unit, it will last up to nine years, and only require an oil change every two years! Working through an automatic transmission, it will drive a pair of powerful 23-in./60-cm-diameter water jets, which will drive it through the water at speed approaching 43 mph/70.5 kph in calm seastates. The propulsion system is so powerful that the twin sets of impeller blades will puree a four-inch-by-four-inch log without a blink. Once it approaches to within a few hundred yards/meters of the beach, the track system will take over, still pushing the AAAV through the surf at around 8 mph/13 kph. Once on dry land, the AAAV will have better mobility than an M1 Abrams tank, while only using about 800 hp from the engine.

- **Retractable Track System—**If the AAAV is to obtain high speeds through the water, the track system must be shrouded from the water flow under the vehicle. To this end, the AAAV's track system retracts into the vehicle, and is dropped when it approaches the beach. All of this takes place in just under twenty seconds. Once on dry land, it utilizes the same kind of hydropneumatic suspension system as the M1, which will give it excellent mobility.

- **Armor Protection System**–The armor-protection system of the AAAV will likely take advantage of the advanced composite armor development being done by United Defense. This will allow the relatively large (27 ft/8.2 m long, 12 ft/3.7 m wide, and 10 ft/3 m tall) AAAV to weigh in at only thirty to thirty-six tons, yet still have better protection than the M2/3A3 variant of the Bradley IFV. In addition, there appears to be some effort to reduce the acoustic, infrared, visual, and possibly even radar signatures of the AAAV.

- **Vehicle Electronics System**–Like the M1A2 Abrams and M2/3A3 Bradley, the AAAV will be equipped to operate on the planned digital battlefield of the 21st century. This will include second-generation FLIR viewer systems for the driver and gunner/commander. In addition, the AAAV will be equipped with the same kind of vehicle electronics as the M1A2 and MZ/3A3, including a digital data bus with onboard diagnostics, GPS tied to a moving map display, a combat identification system to avoid fratricide, and a digital data link fed through three of the new SINCGARS jam-resistant radios. All of this will be controlled by a vehicle software package that will be composed of between 300,000 and 500,000 lines of Ada code, over a Mil. STD-1553 data bus. The driver will even control the throttle, steering, and brakes through a computer, what the Marines call a "drive-by-wire" system!

- **Armament Package** – While this particular item is still being decided upon, current planning has the AAAV equipped with an M242 25mm Bushmaster cannon and a 7.62mm machine gun, like those on the M2/3 Bradley and the LAV-25. There had been plans to perhaps arm the new amtrac with a special 35mm cannon firing time-fused ammunition, but this will probably not happen. But it may carry a twin launcher for the new fire-and-forget Javelin anti-armor missile, if all goes well. All of the AAAV's armament will be usable both in the water and on dry land, and is designed to provide it with the firepower to survive and overmatch other armored vehicles on the battlefield.

- **Payload/Range**–Each AAAV will be capable of transporting a thirteen-man rifle squad and a heavy weapons team—about eighteen personnel plus the three-man crew. Given this load, the AAAV will be capable of swimming up to 65 nm/120 km, or traveling up to 300 mi/483 km on dry land. A normal mission configuration would have the vehicle swimming in from about 25 nm/46 km offshore, moving about 100 mi/161 km to and from the objective, and then swimming back to the mother ship. The minimum high water speed will be 25 mph/40.2 kph, and maximum 43 mph/69.2 km. All of this in seas up to 10 ft/3 m. In the event an AAAV is overturned, it is capable of righting itself automatically in up to Seastate 5.

- **Production Variants**–Current USMC plans call for a total of 1,013 AAAVs to be produced by 2012, the planned termination date of the contract. Of these, there will be 948 transport versions, and 75 configured as mobile command posts. Planned IOC will be in 2006, and the 1,013 AAAVs will replace a force of 1,323 LVTP-7/AAV-7 amtracs. There will not be a recovery version, since it is planned that other chassis (the M88 carried for the M1A1s) can do double duty for this job. Unit cost has yet to be fixed, but will likely be between $2 and $4 million a copy (comparable to the cost of a new M1A2 MBT). Hard work is being done to drive this cost down.

It is likely that the AAAV will be the last armored vehicle procured by the Marine Corps in the foreseeable future. It therefore must be able to survive and dominate its chosen battlespace for most of the first half of the 21st century. It is an ambitious program, though all of the technologies are well proven and understood by all of those involved.

Transportation

While Marine units are anything but "heavy" where vehicles are concerned, they still require their share of trucks and other transportation assets to keep themselves supplied and mobile. For this reason, the Corps has carefully selected a few varieties of transport vehicles to support their expeditionary units, and is generally quite happy with them. A proper complement of transport vehicles is vital for a unit like a MEU (SOC), since there is only so much room aboard its amphibious ships to stow its gear. In fact, while you will find about thirty armored vehicles in such a unit, it will have over one hundred trucks of different types, including those mounting machine guns, mortars, and missiles. Here are the most important of these:

AM General M998 High-Mobility Medium Wheeled Vehicle (HMMWV)

The vast majority of vehicles in the Army and Marine Corps today derive from the classic M998 High-Mobility Medium Wheeled Vehicle (HMMWV). Like the Army, the Corps has embraced the "Hummer," and it has performed nobly in a vast variety of tasks. Produced for over a decade by AM General of South Bend, Indiana, the HMMWV is used for everything from ambulance duty to air defense. Powered by a diesel V-8 engine, it is practically indestructible and can climb anything that a member of any military force in the world might want to take and hold. Today, the Marine Corps' buy of Hummers is pretty much complete, though there will probably be additional buys in the 21st Century as older models wear out. As it is, the M998s that the Marines use are being heavily used, and will probably require a mid-life service life-extension program (SLEP) sometime in the next ten years or so. In the short term, if a Marine offers you a ride, expect it to be in an HMMWV.

M923 5-Ton Truck

No implement of war seems less glamorous than the 5-ton truck, but none is more vital, or causes more sleepless nights for the commander. During World War II, the rapid advance of General Eisenhower's armored spearheads was only made

A Marine Logistics Vehicle System (LVS) transporter truck on maneuvers in Norway in 1994.

possible by a stream of rugged, reliable GM 4X6 trucks. Today's 5-ton trucks are very similar to that 1940s-era design, except that diesel engines replace the old gasoline models. Unfortunately, today's 5-ton trucks are also very old. To put it simply, the Marines' truck fleet is worn out and undersized. With 8,300 vehicles in inventory, you might think there are plenty of trucks to go around, but large numbers of them are tied up on maritime prepositioning ships, in depots, and in support of fixed bases in the rear.

The term "5-ton" describes the nominal cargo capacity, not the empty weight of the vehicle, which is 21,600 lb/9,800 kg. The 5-ton is 25.6 ft/7.8 m long and 8.1 ft/2.5 m wide, and has three axles. The two rear axles are powered and have twin tires on each side, which are tied to a five-speed automatic transmission. The engine is a six-cylinder in-line, liquid-cooled diesel of 250 hp, and the fuel tanks hold 81 gal/306 L, sufficient to take the truck 350 mi/560 km down the highway. The 24-volt electrical system is sufficient to power a radio when one is fitted, and many are also equipped with SLGR GPS receivers. Engineer units are equipped with dump truck and wrecker models, which are subject to particularly severe wear and tear. A major rebuild and SLEP are currently under way to keep the Corps' truck fleet rolling into the 21st century.

Logistics Vehicle System (LVS)

In the category of cross-country heavy military trucks, the Oshkosh Corporation of Oshkosh, Wisconsin, despite stingy and uncertain budgets and extremely stringent requirements, has engineered a line of world-class vehicles. For the Corps, Oshkosh has adapted the Army's HEMTT family of ten-ton 8x8 trucks to produce a transporter good enough for Marines. Known as the Logistics Vehicle System (LVS), it provides the heavy lift capability for expeditionary Marine units. The LVS consists of two units, a standard Mk 48 Front Power Unit (FPU), and a variety of specialized trailers or Rear Power Units (RPUs). The FPU can be attached to any RPU through an articulation joint to produce a flexible 8x8 vehicle. The FPU has

A pair of Marine AV-8B Harrier II Plus's from VMA-542 at MCAS Cherry Point, N.C., on a training mission over the Atlantic. These aircraft are equipped with APG-65 radars so that they can employ the AIM-120 AMRAAM air-to-air missile. *McDonnell Douglas Aeronautical Systems*

a 445-hp liquid-cooled, turbocharged diesel engine. The spacious and fully enclosed cab seats two drivers and provides exceptional visibility. The FPU is 8 ft/2.4 m wide and 8.5 ft/2.6 m high at the roofline, and weighs 12.65 tons/11,470 kg unloaded.

The LVS is equipped with a four-speed automatic transmission, and the vehicle can ford water up to 5 ft/1.53 m deep without special preparations. The fuel tanks hold 150 gal/568 L, providing a nominal range of 450 mi/725 km. The RPUs include cargo trailer, wrecker, crane, and ribbon-bridge variants. The LVS family of vehicles are a critical link in the supply chain that moves bulk fuel, ammunition, and supplies from the beachhead or landing area to the forward combat elements of the landing force. The Marine Corps operates 1,584 of these useful transports, assigned to special combat service support motor transport units. A deployed MEU (SOC) would normally have at least two of these trucks, assigned as diesel fuel carriers.

Marine Corps Aviation

Marine aviation has always had two goals. The first is to support Marines on the ground, and the second is to remain expeditionary, which is another word for mobile and deployable. Today, the Corps deploys one of the most unusual and focused air forces in the world. Its aircraft have been specially selected to support the Marine mission, and this has put the Marines frequently at odds with the leadership of both the nation and the other services. In these conflicts, the Marines have usually won out in the end. In the 1970s, the Administration of President Jimmy Carter killed—several times in fact—the AV-8B Harrier II and CH-53E Super Stallion programs, claiming that they were not necessary or useful. Luckily, the Corps has an awesome Congressional lobby, and was able to sustain the programs until the coming of the 1980s and President Ronald W. Reagan. Today the Marines are winning another battle with the MV-22 Osprey tilt-rotor medium transport aircraft, which then-Secretary of Defense Dick Cheney actually canceled back in 1989. No matter how you look at it, when Marines see something they really want, they will do what is necessary to get it.

McDonnell Douglas/British Aerospace AV-8B Harrier II

Harriers are a species of marsh hawk native to the British Isles that preys on rodents and small reptiles. Not a bad description of the tactical role of this unique British-designed and internationally built aircraft that is now in service with the U.S. Marine Corps. In the 1950s, Sir Sidney Camm of the Hawker Aircraft Company (already a well-respected British aircraft designer) began sketching ideas for a jet plane capable of vertical takeoff and landing (VTOL). The British Government, believing that guided missiles would soon make the manned fighter aircraft obsolete, showed little interest; but the company invested its own funds to build a prototype, the P.1127, which made its first flight on November 19th, 1960, after a series of tethered hovering tests.

Over the years, designers and engineers have proposed many bizarre solutions to the VTOL problem, but the P.1127 used one of the oddest solutions yet, and it proved to be a winner. The key is the Pegasus engine (designed by Dr. Stanley Hooker of the Bristol-Siddeley Engine Company), a turbofan without a tailpipe. The jet exhaust is vented through an array of four nozzles that swivel through an angle of more than 90°. The concept is called "vectored thrust." Point the nozzles straight down, and the plane goes straight up. Point the nozzles aft and the plane zooms off into level flight. To land, reverse the sequence. Sir Sidney observed that, kinetically speaking, it was easier to stop and then land than to land and then try to stop. He was right. Tactically, a VTOL aircraft does not require a ten-thousand foot concrete runway; it can operate from a parking lot, a clearing in the woods, or even a tennis court (if you take down the net). During the Cold War on NATO's Central Front, a Soviet surprise attack might have knocked out most of the concrete runways on Day One, but a force of VTOL fighters, well dispersed and hidden, could have carried on the fight, waging a kind of aerial guerrilla warfare.

The test successes of the P.1127 led to an order in the early 1960s from the Ministry of Aviation for an evaluation unit of nine improved aircraft, under the type designation Kestrel FGA.1 (Fighter, Ground Attack). Pilots from the Royal Air Force (RAF), the U.S. Navy and Air Force (six were shipped to the Navy's flight test center at Patuxent River, Maryland, for evaluation), and the new West German Luftwaffe were invited to test-fly the Kestrel. In February 1965, the RAF ordered the first pre-production batch of VTOL fighters, under the name Harrier, and on August 31st, 1965, the new aircraft made its first flight. (Hawker Siddeley was eventually merged into British Aerospace, while Rolls-Royce took over Pegasus engine production.)

For U.S. naval aviators, wedded to their big deck aircraft carriers, the poky little Harrier (no radar, no afterburner, and look at that cramped cockpit!) was unimpressive in comparison with their mighty new supersonic McDonnell Douglas F-4 Phantom IIs. But for USMC pilots, traditionally committed to delivering close air support that flies really, really close, it was love at first sight. There is a legendary story of how two Marine officers quietly went to the 1969 Paris Air Show (with the backing of the Corps leadership), walked up to the British Aerospace chalet, and told the British representative, "We're here to fly the Harrier!" The rest is history. With the enthusiastic support of the Commandant, the Marines used their considerable political clout to win budget approval for the purchase of a dozen Harriers, modified to carry the AIM-9 Sidewinder missile, and designated AV-8A. By 1977, the force had grown to a total of 110 Harriers, including eight TAV-8A two-seat trainers, equipping four attack squadrons of Marine Air Group (MAG) 32 based at Cherry Point, North Carolina (VMA-223, VMA-231, VMA-542, and VMAT-203). In 1972, the first Harrier detachment went to sea, aboard the USS *Guam* (LPH-7), and proved highly effective. Unfortunately, by 1985, one trainer and 52 single-seaters had been lost in accidents. Like so many early jet designs, the early Harriers were harshly unforgiving of pilot error, especially during the critical transition between vertical and horizontal flight.

One of the lessons learned from the early Harriers was that vertical takeoff was usually both wasteful and unnecessary. A short horizontal takeoff roll saved a great deal of fuel, made it possible to carry a greater payload, and greatly eased the tricky transition from vertical to horizontal flight. In military organizations, every new concept generates a new acronym; hence STOVL, "Short Takeoff, Vertical Landing." For their second-generation Sea Harriers, the British further refined this technique with the development of the "ski jump." Providing an inclined ramp at the bow of a ship, or the end of an expeditionary airfield, gave the aircraft an extra "kick" at the moment of takeoff, and placed it in a safer nose-high attitude in the event of an engine flameout. During the South Atlantic war of 1982, both RAF Harriers and Royal Navy Sea Harriers proved the validity of the concept under difficult combat conditions. Suddenly, the Harrier had become a war-winner. Spain and India ordered various models of Harrier to operate off their small forces of aircraft carriers, and the little aircraft began to develop an international following.

In U.S. naval aviation circles, where doctrine prohibits using the word "small" in the same sentence with "aircraft carrier," the Harrier was regarded as an aberration; and the Marines had to fight a series of bitter budget battles during the late 70s and early 80s to keep the program alive. But they did more than that. In cooperation with British Aerospace, McDonnell Douglas proposed an improved "big wing" version of the Harrier, the AV-8B, Harrier II, which entered service in 1984. The Marines originally hoped to procure 336 of these aircraft to equip every light attack squadron. But by the end of 1993, only some 276 were delivered, including 17 two-seat TAV-8B trainers. At the beginning of 1995, the Marine Harrier force, a small community of eight 20-plane squadrons, was evenly split between the East (Atlantic)

A Marine AV-8B Harrier II of VMA-231, assigned to HMM-264, sits on the deck of the USS *Wasp* (LHD-1). Six of these birds from MCAS Cherry Point, N.C., were assigned to the air component of the 26th MEU (SOC) for their 1995/96 cruise.

John D. Gresham

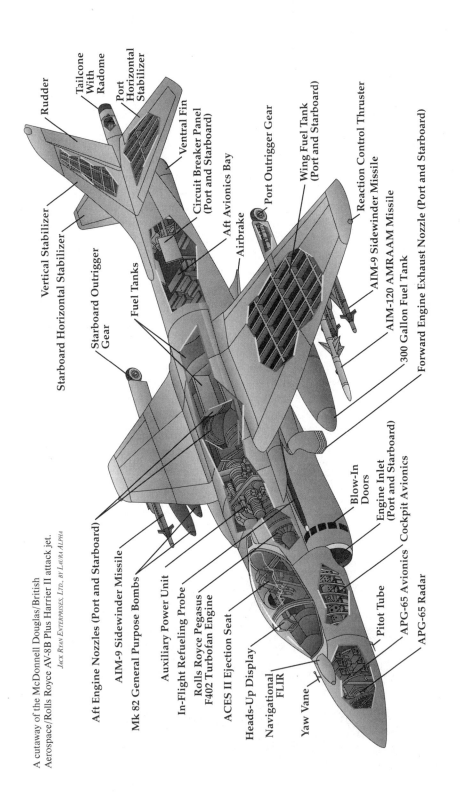

A cutaway of the McDonnell Douglas/British Aerospace/Rolls Royce AV-8B Plus Harrier II attack jet.
Jack Ryan Enterprises, Ltd., by Laura Alpha

Rudder

Tailcone With Radome

Port Horizontal Stabilizer

Vertical Stabilizer

Ventral Fin

Circuit Breaker Panel (Port and Starboard)

Port Outrigger Gear

Wing Fuel Tank (Port and Starboard)

Reaction Control Thruster

Starboard Horizontal Stabilizer

Aft Avionics Bay

AIM-9 Sidewinder Missile

Starboard Outrigger Gear

Airbrake

AIM-120 AMRAAM Missile

Fuel Tanks

300 Gallon Fuel Tank

Forward Engine Exhaust Nozzle (Port and Starboard)

Aft Engine Nozzles (Port and Starboard)

AIM-9 Sidewinder Missile

Mk 82 General Purpose Bombs

Blow-In Doors

Engine Inlet (Port and Starboard)

APG-65 Avionics

Cockpit Avionics

Auxiliary Power Unit

In-Flight Refueling Probe

Rolls Royce Pegasus F402 Turbofan Engine

ACES II Ejection Seat

Pitot Tube

APG-65 Radar

Heads-Up Display

Navigational FLIR

Yaw Vane

and West (Pacific) Coasts. One squadron from Yuma has often been forward-deployed on rotation to Iwakuni, Japan. Squadrons provide detachments of six aircraft for six-month deployments aboard amphibious ships around the world.

The key feature of the AV-8B is an advanced graphite-epoxy composite wing, with integral fuel tankage providing up to 100% greater range than the AV-8A. A built-in-air-refueling probe makes it possible to extend the range even further. The larger wing provides six hard-points, rather than the four on the AV-8A, a 50% increase in armament options. The engine intakes and nozzles were redesigned to reduce drag, and an automatic stability-augmentation system was provided, with small "puffer" jets at the nose, tail, and wingtips, using high-pressure bleed air from the engine. The landing gear is unusual, with a steerable nosewheel and twin-wheel main gear retracting into the fuselage; spindly outriggers at half-span on the wings retract rearward, where the wheels dangle freely in the slipstream.

Visually, the Harrier's most distinctive feature is the sharp angle at which the wings droop downward from root to tip; aeronautical engineers call this "anhedral." This helps to trap a cushion of air under the wing during VTOL operation. The wingspan is 30 ft, 4 in./9.25 m, small enough to fit on shipboard elevators without the added design complexity and weight penalty of folding wings. The Harrier's length is 46 ft, 4 in./14.12 m, and the Harrier does not have (or need) a tailhook. Empty weight is only 13,086 lb/5,936 kg, compared with an F/A-18C fighter-bomber, which tips the scales at 24,600 lb/11,182 kg empty! Maximum vertical takeoff weight is 18,930 lb/8,587 kg, while maximum horizontal takeoff weight is 31,000 lb/14,061 kg, showing the dramatic benefit of a short takeoff roll.

The heart of the Harrier is the Pegasus vectored thrust engine, which gives it such unique qualities. Over the years, the engineers at Rolls Royce have managed to tweak additional thrust out of the Pegasus engine through a series of incremental upgrades. These are shown in the table below:

Engine Model	Aircraft Model	Thrust
Pegasus 101	AV-8A	19,000 lb
Pegasus 102	AV-8B	20,000 lb
Pegasus 104	AV-8B	21,500 lb
Pegasus 11-61/F402-408	AV-8B Plus and Night Attack	23,800 lb

Maximum speed in a "clean" (without external stores) configuration at sea level is 661 mph/1,065 kph. A new bubble canopy greatly improves the pilot's view to the sides and rear. The original twin 30mm ADEN-DEFA cannon (a joint British-French design from the late 1950s) in removable pods under the fuselage have been replaced by the awesome five-barrel rotary 25mm General Electric GAU-12, with the gun in one pod and a three hundred-round ammunition magazine in the other. There are six underwing hard-points, and one on the centerline. The four inboard hard-points have plumbing to accommodate 300-gal/1,135-L drop tanks; and for air-to-air missions up to four AIM-9 Sidewinder or AIM-120 AMRAAM air-to-air missiles (AAMs) can be carried. With regards to air-to-ground ordnance, the following maximum loads can be carried along with the GAU-12 gun pods:

- Up to sixteen Mk 82 500-lb/227-kg general-purpose or Mk 20 Rockeye cluster bombs.
- Up to six Mk 83 1,000-lb/454-kg general-purpose or CBU-87/89/97 cluster bombs.
- Up to four 2.75-in./70mm Hydra 70 Rocket pods (each with ten unguided rockets).
- Up to four AGM-65 Maverick air-to-surface missiles.

Accurate delivery of unguided and laser-guided weapons is ensured by the Hughes AN/ASB-19 Angle Rate Bombing Set (ARBS). In addition, an ALR-67 radar warning receiver and ALE-39 chaff/flare dispensers are fitted in the tail. In high-threat environments the centerline hard-point would be occupied by an ALQ-164 or ALQ-167 defensive-electronics-countermeasures (ECM) pod.

As with so many other weapons systems, the 1991 Persian Gulf War gave the Marines and the Harrier II a chance to prove themselves in combat. Only seventeen days after Iraq invaded Kuwait in 1991, forty AV-8Bs of Marine Attack Squadrons VMA-311 and VMA-542 arrived at Sheikh Isa Air Base (also known as "Shakey's Pizza") in Bahrain after a grueling trans-Atlantic flight. An additional twenty aircraft arrived with VMA-331 aboard the USS *Nassau* (LHA-4). And at the end of August 1990, VMA-311 moved up the Saudi coast to King Abdul Azziz Air Base. By late December, another squadron had arrived, VMA-231, flying eighteen thousand miles—more than halfway around the world—from Iwakuni, Japan, across the Pacific, the United States, and the Atlantic. As the start of the air war approached, in order to get really close to the action, a forward operating location was established at Tanajib, a helicopter field only 40 mi/64 km south of the Saudi/Kuwait border. The narrow 6,000 ft/1,828-m runway provided space for about a dozen Harriers at a time, but a good truck road allowed continuous delivery of fuel and ordnance. The Desert Storm air campaign plan envisioned holding the Harriers in reserve until they were needed for direct support of Marines in the ground war. But early on January 17th, 1991, Iraqi artillery batteries fired on Marine positions near the Saudi coastal town of Khafji, and the Harriers were called in to deal with the situation:

> *"We launched four aircraft. They made two passes each, releasing the one-thousand-pound bombs right onto the artillery pieces themselves. We watched the video of the sortie, and you could actually see the big 122mm guns going end over end as though they were toys."*

—Lieutenant Colonel Dick White, USMC, VMA-311

You can appreciate the skill of the Marine pilots when you remember that these were unguided, "dumb"- bomb attacks. To avoid Iraqi SAMs and gunfire, Harriers tried to stay above 10,000 ft/3,048 m, making targets relatively difficult to spot. The typical attack profile was a 45° jinking dive at 525-kt/960-kph airspeed, with bomb release at between 10,000 and 7,000 ft/3,048 and 2,134 m. Chaff would be dispensed on the way in to confuse enemy radar, and flares would be dropped on the way out to decoy heat-seeking SAMs. By the end of the war, Harriers were ranging up to 210 mi/338 km deep into Kuwait to find targets. Pairs of aircraft would attack

A pair of Marine AV-8B Harrier IIs operate on the deck of the USS *Wasp* (LHD-1) during operations in a Norwegian fjord in 1994.

from different directions, often relying on targeting information from a forward air controller in a low-flying Marine OV-10 Bronco or Navy/Marine F/A-18 Hornet.

During the first week of the air war, Harriers carried one or two Sidewinders for self-defense, but the Iraqi Air Force was neutralized so quickly that no Harrier pilot even saw an enemy aircraft. Of eighty-six Harriers that operated over Kuwait, five were lost to enemy ground fire during the war, and one to a non-combat accident. Since they had experienced the joys of Yuma, Arizona, and Cherry Point, North Carolina, the desert heat of Saudi Arabia was nothing special to the Harrier squadrons, and there were remarkably few problems caused by the blowing powdery sand. In total, Harriers flew 9,353 sorties during Desert Shield and Desert Storm, including 3,380 combat missions, which delivered almost six million pounds of ordnance onto enemy targets. During the war, Harriers rarely flew more than two missions in a day, due to the bad weather.

During Desert Storm, the Harrier was largely limited to its designed role as a daylight/clear-weather aircraft, due to its lack of radar or precision-targeting electro-optical systems. Since wars don't stop at night or take breaks for bad weather, this was a serious limitation. Beginning in mid-1987 (with initial deliveries in September 1989), sixty AV-8Bs have been converted to Night Harriers through the installation of an FLIR sensor and new cockpit lighting compatible with night-vision goggles. The FLIR, mounted in a fairing above the nose of the aircraft, projects a green-and-white video image on the pilot's heads-up display (HUD). A color digital moving map display, using data stored on a laser disc, eliminates the hassle of fumbling with paper charts in a dark cockpit.

Even better things were to follow. With the Sea Harrier, the Royal Navy had already demonstrated that it was possible to fit a radar in the Harrier's nose. With the Harrier II Plus, McDonnell Douglas engineers did not just settle for a simple range-only or air-search radar. They essentially redesigned the airframe to accommodate the powerful Hughes APG-65, the same multi-mode radar used on the F/A-18 Hornet. This means that in the fall of 1996, the Harrier force will add the mighty AIM-120 AMRAAM missile to their weapons suite, making it one of the most dangerous birds

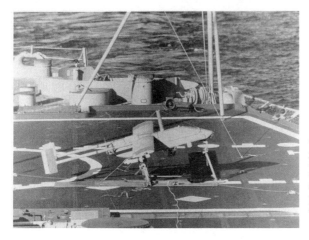

A Pioneer Unmanned Aerial Vehicle (UAV) just before launch from the deck of a U.S. Navy ship. A small rocket motor ignites, powering the craft until the cruise motor takes over.

OFFICIAL U.S. NAVY PHOTO

in the sky. Since the radar adds some 900 lb/408 kg of weight, and extends the airframe by 17 in./43 cm, a completely new fuselage was fabricated, and a new engine installed. The last twenty-four production Harrier IIs were built to the Plus standard. After that, additional aircraft will be "remanufactured." To save money, the wing, tail surfaces, landing gear, ejection seats, and other major components of existing AV-8Bs are being recycled to produce a new aircraft, at about two-thirds the cost of manufacturing a completely new aircraft. Italy (sixteen aircraft) and Spain (eight aircraft) are sharing in the development cost and production of the Harrier II Plus, under an agreement signed in September 1990. The Marine Corps plans to remanufacture seventy-three airframes to the II Plus configuration. The first Harrier II Plus made its inaugural flight on September 22nd, 1992.

Harriers will remain in service with the Marines well into the 21st century. Most likely, they will gradually be replaced sometime after 2010 by a variant of the Air Force/Navy STOVL joint strike fighter (JSF), which is currently in the early stages of development. Between now and then, the variety of weapons loads and mission capabilities are due to greatly increase. For example, there will soon be a competition for a laser targeting/designation pod for the centerline stores station, which will allow the Harrier to employ laser guided bombs and missiles by itself.

Pioneer Unmanned Aerial Vehicle (UAV)

They used to be called "drones" or "remote controlled" (R/C) aircraft. Today we call them UAVs ("unmanned aerial vehicles") to emphasize that they operate without a human pilot on board. The idea of a pilotless aircraft makes many pilots feel uneasy. ("This machine wants your job . . . and it might cause a mid-air.") Since pilots become the Generals and Admirals who call the shots in military aviation, UAVs have had to overcome deeply entrenched institutional resistance to win acceptance. All the same, the advantages of a UAV are obvious. For one thing, compared to a manned aircraft, it can be made very small and cheap. For another, advances in software and miniaturized electronics have made it possible to provide relatively "intelligent" autopilots. And the development of miniaturized video cameras in stabilized mountings ("steadicams") provides high-resolution imagery, day or night. Even if the enemy manages to shoot one down, it makes a lousy hostage.

In early 1996 the Pioneer is the only UAV operational with the U.S. Navy, Army, and Marine Corps. Pioneer was developed in the 1970s by Israel Aircraft Industries (IAI), and it played a key role in the 1982 Bekaa Valley air campaign, in which the Israeli Defense Forces utterly smashed Syria's advanced Soviet-made integrated air-defense system. In 1985, following our own miserable experience in Lebanon, Secretary of the Navy John Lehman ordered the immediate procurement of an off-the-shelf UAV, to be carried on board the newly reactivated and modernized Iowa-class battleships, where they were to be used for gunfire spotting, reconnaissance, and battle-damage assessment, which had so far been impossible in Lebanon. Pioneer won the competition, and entered service with the fleet late in 1986. The following year, the Marine Corps procured additional Pioneers to operate from LPDs or mobile ground bases. In 1991, during Operation Desert Storm, six Pioneer units deployed to the Persian Gulf, flying some 523 missions. One of these unmanned aircraft earned a unique place in aviation history when an Iraqi unit attempted to surrender to it.

Pioneer has a wingspan of 17 ft/5.2 m, and a length of 14 ft/4.3 m, Empty weight is only 264 lb/120 kg, and maximum takeoff weight is 429 lb/195 kg. A 26-hp 2-stroke piston engine drives the pusher-type wooden propeller, located between twin tail booms. The engine also drives an electrical generator to power the sensor package, flight controls, and data link. Pioneer can reach a ceiling of 15,000 ft/4,600 m, but missions are generally flown at 3.280 ft/1,000 m or less. Top speed is 110 kt/204 kph, but the normal cursing speed is 65 kt/120 kph. Mission endurance is around five hours, allowing a tactical mission radius of about 100 nm/185 km. Fuel capacity is 12 gal/49 L of 100-octane aviation gasoline, mixed with a small amount of motor oil. Pioneer breaks down easily into modular components for storage in rugged shipping containers, which the crews call "bird boxes." For shipboard operations, Pioneer requires a rocket-assisted takeoff, which needs very little deck space. For ground operation, there is a truck-mounted pneumatic catapult. At the end of a shipboard mission, Pioneer is flown into a nylon recovery net rigged on the fantail of the ship, like a big volleyball net. When a runway is available, it can make a normal takeoff or landing on its fixed tricycle landing gear.

Pioneer can carry one of two standard electro-optical payloads, either of which can be swapped out in about an hour. The day package includes a stabilized, turret-mounted monochrome video camera with a full zoom. A full-color camera has been proposed as an upgrade, trading off contrast for color information. Color might also require a data link with higher bandwidth. The night package includes a high-resolution FLIR system, which can zoom to fixed lengths, and can be switched between "white hot" and "black hot" display modes. The radio command and data link uses the spread-spectrum technique, which is highly resistant to jamming. Since Pioneer is constructed from lightweight composite materials, it has a very low radar cross-section. It is equipped with a standard Mode 3 IFF transponder, allowing friendly aircraft to track it and avoid airspace conflicts. The system software automatically displays the time and date, geographic coordinates, and range to target on

A Marine UH-1N assigned to the 26th MEU (SOC)'s HMM-264. The "IFOR" markings indicate that the unit was recently part of NATO's Bosnia "Implementation" Peacekeeping force.

JOHN D. GRESHAM

the imagery transmitted over the data link. It also generates symbology showing the aircraft flight direction and attitude, similar to the HUD (Heads-Up Display) of a fighter aircraft, but much simpler.

Four Landing Assault Ships (LPDs) are currently equipped to operate Pioneer. A UAV detachment consists of about thirty personnel and five air vehicles. The control station is an air-conditioned shelter with separate consoles for the flight operator and the sensor operator, who work under the supervision of a mission commander. The flight operator hands off control of an aircraft to a remote Portable Control Station for landings and recovery. A tracking technician operates the tracking and communication system, which requires a pole antenna and a steerable dish antenna, which may be installed on the ship, or mounted on a light truck. A recording technician operates the videocassette recorders, which can feed their signals to other ships and ground stations.

Pioneer has suffered some reliability problems due largely to insufficient procurement of spare parts. In operation, Pioneers often suffer minor damage when they hit the recovery net, and the complex sensor packages demand highly skilled maintenance. Nevertheless, they have proven to be invaluable national assets. So much so that additional vehicles are about to be procured. The Pioneer system will continue to serve well into the 21st century. The prime contractor is Pioneer UAV, Inc., a joint venture of Israel Aircraft Industries and AAI Corporation, located in Hunt Valley, Maryland.

Bell-Textron UH-1N Twin Harvey

Every American war has its distinctive icons in our collective historical imagination. For the Civil War, it's the forage cap and the 12-pounder bronze smoothbore "Napoleon" cannon. For the Second World War, it's the Sherman tank and the GI helmet. For Vietnam it's the "boonie hat" and the Bell UH-1 helicopter. Officially it's called the Iroquois, because the Army insists that helicopters should be named after Indian tribes. But to the troops, it will always be simply the "Huey." Based on a 1955

Army design competition, the UH-1 made its maiden flight on October 22nd 1956. Over eleven thousand have been produced in a dozen major models and countless variants. In 1996 it remains in production around the world.

A major factor in the longevity of an aircraft design is the ability of the airframe to accommodate more powerful engines. No pilot worth his or her wings ever thinks an aircraft has enough thrust or lift. The initial batch of production Hueys had an anemic (by current standards) 700-hp Lycoming turboshaft engine. The current model has a pair of Pratt and Whitneys, each rated at 900 hp each, but with burst transmission power rating of up to 1,290 hp.

Originally intended as an angel of mercy for battlefield casualty evacuation, the Huey proved to be a jack-of-all-trades, providing a bird's-eye view of the battlefield for commanders and forward observers, ferrying troops in and out of hot landing zones, hauling cargo to mountaintop fire bases, and serving as a platform for door-mounted machine guns and rockets. Hueys are currently the only aircraft being used by all four services—the USAF still uses a small number for VIP transport, missile range safety, and support of remote missile silos. The first Huey designed for the Marine Corps was the UH-1E, which entered service with MAG-26 in February 1964. It was equipped with an uprated 1,400-hp engine, a rescue hoist, improved electronics, and a rotor brake (to lock the rotor in position, fore and aft, for shipboard parking).

The current Marine version is the UH-1N, which was introduced in 1971, of which 111 remain in inventory. The pilot and copilot are supplemented for combat missions by a pair of door gunners manning 7.62mm or .50-cal. machine guns. Their primary mission is to act as a command and control platform for MEF and MEU (SOC) commanders. To this end, a special communications package can be fitted to the Marine Huey for use by a task force commander. The Marines figure the current upgrade cost at $4.7 million. The big news about the Huey these days is the planned upgrade program, which will be combined with a similar upgrade for the AH-1W Cobra attack helicopter. Beyond that, current plans have the UH-1N serving until about 2020, when a command and control version of the new V-22 Osprey will probably take over the job.

Bell Textron AH-1W Cobra Attack Helicopter

> *"There were many airplanes, but it was the skinny bird that scared us the most."*
>
> —*Iraqi POW Debriefed after Desert Storm*

The Iraqis called it the "skinny bird." The Marines call it "Whiskey Cobra." "Whiskey" is the military phonetic code for the letter W. Whatever you call it, it's one of the most lethal and versatile flying machines on the battlefield, the Bell Textron AH-1W Cobra. The origins of the attack helicopter can be traced back to the long, bloody colonial war in Algeria in the 1950s, where the French Army experimentally rigged guns up to 20mm to their light Alouette helicopters. In Vietnam, the U.S. Army carried out similar experiments with automatic weapons and rocket pods on various models of the Huey. It soon became obvious that hitting a moving target from a moving helicopter required some kind of fire-control system more sophisticated than the Mark 1 human eyeball. It was also clear that the workload of flying a

A Marine AH-1W
Cobra attack helicopter
of HMM-264 conducts
a low-level run during
an exercise at
Camp Lejeune, N.C.
JOHN D. GRESHAM

helicopter, especially when people on the ground were shooting back, made it necessary to divide the combat tasks between a pilot and a gunner. As helicopter losses mounted, it was also clear that to survive, a gunship would need to present the smallest possible target, and carry as much protective armor as the engine(s) could lift.

The result was the Army's original AH-1G Cobra (Army aviators call it "The Snake"). This used the engine, transmission, and rotor of the Huey, installed in a very narrow fuselage, with a gunner seated in the forward cockpit and the pilot seated behind and above him. Two stub wings provided mounting points for rockets and machine-gun pods, and a nose-mounted turret provided room for a machine gun, or 40mm grenade launcher. The Marines were sufficiently impressed with the new birds to ask for the loan of thirty-eight Army Cobras, which were pressed into service for Vietnam. Experience with these early Cobras convinced Marine aviators that they needed more power, which meant a second engine. Shipboard operation also required adding a rotor brake, which locked the rotor in the fore-and-aft position for reduced stowage space. Designated the AH-1J Sea Cobra, the aircraft was upgunned with a three-barrel rotary 20mm cannon mounted in a power-driven chin turret, allowing the gunner to fire on targets up to 110° off the nose.

The Sea Cobra entered service in 1971 with HMA-269, and sixty-nine aircraft were eventually delivered. An improved version, designated AH-1T was stretched 3 ft, 7 in./ 1.1 m to provide additional internal fuel. It was also equipped to launch the TOW anti-tank missile. This led to the ultimate Cobra design, the AH-1W "Super Cobra," which entered service early in 1986, powered by two GE T700 engines rated at 1,690 hp each. Maximum level speed is 175 kt/320 kph, and the maximum range with internal fuel is 395 mi/636 km. The Whiskey Cobra has a laser range finder and stabilized optical system mounted in the nose, carries chaff and flare launchers, and has a "Black Hole" IR signature-suppression system that mixes outside air with the hot engine exhaust. Up to eight TOW or Hellfire missiles can be carried. The stub wings can even be fitted with launch rails for the AIM-9 Sidewinder, enabling Cobra to engage enemy helicopters or aircraft. By 1996, over one hundred new aircraft had been delivered, while more than 42 older "-1T" birds have been upgraded to the AH-1W configuration. They serve with six operational squadrons and a training unit, HMT-303 at Camp Pendleton, California.

A Marine CH-46E Sea Knight transport helicopter of HMM-264 gets ready for engine start on the deck of the USS *Wasp* (LHD-1). Also known as the "Bullfrog," this elderly bird will be replaced in the 21st century by the MV-22B Osprey tilt-motor transport.

JOHN D. GRESHAM

During Desert Storm, the typical weapons load was a pair of LAU-68 rocket pods on the inboard pylons, with anti-tank missiles outboard. Marine Cobras played a key role in the battle of Khafji, decimating Iraqi armor. One Marine commander watched in amazement as an Iraqi artillery round detonated directly underneath a hovering Cobra. The helicopter shuddered and continued its mission. Despite sandstorms and salt fog, the Super Cobra maintained a 92% mission-readiness rate, 24% better than the Army's more complex (and much better publicized) AH-64A Apache, which required continuous support by civilian contractor technicians.

Current plans for upgrading the Whiskey Cobra will extend the service life of the fleet until at least 2020. One goal is to achieve commonality of engine, transmission, and other systems between the AH-1W and the UH-1N, thereby reducing maintenance costs and spare parts inventories. Key changes will include a new composite four-bladed rotor for improved agility and lower noise and vibration levels, an improved night-targeting system (NTS) based on an Israeli design, and numerous digital cockpit display improvements to reduce the pilot and gunner workload. The NTS system is designed to provide Marine Cobra crews with the same kind of FLIR and laser-designation system that is carried by the AH-64A Apache and OH-58D Kiowa Warrior. This means that it will be able to self-designate for delivery of Hellfire missiles, or even Paveway laser-guided bombs. By the time the program is completed, it will mean that the Cobra fleet will remain viable into the second decade of the 21st century. By that time, an attack version of the V-22 Osprey is a likely development, and may finally replace this classic warbird.

Boeing Vertol CH-46E Sea Knight

In the late 1940s, a visionary group of young Marine officers began to explore the possibilities that rapidly evolving helicopter technology offered for amphibious assault. They called the new concept "vertical envelopment." As the main landing force came ashore over the beach, small helicopter-borne detachments would seize key terrain and blocking positions deep behind the enemy's coastal defenses.

Something like this had been tried with parachute and glider-borne infantry in the Normandy invasion, but the confused and scattered night drop had nearly turned into a disaster. During the Korean War, the small numbers of fragile piston-engined helicopters available had proved their value in medical evacuation of the injured and battlefield observation for commanders. But it took the development of helicopters powered by turbine engines in the early 1960s to make the dream of vertical envelopment a reality.

Forbidden to operate its own fixed-wing armed aircraft, the U.S. Army adopted helicopters enthusiastically, developing a doctrine called airmobile warfare. It was an expensive way to fight a war, though. By one estimate, over four thousand American helicopters were shot down in Vietnam while practicing airmobile warfare. One of the helicopters very much present in Vietnam was the CH-46E, the now-aging workhorse of Marine helicopters. "Sea Knight" may be the official nickname, but Marines call them Bullfrogs. The aircraft entered service with Marine Medium Helicopter (HMM-265) in June 1964. The Navy and Marine Corps procured a total of 624 units, which served through the Vietnam War and in every Marine operation since then. Production ended in 1977, and the current inventory is 242 aircraft. Despite the best maintenance and several service-life extensions, these machines are quite simply worn out. They continue in service today with 15 HMMs, for lack of any replacement. However, when the V-22 Osprey finally enters service they will be retired rapidly.

The CH-46 is a twin-engine, twin-rotor design, which eliminates the need for a tail rotor. The three-bladed fiberglass rotors rotate in opposite directions, and are designed to fold for shipboard storage. Each General Electric T-58-16 turboshaft engine is rated at 1770 horsepower. Both engines are mounted side by side above the tail, leaving the cabin relatively unobstructed, and incredibly noisy. The transmission is cross-connected, so that in case of damage or failure on one engine, the remaining engine can drive both rotors, albeit with vastly less performance. Marines enter and exit through a loading ramp at the rear, or forward passengers doors on either side. Maximum speed is 161 kt/259 kph, and since the fuselage is unpressurized, the maximum practical altitude is about 14,000 ft/4,267 m. The cabin is watertight, and can safely land in choppy seas, but this is an emergency procedure, not a normal operational technique.

A normal flight crew includes pilot, copilot, crew chief, and mechanic. On combat missions, the mechanic is replaced by two door gunners, and up to twenty additional troops can theoretically be carried. The gradual increase in overall weight, due to the addition of defensive electronic countermeasures, armor, and reinforced structure, has seriously reduced the actual carrying capacity of the surviving aircraft. In matter of fact, only eight to twelve loaded troops can be carried. For medical evacuation missions, the capacity is fifteen litters and two corpsmen. Up to 5,000 lb/2270 kg of cargo can be carried as an external sling load. Officially, the combat radius is given as 75 nm/139 km, but in practice the aircraft are limited to about 50 nm/91 km from their mother ship. As for the future, there will be one more planned upgrade of the Bullfrog fleet to keep it going until the MV-22 Osprey arrives in the early 21st century. Only then will the noble CH-46 take its place as a "gate guard" for Marine bases around the world.

A CH-53E Super Stallion heavy transport helicopter of HMM-264 pulls up and away after takeoff. The CH-53E is the fastest and most powerful helicopter currently in Marine service.

JOHN D. GRESHAM

Sikorsky CH-53E Super Stallion Helicopter

"When the balloon goes up, commanders turn to the CH-53 to get the job done. We have seen this in the Gulf War, Somalia, Rwanda, and most recently with the rescue of Captain Scott O'Grady in Bosnia."

—Marine Officer's Letter in U.S. Naval Institute *Proceedings,*
February 1995

One of the star aerial performers in the Vietnam War was an Air Force adaptation of a big Navy helicopter, Sikorsky's HH-3 "Jolly Green Giant." These served with units like the 37th Aerospace Rescue and Recovery Squadron, flying deep into enemy jungle and mountain areas to rescue crash survivors, often under fire. Apparently, to survive on the battlefield, it isn't enough just to be agile and smart; a helicopter needs to be big and tough. The Marines were impressed enough with the HH-3 to order a new heavy assault helicopter, the CH-53A "Sea Stallion," which combined the Jolly Green Giant's fuselage and basic design with the twin engines and heavy-duty transmission of the Army's monster CH-54 Tarhe "flying crane." The Sea Stallion first flew on October 14th, 1964, and entered service with Marine Heavy Helicopter Squadron (HMH) 463 in November 1966. When production of the basic Stallion ended in 1980, the Navy and Marine Corps had taken delivery of 384 aircraft, and additional Stallions were serving with the U.S. Air Force, U.S. Navy, Austria, Germany, Iran, and Israel. By that time though, a second-generation Stallion was in the works, and was ready to enter production, the CH-53E Super Stallion.

The Sikorsky CH-53E is both big and tough. You want redundant systems? How about three engines? And how about seven rotor blades, with main spars forged from titanium? You need to fit a big helicopter on a small deck? How about folding rotor blades and a hinged tail boom, which together reduce the overall length (including rotors) from 99 ft/30.2 m down to 60 ft, 6 in./18.4m! The landing gear is fully retractable and the fuselage is watertight, in case of an emergency landing at sea. An in-flight refueling probe provides almost unlimited potential range, as long

An HMM-264 CH-53E sits fully folded on the port elevator of the USS *Wasp*. Most Marine helicopters have some capability to fold their rotors to save space on ship.

JOHN D. GRESHAM

An artists concept of the McDonnell Douglas/Northrup Grumman/British Aerospace Marine STOVL Joint Strike Fighter entry. This aircraft is designed to replace both the AV-8B Harrier and F/A-18 Hornet for the Marine Corps, as well as the Royal Navy Harrier FRS.2s.

MCDONNELL DOUGLAS
AERONAUTICAL SYSTEMS

as an appropriate tanker aircraft (such as a KC-130 Hercules) is available. The cargo hook can handle an external sling load of up to 36,000 lb/16,330 kg, which means that a LAV or M198 howitzer can be delivered by air. With a sixteen-ton load, the combat radius is 50 nm/92.5 km, though this increases to 500 nm/926 km with a ten-ton sling load. No radar or FLIR is fitted, but the crews train to operate with night-vision goggles. In addition, no armament is permanently fitted, though machine gunners can easily rig machine guns to fire from the forward crew door and either side of the open rear loading ramp. The normal crew consists of a pilot, copilot, and crew chief. Up to fifty-five fully loaded troops can be carried in reasonable discomfort on folding canvas seats. A passenger tip, though: Don't sit directly under the rotor head, where the transmission tends to drip hot hydraulic oil.

The Marines have a requirement for 183 of these mission-critical birds, of which 155 had been delivered by the end of 1995, and production continues at a low rate of four per year. Eleven also have been built for the Japanese Maritime Self-Defense Force under license by Mitsubishi. With a SLEP underway, the Super Stallions are expected to serve until about 2025. By any measure—range, payload, speed, or survivability—the CH-53E is an awesome hunk of aeronautical technology. Back in the days when money was no object, the Soviet Union managed to produce a bigger troop-carrying helicopter, the Mi-26 "Halo." But nobody has ever built a better one.

The Future: Joint Strike Fighter (JSF)

During the 1950s the United States built over a thousand B-47 medium bombers. During the 1990s, the most bitter and protracted budget battles managed to provide only twenty-one B-2A stealth bombers, each costing more than a billion dollars. More aircraft have been killed by cost overruns in the design and development stage than have ever been downed by enemy guns and missiles, pilot errors, or engine flameouts. Projecting the trend into the 21st century, industry observers sometimes joke about a future when the entire defense budget will only suffice to purchase one aircraft; Air Force pilots will be allowed to fly it Monday through Thursday, Navy aviators on Friday and Saturday, and the Marines on alternate Sundays, if it isn't down for maintenance.

With these depressing realities in mind, there are two technical approaches to making a high-performance aircraft affordable. First, make it light. Every non-essential pound/kilogram imposes severe cost and performance penalties. The best example of how to make an aircraft light, simple, and advanced is the Douglas A-4 Skyhawk—Ed Heinemann's classic 1951 design—a five-ton airplane designed to deliver a one-ton nuclear bomb with a single engine of 7,700 lb of thrust. Second, make it generic. That is, make a single basic airframe design serve the widest possible range of roles and missions. Beginning in the late 1980s, under the acronym JAST (Joint Advanced Strike Technologies), a Defense Department Program Office made a serious effort to push these approaches right to the "edge of the envelope." The aerospace industry, seeing the only opportunity for a major new program in the opening decades of the next century, responded with enthusiasm. Now called the Joint Strike Fighter (JSF), the program office is headed by a rear admiral, who reports to an Air Force Assistant Secretary.

Pilots tend to be exceedingly suspicious of anything with the word "joint" attached to it, unless they are talking to an orthopedic surgeon about a sports injury. From an aviator's perspective, a "joint" aircraft is likely turn into a camel (i.e., a "horse designed by a committee"). The three services have radically different tactical doctrines and tribal cultures, and even the most brilliant design team will face a thicket of compromises in trying to fit one airframe to such widely different customers. If you are flying an aircraft into combat, you want that feeling of confidence that only comes from knowing that the designer made no compromises with anything, including the laws of physics. JSF's program managers, aware of this issue, are striving for a relatively modest goal—80% "commonality" of major structural components and systems.

JSF is actually three aircraft. A conventional takeoff and landing (CTOL) model will replace the Air Force's F-16 Fighting Falcon, with a unit price target of $28 million, and an awesome procurement target of 1,874 units. A Navy version to replace the aging F-14 Tomcat and early F/A-18 Hornet types will have a strengthened fuselage structure and special landing gear for carrier operations, raising the unit cost to between $35 and $38 million, with a requirement for at least 300 units.

The Marine version, to replace the Harrier, will be capable of short takeoff and vertical landing (STOVL), at a unit price of $30-32 million. The Marines want 642 units. Three industry teams are competing for the contract. They include Boeing, McDonnell Douglas/Northrop Grumman/British Aerospace, and Lockheed Martin.

All the designs show a strong influence from the F-22 and F-23 advanced stealth fighter designs, with widely separated twin tail fins, splayed out at a sharp angle. The Boeing design has a hinged air scoop under the nose, which gives the aircraft an uncanny resemblance to a gasping fish. The inlet swings down to increase airflow to the engine at low speeds, and swings up to reduce overall drag at high speed. The twin exhaust nozzles rotate, just like on the Harrier. The McDonnell design looks like a slimmed-down F-23, with sharply swept wings mounted well aft. The Lockheed Martin design has a vertically mounted lift fan, driven during takeoff by power from the main engine, just behind the cockpit. There are small canards (auxiliary wings) mounted just forward of the main wings, which closely resemble the diamond-shaped planform of the F-22. In effect, there will be as many as nine different prototype designs, all using the Pratt & Whitney F119 turbofan developed for the F-22. This was the first turbofan capable of supersonic cruise without use of a gas-guzzling afterburner. General Electric will also continue development of its F-120 engine, which was not selected in the F-22 competition, but represents a viable alternative if F-119 development runs into difficulties.

A Bell Boeing MV-22 Osprey tilt-rotor transport begins the transition to forward flight following takeoff. The engines are beginning to tilt forward and the landing gear are retracting.

BELL HELICOPTER-TEXTRON

A prototype MV-22 Osprey on the USS *Wasp* (LHD-1) during compatibility trials. This aircraft has the wing and rotor blades fully folded in the fore-and-aft position to save stowage space aboard ship.

BELL HELICOPTER-TEXTRON

For the JSF program, failure is not an option. Low-rate initial production is scheduled to begin in 2005, and deliveries to operational units are pegged for 2007. By that date, several generations of combat aircraft will be facing block obsolescence, even if we are lucky and no unexpected new threats emerge. Many types that are familiar sights in 1996 will have been prematurely retired, due to escalating maintenance and support costs, airframe fatigue, and normal peacetime attrition.

The Future: The Bell Boeing MV-22 Osprey

We call it a helicopter only because it takes off and lands vertically, but the V-22 Osprey really performs like a small C-130 Hercules transport. As for the importance of the program, the Osprey is designed to replace the entire fleet of CH-46 Sea Knights, which will be entering their fifth decade of service by the time that the V-22 arrives on the scene. It also represents the single biggest technological gamble in the history of the Marine Corps. On the strange wings of the Osprey, the Marines have bet not only their ability to conduct vertical envelopment assaults, but the whole future of over-the-horizon/standoff amphibious warfare.

Ever since the Wright brothers began to fly heavier-than-air vehicles on the Atlantic shore at Kitty Hawk, North Carolina, there has been a dream that you could build an airplane that would take off vertically like a helicopter and still fly like a conventional airplane. The Harrier represents one set of engineering compromises to achieve this, though at a high cost in range and payload. But even tougher to build than a fighter/bomber is a medium lift transport aircraft with the lifting performance of a CH-46 and the speed and range of a C-130 Hercules. Back in the 1950s the idea was put forth that perhaps you could place the engines of such an aircraft out on the ends of the wings, then tilt the engines in much the same way that the vectored thrust nozzles of the Harrier's Pegasus engine rotate. The first aircraft demonstrates this was the Bell XV-3, which flew in 1955, and spent eleven years testing out the tilt-rotor concept. Following this, NASA had Bell build a more advanced aircraft, the XV-15, which first flew in 1976. This incredible experimental aircraft's achievements are still legend in the flight-test world. It proved once and for all that a tilt rotor transport aircraft was not only possible, but would have some very desirable qualities.

Next came the multi-service Joint Vertical Experimental (JVX) requirement, for over five hundred tilt rotor transport aircraft for combat search and rescue (CSAR), special operations (SPECOPS), medical evacuation (MEDEVAC), and replacement of the entire fleet of CH-46 Sea Knights and CH-53D Sea Stallions. In 1983, a team of Bell-Textron and Boeing Vertol won the JVX contract for design and development of what would become known as the V-22 Osprey. Development continued throughout the 1980s, and appeared to be going well despite the usual glitches associated with any new aircraft. Then, as a cost-cutting move, Secretary of Defense Dick Cheney abruptly canceled the entire program in 1989, leaving Bell Boeing with a big nothing for all their work, and all four of the services scrambling to find replacements for the Osprey. As it turned out, they never did, and this caused a small guerrilla movement to break out among the services to revive the V-22. As if this was not enough of a challenge, there were a pair of crashes by prototype V-22s (neither of which was design-related), which gave opponents lots of ammunition for keeping the

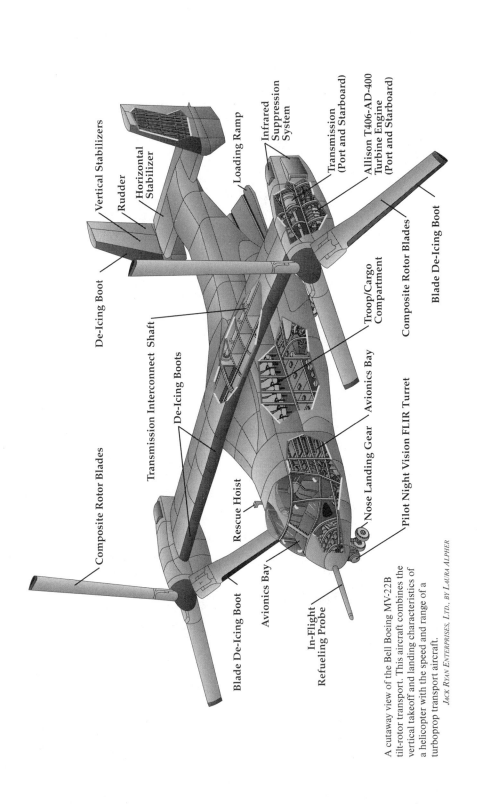

Composite Rotor Blades

Vertical Stabilizers

Rudder

Horizontal Stabilizer

De-Icing Boot

Loading Ramp

Infrared Suppression System

Transmission (Port and Starboard)

Allison T406-AD-400 Turbine Engine (Port and Starboard)

Transmission Interconnect Shaft

De-Icing Boots

Rescue Hoist

Troop/Cargo Compartment

Avionics Bay

Composite Rotor Blades

Blade De-Icing Boot

Nose Landing Gear

Pilot Night Vision FLIR Turret

Blade De-Icing Boot

Avionics Bay

In-Flight Refueling Probe

A cutaway view of the Bell Boeing MV-22B tilt-rotor transport. This aircraft combines the vertical takeoff and landing characteristics of a helicopter with the speed and range of a turboprop transport aircraft.

JACK RYAN ENTERPRISES, LTD., BY LAURA ALPHER

program canceled. Though no one was lost in the first accident, in the second, all seven aboard were killed; and things looked bleak for the Osprey and those who had backed it.

Then, in 1993, good things began to happen for the V-22. The coming of a new Administration allowed the Department of Defense to take a fresh look at the aircraft and the requirements it was meant to fulfill. After a small mountain of studies, the Clinton Administration decided to restart the Osprey production program, and began to work towards a planned initial operational capability for the first squadron of 2001. Since that time, the first new production Ospreys, officially designated MV-22B (this is the Marine variant), have been mated and are moving towards final assembly. The first flight is scheduled for 1996, and the program is moving along well; it's on time and on budget. As an added bonus, the other three services have reevaluated their requirements and are beginning to get back into the V-22 program, with the Air Force's SPECOPS program first among the newcomers. Current program production plans have the USMC buying 425, the Air Force 50 for special operations, and the Navy 48 for CSAR, for a planned total of 523 units. Current cost estimates place the average flyaway cost (including non-recurring R&D costs) of around $32 million a copy, though Bell-Boeing thinks that they can get that down to under $29 million.

As currently planned, the MV-22 will be about 57 ft, 4 in./25.8 m long with a wingspan of 50 ft, 11 in./15.5 m, and a height of 22 ft, 7 in./6.9 m. It will weigh in empty at 31,886 lb/14,463 kg, and will have a maximum takeoff weight of 60,500 lb/27,947 kg in a STOVL mode. Maximum payload will be twenty-four fully loaded troops or 20,000 lb/9,072 kg of cargo. Performance will include a top-level flight speed of 314 kt/582 kph and a maximum ferry range of around 2,100 nm/3,829 km, and a tactical range of around 1,800 nm/3,336 km. These are impressive numbers for an aircraft with roughly the same folded dimensions as the CH-46. Inside the MV-22B will be a cockpit that is arguably the most advanced of any aircraft in the world. Based on the cockpit of the Air Force's MH-53J Pave Low III SPECOPS helicopter and the MC-130H Combat Talon II aircraft, it has undergone many improvements in the years that the program has taken to mature. This is a good thing, because a few years back, I nearly killed myself and a few other folks in a full-motion V-22 flight simulator, trying to fly the thing like a normal helicopter. Today, the MV-22's two man cockpit looks a lot like a normal military cockpit, with a control stick, left-side thrust control lever, and a whole panel of flat multi-function displays (MFDs) to show them all of the vital flight data. This includes a moving-map display tied to a GPS aided inertia navigation system, so that pinpoint, split-second landing operations can become the rule rather than the exception. There is also a FLIR pilotage system to allow enhanced night operations. The entire aircraft is sealed against chemical, nuclear, and biological threats by an overpressure/filter system.

Flying this new bird is, to say the least, a bit strange. I got to try it on the new mission simulator at Bell's Plant in Fort Worth, Texas, and it was an eye-opener. To lift off, you advance the thrust control lever on your left forward, and the MV-22 lifts off smoothly. To transition to high-speed level flight, you push a small thumb-wheel on the thrust control forward, and the engines rotate down in 3° increments.

Once they are in the full "down" position, you are essentially flying a high-performance turboprop transport, which is actually quite agile and comfortable. To land, you begin to pull back on the thumbwheel, causing the engines to rotate back to the vertical. The fly-by-wire system makes this very comfortable, and your eyes begin to transition to the MFD, which tells you the sink rate towards the ground. This is the critical condition to watch, because you need to keep this fairly low. Tilt-rotor aircraft cannot apply power quite as fast as normal helicopters, and you have to think a little "ahead" to make this go smoothly. If you've done it right, you should feel a gentle "thump," and you are down.

Right now, the biggest problem facing the Osprey program is the planned rate of procurement. Originally, the Clinton Administration had planned to buy less than two dozen a year. This meant that the buy would run out to the year 2025. General Krulak is planning to speed this up to around thirty-six a year, so that the procurement of MV-22B will be completed before 2010. In this way, he hopes to avoid a funding conflict between Osprey and the planned JSF buy.

Getting There:
The 'Gator Navy

Amphibious warfare is one of the most expensive and risky forms of combat ever devised. You have to move difficult and unruly cargo (combat troops), feed and care for them, and safely bring them through hostile waters to an enemy shore. You have to then deliver them, with all of their equipment and supplies, onto a beach to fight their way inland. And then they have to wait for follow-on forces or evacuation at the end of the mission. Today, most nations with coastlines have radar-equipped planes and patrol boats to locate an incoming force over the horizon. They are armed with guided missiles, coastal artillery, and mines.

When they were planning the Normandy invasion, General Dwight D. Eisenhower and his staff in 1944 faced this problem. But things have changed a lot since then. The weapons of our time are far more lethal than those of World War II; and General Eisenhower had the unlimited resources of American, British, and Allied industry to build over five thousand ships and landing craft to "kick-in-the-door" of Nazi-occupied France. Today, a theater commander in chief (CinC) might be lucky to have a dozen such craft within a single amphibious ready group (ARG). Eisenhower could land five divisions with over 100,000 men on D-Day (June 6th, 1944). Today's CinC might have only 2,500 fighting men and women to throw onto a hostile coast. Clearly, in the fifty years since we invaded our way to victory in Europe and the Pacific, the problem has become more difficult.

The drawdown of amphibious shipping and landing craft by the U.S. over the last few decades has been so precipitous that it has occasionally destabilized the global balance of power. When the Royal Navy announced plans in 1982 to decommission its tiny amphibious force—two Assault Ships (LPDs), and six Landing Ship, Tanks (LSTs)—Argentina promptly invaded the Falkland/Malvinas Islands. Similarly, the perceived inability of the United States to project power into the Persian Gulf in 1979 encouraged the Soviet invasion of Afghanistan and the takeover of the U.S. Embassy in Tehran by Iranian militants. By early 1996, our amphibious shipping force had fallen to its lowest level since before Pearl Harbor. This leaves the United States and her allies with just two options. One is to simply abandon the ability to influence events in global crisis areas beyond our shoreline. The other choice is to make the best use of the limited assets we retain. Luckily, we have adopted that one. This is the core of *From the Sea* and *Forward from the Sea*. The concept of operations outlined in these documents allows the U.S. to maintain a "kick-in-the-door" capability, without bankrupting the treasury or compromising other commitments.

We don't yet have all the tools to accomplish the missions spelled out in *From the Sea/Forward from the Sea*. U.S. amphibious forces during the next decade or so will be a mix of older equipment and ideas and newer "over-the-horizon" (OTH) concepts. As older ships retire, a limited building program will eventually stabilize the amphibious fleet at about thirty-six ships. There will be several hundred landing craft of various types, three Maritime Prepositioning Squadrons (MPSRONs) with a dozen or so ships, and a few older ships in the Ready Reserve Fleet (RRF). And that will be it. Anything else we need will have to be borrowed from the British or another ally, or chartered from commercial shipping.

The good news is that it will all probably work, at least under the current world order, or rather, disorder. The key is a new view of amphibious warfare that has quietly taken hold within the military over the last twenty years or so. This is the OTH concept. Instead of closing within a few thousand yards/meters of a beach to unload troops and equipment, the big ships will stay between 25 and 250 nm/46 and 457 km offshore, out of range of enemy sensors and weapons. High-speed vehicles like the Landing Craft, Air Cushioned (LCAC), the new Advanced Amphibious Assault Vehicle (AAAV), the MV-22 Osprey tilt-rotor aircraft, and the CH-53E Sea Stallion helicopter will deliver the assault forces to their assigned targets. With these vehicles and aircraft, there will be less need to be so picky about beach topography (sand, shale, beach incline, etc.) or oceanographic conditions (tides, seastates, etc.). This will mean that the area of operations (AOR), or "battlespace," can be vastly expanded, making the problems of defending a coastline more difficult. The result of all this will be to increase the value of our limited amphibious forces, while decreasing the risks they face. Meanwhile, those thirty-six amphibious ships will be the most capable and powerful ever built.

This chapter will introduce you to the Navy's amphibious vessels. It will give you some feel for how the men and women of the 'Gator Navy live, as they do their hard, dangerous jobs in the "littoral regions" of the world.

Amphibious Shipping/Landing Craft Development

The fragile, lightweight oared warships of antiquity could be hauled up on a beach, but they were awkward platforms for amphibious assault. Alexander the Great's siege of the island fortress of Tyre on the Lebanese coast in 332 B.C. saw early examples of ingenious improvisation on both sides, with ships lashed together to provide platforms for siege towers and battering rams. The Viking longships of the Dark Ages demonstrated amazing seaworthiness and adaptability—the amphibious raiding strategy of the Norsemen dominated Europe for centuries. During the age of wooden sailing ships, various nations built landing barges with assorted fixtures (ramps, cranes, etc.) to load and unload troops, horses, and equipment. This is all well and good, but having a big navy and lots of troops does not guarantee a successful amphibious assault. The Spanish Armada in 1588 and Napoleon Bonaparte's aborted invasion of England in 1805 are classic examples of failures. The land-oriented military doctrine of continental empires could never quite solve the problem of crossing even the 30 nm/55 km of English Channel. In 1940, German General Staff planners thought crossing the Channel would simply be a "river crossing along a wide front." Wrong!

Many factors go into the execution of a successful amphibious assault, including air supremacy and sea control. But crossing the interface of land and water, known to most of us as "the beach," is the most difficult part, technologically and militarily. The beach or littoral zone can be a dangerous place, even if you just want to swim and sun yourself. Now try to move thousands of troops, hundreds of vehicles, and thousands of tons of equipment and supplies across it. It takes a lot of horsepower and engineering to create machines that enable men to do the job, and more than a little political capital. That is where our story about landing craft and amphibious ships starts. During the period between the World Wars, the problem of beach landing obsessed several groups of officers and engineers on both sides of the Atlantic. In America, Marines searching for a new mission to justify their continued existence saw amphibious assault as their future. During the 1930s they observed with interest a series of small operations by Japanese naval landing forces in China, utilizing specialized landing barges.

On the other side of the ocean, British officers, studying the failure of their 1915 invasion at Gallipoli, looked for ways to cross the beach rapidly to conduct mobile operations inland. The Gallipoli landing was the idea of the former First Lord of the Admiralty, Sir Winston Churchill; and when it bogged down into a bloody stalemate, it nearly ended his political career. These problems became even more urgent for Churchill in World War II, after another disastrous landing in Norway and the Nazi conquest of Europe in 1940. For all of his many shortcomings as a strategist, Churchill clearly saw the need to build ships and landing craft in vast numbers if Europe was to be liberated from Hitler.

Even as the Battle of Britain was being fought in 1940 to fend off German invasion, the British were designing their first purpose-built landing craft, the Landing Craft, Assault (LCA, the American designation when we built them from the British design). Just over 40 ft/12.2-m long and powered by a pair of 65-hp Ford V-8 gasoline engines, they could haul thirty-five troops and 800 lb/364 kg of equipment some 50 to 80 nm/91 to 146 km. The open-topped LCA had a long, flat bottom suitable for beaching, an armored front to protect the embarked troops, and a bow ramp for rapid off-loading. LCAs could hang on a transport ship's davits, like large lifeboats. Assault troops boarded them by climbing down rope ladders and nets. The same features would appear on almost every landing craft, including the Landing Craft, Utility (LCUs), and Landing Craft, Medium (LCMs), still in use today. From the LCA design came literally dozens of specialized landing craft that would be used for the next half century. At the same time, American engineers were coming up with their own designs, such as the famous "Higgins" boat, which was based on a surf-rescue craft. Evolutionary improvements led to standard designs like the Landing Craft Vehicle, Personnel (LCVP), built in the thousands as the backbone of the landing craft fleet that helped win World War II.

Once the landing craft had been developed, the next problem was getting the frail little boats across the oceans. Amphibious operations are fought against the elements of the ocean and the shore, as well as the enemy's defenses. The flat-bottomed assault boats, while handy in the shoal waters of beaches and atolls, needed larger "mother" vessels to move them close to their objectives. This requirement led to specialized attack transports, grouped into amphibious "tractor" groups. Early attack transports were converted freighters and passenger liners. They lacked cranes and other handling gear for hoisting out and loading embarked landing craft and troops. Later in the war, purpose-built ships were significant improvements, but they still had to run in close to the beaches to unload; and they were vulnerable to enemy coastal artillery, mines, and aircraft.

An important development was the Landing Ship, Tank (LST—their crews said it stood for "large, slow target"). This was an oceangoing vessel that could beach itself, open its bow doors, drop a ramp, and then off-load vehicles up to the size of heavy tanks directly onto the beach. The last U.S. Navy LSTs (built in the 1960s) only recently retired from active service. Another special-purpose amphibious ship was the Landing Ship, Dock (LSD), equipped with ballast tanks and an interior well deck that allowed landing craft to load in relative safety. By flooding the well deck, the landing craft could easily float out, without the need for hoists or cargo nets to load the boats. The well deck was so successful that all thirty-six of the U.S. Navy's amphibious ships in the 21st century will have one. Other specialized amphibious ships built during World War II included amphibious command ships and fire-support vessels carrying rockets and guns.

These bizarre craft provided the sealift to liberate North Africa, Europe, and the Pacific. Soon after the victory they helped to win, virtually all of the landing craft and amphibious ships were sold for scrap or mothballed. The atomic bomb seemed to signal an end to amphibious warfare. This attitude would not last. The Korean War marked the rebirth of amphibious operations. Recalled from the mothball fleets, World War II amphibious ships provided General Douglas MacArthur with the lift for his brilliant landing at Inchon in the fall of 1950. Some of these same ships served off Lebanon when that troubled land erupted in 1958. While the amphibious vessels of the Second World War held the line in the 1950s, the U.S. Navy began to design new amphibious ships, suitable for the atomic age. The most important of these was the assault helicopter carrier (LPH), designed to carry a Marine battalion and land it by helicopter onto an enemy shore. The first LPHs were converted World War II aircraft carriers, but the purpose-built *Iwo Jima* class (LPH-3) was in production by the early 1960s. By the end of the decade, in addition to the LPHs, new classes were in production—the *Newport* class (LST-1179), the *Charleston* class amphibious cargo ships (LKA-113), and the *Anchorage* class (LSD-36), as well as new designs like the *Austin* class (LPD-4), which was equipped with a well deck. These ships maintained a credible amphibious lift capability through the Cold War years. Despite all this building, the tactics of assault with landing craft through the surf-line from a few thousand yards offshore had changed little since World War II. Landing craft themselves had changed little, with conventional medium (LCMs) and utility landing craft (LCUs) constructed as late as the 1980s.

While the technology of amphibious assaults had not changed much by the close of the 1960s, the soldiers they carried would. After the experience of Vietnam, with its conscripted combat troops, military leaders were forced to accept an all-volunteer force as the basis for a new, professional military in the 1970s. This change had many consequences. One not often noted affected the U.S. Navy: Realizing that it would have to take better care of all-volunteer crews, the Navy began to improve the habitability of warships. In the 18th century, Samuel Johnson observed that serving on a warship was like being in jail, with an added chance of drowning. This was not quite true on the World War II-vintage ships of the Vietnam era, but they were hardly designed for comfort. Naval architects try to pack as many men as possible into a warship. Personnel are needed to operate a maximum of weapons, sensors, and other systems. The emergence of the all-volunteer Navy in the 1970s meant that future warship designs would need improved habitability standards.

Another Navy goal in those days was to make warships capable of accomplishing more various missions. The results were seen in the *Spruance*-class (DD-963) destroyers and the *Tarawa*-class (LHA-1) helicopter assault ships. The LHAs were revolutionary; they were capable of operating both landing craft and helicopters, plus the new AV-8 Harrier V/STOL fighter bombers. Lessons learned from the *Spruances* and the *Tarawas* were applied to every future class of U.S. warship. Unfortunately, both types suffered rapid procurement-cost growth. The *Tarawas* were originally priced as a class of nine, but only five could be bought during the double digit inflation of the 1970s. The late 1970s were a bad time for the Navy in general, and amphibious forces in particular. The Administration of President Jimmy Carter took an axe to the Navy budget, particularly in shipbuilding, operations, and maintenance. And by 1979, when a series of crises broke out in Southwest Asia, the U.S. had only a minimal amphibious capability. Amphibious forces are expensive to build and tough to maintain. They are often among the first items cut in times of austerity.

As a result of the Carter austerity, planners reconsidered the capabilities of merchant shipping to supplement the specialized 'Gator ships. The first use of containerized merchant ships for amphibious forces was to be seen in the creation of Maritime Prepositioned Squadrons to provide a mobile, floating base for Marine task forces. Three such squadrons would be created, with additional units for the U.S. Army and Air Force. During the 1982 South Atlantic War, the British employed "Ships Taken Up From Trade" (STUFT) to transport the bulk of their landing force and supplies. Both programs showed the limitations of civilian ships to support military operations.

The inauguration of President Reagan in 1981 led to Secretary of the Navy John Lehman's ambitious plans for a six-hundred-ship Navy. This included a follow-on class of LHAs, the *Wasp* (LHD-1) class, and a new class of LSDs, the *Whidbey Island* (LSD-41) class. And procurement of a radical new landing craft began, the LCAC (Landing Craft, Air Cushioned). LCAC was the first new technology for amphibious warfare since the helicopter; its introduction allowed the big ships for the first time to stand away from coastal landing zones. Meanwhile, amphibious warfare capabilities that had been lost after Vietnam were slowly rebuilt. Unfortunately, building ships takes time. The Reagan Administration was history, and the Bush Administration was well along before the new ships began to join the fleet. In fact, the LHD and the LSD-41 building programs continue, more than fifteen years after they were started.

In the 1990s the amphibious forces of America and her allies have been busier than at any time since World War II. In addition to supporting the liberation of Kuwait in 1991, amphibious forces have been constantly engaged in crises and contingencies from Haiti to Somalia. The future of amphibious shipping is of interest to everyone from Marine privates to the President of the United States.

The 'Gator Navy

The U.S. Navy is divided into three distinct communities. There is a submarine navy, with its nuclear attack and ballistic missile submarines. There is naval aviation, with its carriers and aircraft. And last, but not least, is the surface navy, with squadrons of cruisers, destroyers, and frigates, to escort carrier battle groups and vital supply ships. Shoehorned into a corner of the surface navy are a few dozen ships and few hundred small boats and landing craft called the 'Gator Navy. 'Gator refers to the alligator-like ferocity of the Marines when their combat power is combined with the mobility of the Navy. Like their reptilian namesakes, 'Gators can give you a nasty bite, in the water or out.

Command of amphibious shipping was once viewed as a second-class assignment, with less prestige than command of a real warship like a cruiser or destroyer. No more. Today, officers who command amphibious ships and ARGs hold some of the most coveted assignments in the Navy. *Wasp*-class (LHD-1) helicopter assault ships are the largest vessels that a non-aviator can command in the U.S. Navy (only aviators can command big-deck aircraft carriers). At over forty-thousand-tons displacement, with a crew of more than 1,100, carrying almost 1,900 Marines with all of their gear, as well as over forty aircraft and helicopters, an LHD is a major warship! The other new amphibs, like the *Whidbey Island/Harpers Ferry* class (LSD-41/49), are also very large vessels. For comparison, the biggest amphibious ships built by the former Soviet Union were three eleven-thousand-ton *Ivan Rogov*-class LSDs.

Navy plans envision a force of thirty-six vessels of three different types (LHD/LHA, LSD, and LPD), organized into twelve amphibious ready groups (ARGs). These ships could deliver twelve reinforced battalions, each about 1,600 Marines. This would represent about 2.5 Marine Expeditionary Brigades (MEBs) if every ship could be deployed at one time. Unfortunately, ships that stay in the fleet for thirty to forty years need periodic overhaul and maintenance. Large warships spend about one year in four out of service, "in dockyard hands." So, only about three quarters of our amphibious shipping will be available at any time. These ships are split between the Pacific and Atlantic fleets. Not much strength for any particular crisis when you consider the thousands of miles/kilometers of hostile shoreline the U.S. might have to face. For example, during Desert Storm, the Navy assembled four ARGs with a single afloat brigade from both fleets. The arithmetic demands that each and every amphibious ship constructed for the Navy must be highly mobile and sustainable. The "amphibs" are the high-value units in any naval task force—sometimes even more valuable than the big-deck carriers which often accompany the ARGs these days.

Amphibious ships are evaluated by five different capacities or "footprints" as they are known. These include:

- **Troop Capacity**—The number of Marines the ship can comfortably berth, feed, and support.
- **Vehicle Space**—Called cargo,[2] this is measured in square feet of vehicle storage, along with a little extra room for maneuvering vehicles in and out (called "turnout" space). Total area can be converted to standard vehicle dimensions, based upon the footprint of an HMMWV.

- **Cargo Space**—This is a measure of storage space for packaged cargo, supplies, and equipment. Called cargo,[3] it is measured in cubic feet (ft^3).

- **Landing Craft Capacity**—This footprint indicates how many LCAC landing craft can be carried in the vessel's well deck.

- **Aircraft Capacity**—The number of aircraft that can be operated, stowed, and maintained on deck and in the hangar. The capacity is based on the CH-46E Sea Knight helicopter. An AH-1W Cobra attack helicopter only occupies about .5 units of deck space, whereas the new MV-22B Osprey will have a 1.4 equivalent.

These five measures tell you how valuable a particular ship is to an ARG. For example, the new LPD-17 will replace four different ship classes (the LST-1189, LPD-4, LSD-36, and LKA-113) in the ARG. You can see how critical this one ship must be to future ARG commanders.

Amphibious ships are nothing without people. Life for the sailors in the amphibs is a mix of high technology (like satellite communications and navigation) and old-style seamanship (like small-boat handling and the ancient skills of knotting lines). It is also long, hard work. Marines love to practice their exciting tasks in the wee hours before and around dawn. So, whenever the ARG is conducting operations, the ships' crews go on a fatiguing round-the-clock schedule. The work *is* hard; but when you talk to the sailors, they tell you that it's exactly what they joined the Navy to do. 'Gator sailors love their jobs. Senior chiefs tell you it's like the "old" Navy they grew up in. They frequently see the 'Gator Navy as a refuge from the "political correctness" that seems to infect today's U.S. Navy. For officers, life in the amphibs is a chance to truly test themselves in their chosen profession.

Navigation and warfighting in the littoral zones is demanding and dangerous. Inshore operations present all kinds of natural and man-made hazards to the sailor. Consider the cruise of the assault carrier *Tripoli* (LPH-10), which was mined while operating in the Persian Gulf during Desert Storm. The ship survived, albeit with heavy damage. As we have seen, the British Royal Navy learned even harder lessons during the invasion of the Falkland Islands in May 1982. Also, nature is not kind to

The USS *Wasp* (LHD-1) at anchor in Onslow Bay on August 29th, 1995, just prior to her Mediterranean deployment. She is operating helicopters and taking landing craft aboard simultaneously.

Official U.S. Navy Photo

sailors working near shore. Everything from rogue waves to hurricanes can foil an amphibious assault. D-Day, originally planned for June 5th, 1944, had to be delayed twenty-four hours because of storms in the English Channel. Like flying, amphibious landings are unforgiving, and only a complex combination of planning, skill, experience, and equipment can make them successful.

One quick note before we begin. There are many different ways to interpret warship specifications and statistics, and "official" sources often disagree. On matters of fact, I defer to A.D. Baker III's superb biannual work, *Combat Fleets of the World* (U.S. Naval Institute Press). For over two decades, Dave Baker has made this book his life's work, and all of us who write about defense matters are in his debt. I ask your patience in my use of tables. Amphibious ships are number-intensive! Now, let's go aboard!

USS *Wasp* (LHD-1)

It is the largest and mightiest amphibious ship ever built. At over forty thousand tons, it is the largest man-made object to ever move across the land (so says the *Guinness Book of World Records*). The landing helicopter dockship USS *Wasp* (LHD-1) is the lead ship of a seven-ship class that represents the best America's shipbuilding industry can produce. The largest combatant in the U.S. Navy aside from the supercarriers, it is a virtual one-ship task force that can probably take down a small nation by itself. The story of *Wasp* and her sisters is the story of the Navy's amphibious force after the blight of the Vietnam War and the move to an all-volunteer force. It is also the story of a contractor that saw the future and decided to remake itself.

At the end of World War II, the Marine Corps began to examine ways of avoiding amphibious frontal assaults against fortified enemy shores. The losses suffered in taking Japanese island fortresses like Iwo Jima and Peleliu left a lasting impression on Marine and Navy leaders. Out of all this thinking came the concept of vertical envelopment using the new technology of the helicopter. The father of the current Commandant, Victor "Brute" Krulak, was quick to support the concept. And by the mid-1950s, several World War II aircraft carriers had been converted into experimental helicopter assault carriers. Designated LPH (for "Landing Platform, Helicopter"), they proved successful, though their size and large crews made them expensive to operate. The first conversion, USS *Block Island* (LPH-1, ex-CVE-106), was never completed. But several others, including USS *Boxer* (LPH-4, ex-CV-21), USS *Princeton* (LPH-5, ex-CV-37), USS *Thetis Bay* (LPH-6, ex-CVE-90), and USS *Valley Forge* (LPH-8, ex-CV-45), were converted from surplus aircraft carriers during the 1950s and 60s. Even before these conversions were completed, plans were underway for an LPH designed from the keel up. The idea was to pack a Marine battalion and a reinforced helicopter squadron into the smallest hull possible, so that the ship would be cheap to build and efficient to operate. Crew and passenger (i.e. Marine) comfort would be minimal.

The result was the *Iwo Jima*-class (LPH-2) assault carriers, of which seven were eventually built. Designed around the hull form and engineering plant of a World War II escort carrier, they were built for maximum storage density of aircraft, equipment, supplies, and Marines. Ingalls Shipbuilding (now Litton Ingalls Shipbuilding) of Pascagoula, Mississippi, and a pair of government shipyards built the LPHs, and they proved highly successful. Displacing only 18,300 tons (compared to almost 29,000 tons for the *Essex*-class LPH conversions) and powered by a pair of steam boilers driving a single screw, the LPHs were everything that their designers hoped. Over the thirty-five years since *Iwo Jima* was commissioned, they have been in the front lines of almost every major American military action. They have also served as rescue vessels during the Apollo space missions, trials ships for the deployment of Harrier V/STOL fighter bombers, and as command ships for minesweeping during Desert Storm. This was how USS *Tripoli* (LPH-10, now MCM-10) wound up being mined in the northern Persian Gulf in 1991. America has gotten its money's worth from the LPHs, several of which will serve for a few more years. By the early 21st century these hard-working carriers will go to a well-earned retirement.

The success of the LPH in the 1960s might have led to a follow-on class but for the Vietnam War and the coming of an all-volunteer Navy. And then requirements for more capability and habitability caused a rewrite of the specifications for new warships that would be built in the 1970s. Whatever would replace the LPHs in production would be larger, more comfortable, and more capable. The downsizing of the Navy by the Nixon Administration in the late 1960s also meant that future ships would have "doubled-up" functions. The ideal was a ship that could be both a helicopter carrier and an amphibious dockship, but the Navy only had to pay for one set of engines and a single crew to man it. Thus the stage was set for the Landing Assault Ship, known as the LHA.

There were a number of innovations planned for the LHAs. The entire class was to be built by a single yard under a "fixed price" contract. By awarding the entire program to one shipyard at a "fixed" price, the Government would get a better deal, because of assumed economies of scale. This was a good idea at the time, but problems emerged that neither the Government nor contractors foresaw. Meanwhile, the planned class of nine LHAs represented a huge pool of work for a shipbuilding industry that was already feeling the pinch of declining military orders and competition from overseas. This meant that every major construction yard on both coasts was prepared to fight like hell to win a contract that would be worth over a billion dollars in the 1970s. Down at Pascagoula, Mississippi, Ingalls Shipbuilding (which merged with Litton in 1961 to form Litton Ingalls Shipbuilding) had come to a startling conclusion: The traditional manner of building ships on slipways was both inefficient and overpriced. If a ship could be built in modules, like the sub-assemblies of an automobile, and then put together on an assembly line, cost and building time could be slashed. Now, you have to remember that they were doing all this thinking in the 1960s when gasoline was $.20 a gallon, love was still "free," and a "throwaway" society devalued "quality."

Ingalls has always been a forward-thinking, innovative place, having built the first all-electrically-welded ship, the C3 cargo ship SS *Exchequer* in the 1930s. They worked hard to stay competitive in a business dominated by overseas yards operating with government subsidies (as in Europe), or with incredibly cheap labor (as in Asia). In 1967, they made the decision to construct a new kind of shipyard, across the river from their existing yard in Pascagoula, Mississippi. The new facility would use modular construction techniques and would take advantage of the newest technology for computer-aided design and automated inventory tracking. The idea was that Ingalls could build the same warship as any other yard, but with a competitive price advantage that nobody would be able to touch without making the same investment. At the time, their competitors made fun of the millions of dollars poured into the new facility on the Gulf Coast. But Litton Ingalls stayed the course, and submitted bids for both the LHA and *Spruance*-class (DD-963) programs. Incredibly, amid a howl of protests, they won both contracts.

The *Tarawa*-class (LHA-1) assault ships were 820 ft/249.9 m long, weighing 39,967 tons (fully loaded), and looked a lot like a straight-decked *Essex*-class (CV-9) carrier from World War II. Powered by a pair of large Combustion Engineering boilers feeding twin Westinghouse steam turbines driving two screws with some 70,000 shp, the new ship was capable of a maximum speed of 24 kt/43.9 kph and a sustained speed of 22 kt/40.2 kph. Their broad beam of 106 ft/32.3 m and draft of 26 ft/7.9 m would just fit through the locks of the Panama Canal, so that they could switch between the Atlantic and Pacific Fleets in a hurry. They were long and slab-sided, their dominant feature a huge island structure along the starboard side amidships. This island contains command, flag, and navigational bridges, along with planning and command spaces for embarked Marine units. The hull of an LHA consists of five zones, each with a different function. They include:

- **Flight Deck**—This runs the full length of the LHA; it has nine helicopter landing spots and two aircraft elevators to the hangar deck. There is access to the interior of the ship through the island structure. While there is no "ski-jump" to assist in launching V/STOL aircraft like the Harrier (as found on British, Italian, Spanish, and Russian carriers), there is enough length for a normal takeoff run.

- **Hangar Deck**—Directly below the flight deck in the after half of the ship, this enclosed hangar holds a reinforced squadron of medium lift helicopters. Between the flight deck and the hangar deck, there is room to stow and operate roughly forty-two CH-46-sized aircraft.

- **Well Deck/Vehicle and Cargo Stowage**—Directly below the hangar deck and extending forward is the well deck for launching and retrieving landing craft, as well as the stowage areas for Marine vehicles, equipment, and supplies. The well deck was originally configured for four LCUs, or seven LCM-8s (described shortly). To operate landing craft, ballast tanks at the aft end of the ship are flooded, giving a slight "tip" to the LHA and creating an artificial "beach" for landing craft. Then the tanks are pumped out, and a large stern gate is raised to protect the landing craft and the well deck from the elements.

- **Engineering**—Located amidships below the vehicle and cargo stowage is the engineering plant. This area contains boilers, turbines, generators, and heavy equipment—everything from the engines to the air-conditioning and electrical systems. From here, the exhaust from the boilers and other equipment runs through uptakes on the starboard side, where it is vented through the top of the island structure.

- **Crew/Troop Accommodations**—Most of the forward half of the ship contains berthing, mess, and other spaces for the crew of 925 sailors and 1,713 Marines. Accommodations on the *Tarawa* were considered lavish by contemporary standards, with air-conditioning in all berthing compartments, enlarged bunk and personal stowage space, and a climate-controlled conditioning room for the embarked Marines (now converted to a gym for the entire ship's company).

Compared with earlier amphibious ships, the *Tarawas* were armed to the teeth. In addition to a pair of launchers for the new RIM-7 Sea Sparrow Surface-to-Air Missile (SAM), there were a pair of new lightweight Mk 45 5-in./127mm 54-cal. guns, to provide naval gunfire support, and mounts for six Mk 67 20mm cannons, for protection against enemy patrol boats and other threats. All of this firepower was backed up by a combination of air, surface search, and fire-control radars, as well as by a low-light television camera. *Tarawa* and her sisters were at the time the largest, most powerful amphibious ships ever built. They combined the best features of an LPH, LKA, LSD, and LPD, all in a single, highly survivable hull. Sailors and Marines lined up to get duty assignments to the new "king of the 'gators."

While the new ships were everything the Navy and Marines wanted, they came at a high price, and with a lot of teething problems. The fixed-price contracts had assumed that inflation of construction costs (labor, energy, materials, etc.) would remain stable through the early 1970s. Unfortunately, the 1970s were anything but stable. Several bouts of double-digit inflation, a five-fold increase in the cost of energy, and a huge increase in labor rates caused the construction cost of the LHAs (and everything else!) to skyrocket beyond the expectations of either Litton Ingalls or the Navy. The original plan was that $1.2 billion would buy nine *Tarawas.* The government wound up paying $1.6 billion for five: *Tarawa* (LHA-1), *Saipan* (LHA-2), *Belleau Wood* (LHA-3), *Nassau* (LHA-4), and *Peleliu* (LHA-5). Nobody had seen price inflation like that of the 1970s in over a generation, and it simply was not taken into account when the contracts were written. Since there was no "fault" on the part of either Litton Ingalls or the Navy, the two sides agreed to an additional $400 million for completion of five units. After this forecasting breakdown, Navy contracting was changed forever. Today, contracts have a built-in growth factor to adjust for

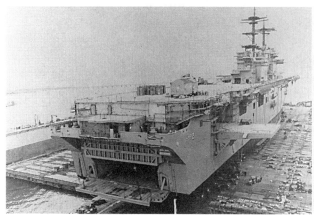

The USS *Essex* (LHD-2) is moved from its final assembly area to a floating barge for launching at the Litton Ingalls production facility at Pascagoula, Miss., on January 4th, 1991. Ships of this class are the largest man-made objects to be moved across the earth.

OFFICIAL U.S. NAVY PHOTO

inflation (determined by the government). This "cost-plus" contract lets the contractor and the government split cost overruns, reassuring contractors who take huge risks on billion-dollar projects that they have a chance to turn a profit someday.

Meanwhile, there were problems at the new Litton Ingalls yard with modular construction. Until engineers realized that they had made the tolerances too tight, the pre-assembled modules wouldn't fit together. They had failed to allow for the normal metal expansion and contraction that might occur between cool Mississippi mornings and the blazing heat of summer afternoons. Simply adding a little extra "meat" to joints between modules and trimming it as they were assembled solved this problem. Another problem developed out of the LHA design itself, which tried to trim top weight by thinning down structural assemblies topside. Unhappily, the strength of the ocean sometimes exceeded the expectations of engineers. The fix for this—structural stiffening—was made when the ships came back in for refits. But generally, the new concept worked, keeping Litton Ingalls the most profitable and busy shipyard in America. As the U.S. shipbuilding industry has crumbled (in 1996 we're down to just five yards capable of building major combatants), they have remained competitive, branching out into building railroad cars and oil platforms.

While the Navy and Litton Ingalls were sorting out financial and engineering problems, the five LHAs were making their presence so much felt around the world, that the Navy and Marines soon realized they should have bought more of them, whatever their cost. While the policies of the Carter years prohibited this, the coming of the Reagan Administration changed everything. John Lehman's planned six-hundred-ship Navy included funding for new amphibious vessels and landing craft. First on the wish list was a batch of new big-deck amphibious assault carriers, based on the LHA design. The new class, designated Landing Helicopter Dockships (LHDs), would consist of five units. By 1996, seven LHDs had been contracted, with possible extra units to replace retiring LPHs. The LHDs would bear the proud names of World War II aircraft carriers. The lead ship was christened USS *Wasp* (LHD-1) after two carriers (CV-7 and CV-18) that served in World War II and the Cold War. *Wasp* was a traditional name dating back to the American Revolutionary War.

The LHD is based upon the LHA design, with significant new features. These included:

- **Standoff Capability**—The ability to support amphibious operations from over the horizon (OTH), utilizing the new LCAC, MV-22B Osprey, CH-53E Super Stallion, and AV-8B Harrier II V/STOL fighter bomber.

- **Survivability**—The capability to fight in environments contaminated by nuclear fallout, chemical agents, or biological weapons. Survivability includes active defense against patrol boats or suicidal small craft, and the ability to avoid, withstand, or repair damage from mines, bombs or cruise missiles.

- **Sea Control Ship Convertibility**—During the 1970s, several CNOs, including Admirals Elmo Zumwalt and James Hollaway, tried and failed to build small aircraft carriers, with up to twenty V/STOL fighter/bombers and eight to ten antisubmarine (ASW) helicopters to

USS *Wasp* (LHD-1)

Profile and interior views of the
USS *Wasp* (LHD-1), the lead ship of
a seven ship class of big-deck amphibious
ships. The *Wasp*-class ships are the
largest amphibious ships in the world.

JACK RYAN ENTERPRISES, LTD. BY LAURA ALPHER

SPS 48E Radar

Bridge

Mk 16 Phalanx CIWS (Port & Starboard)

Mk 29 Sea Sparrow SAM Launcher
(Port & Starboard)

SPS 49 Radar

Life Raft Containers

Mk 29 Sea Sparrow
SAM Launcher

Mk 16 Phalanx CIWS

Mk 91 Fire
Control Radar

Propeller Shaft

Aircraft Elevator
(Port and Starboard)

Stern Gate   Rudder

Island/Bridge (Command, Control,
and Communications)

Hospital/Medical   Accommodations
Facilities

Cargo   Stores

Vehicles

Propulsion

Maintenance Shops

Aircraft Hangar Deck

Office Spaces

Cargo

Landing Craft
Well Deck

escort convoys and amphibious forces. These "Sea Control Ships" would resemble the British Invincible class. The very successful Spanish light carrier *Principe de Asturias* is based on a U.S. design from this period that never got off the drawing board. By simply embarking with a suitable air group, the LHD could perform Sea Control missions in addition to its amphibious role.

While *Wasp* would be based on the good basic design of the *Tarawa* class LHAs, it would be a greatly improved and more capable vessel. One way to compare the two classes is to consider the five critical payload footprints discussed earlier in this chapter:

LHD vs. LHA/LPH Payload Footprints

Class	Troops	Cargo^2 (Ft^2)	Cargo^3 (Ft^3)	LCACs	Helicopters
LHD-1	1,686	20,900	125,000	3	45
LHA-1	1,713	25,400	105,900	1	42
LPH-3	1,489	3,400	40,500	N/A	26

As can be clearly seen, with the exception of vehicle space (Cargo^2), the LHD is superior to each of the ships it replaces. The Navy decided to trade additional Cargo^3 space for Cargo^2. Provision of a chemical/biological/nuclear Collective Protection System (CPS) took up a lot of internal volume in the LHD design, but was considered essential to the ship's mission. Like any warship, the LHD is a set of design compromises. The design of *Wasp* has the advantage of a more spacious well deck for the new LCACs, plus more room to operate aircraft.

To better understand how these huge ships are put together, I visited the Litton Ingalls Shipbuilding plant in Pascagoula, Mississippi, on the Gulf Coast. Pascagoula is a shipbuilding town, with a bit of roughneck, wildcat spirit still left. Litton Ingalls is the largest employer in the area, which butts up against Mobile, Alabama, and Pensacola, Florida, to the east. The West Bank facility, where they build the LHDs, is a joint venture of Litton Ingalls Shipbuilding and the State of Mississippi, which issued state bonds to finance construction of the world's most advanced shipyard. It is the only new shipbuilding yard built in the last thirty years in the U.S. Other yards still build ships on slipways carved into the banks of rivers. Litton Ingalls builds them in a vast open space, where ships move along a production line of mammoth proportions. Over the past few years, Ingalls has built four different classes of warship here, including *Ticonderoga* class (CG-47) cruisers, *Arleigh Burke* class (DDG-51) destroyers, *Sa'ar V* class corvettes for Israel, and *Wasp*-class (LHD-1) amphibious carriers.

The best place to get a feel for how Ingalls works is the control tower in the middle of the facility. From over twelve stories up on the observation platform, you can see the work flow around the 611-acre yard, and it is fascinating to watch. From the railroad and truck receiving areas on the north side, raw materials and equipment feed into fabrication shops. From the moment it hits the receiving dock, every metal plate, wire spool, or equipment crate is tagged with a bar code for computerized tracking in nearly real time. This lets Litton Ingalls order materials and equipment for "just in time" delivery, which reduces inventory costs.

Assembly takes place in five work "Bays," which are open areas of concrete pads overlaid with a grid of railroad tracks, surrounded by mobile cranes to lift and position ship modules as they are assembled. At the time of my visit, the *Arleigh Burke*-class destroyer construction occupied Bays 1 through 3 on the Eastern side of the yard. Litton Ingalls calls them the Barry class, after the first unit that they built (DDG-52). Bays 4 and 5 are assigned to work on the LHDs. The massive vessels are assembled much the way that a sandwich shop stacks a "hoagie." Each module is "stuffed" with electrical, water, hydraulic, steam, and cable "runs," reducing the need to work deep inside a dark, partially completed ship. It also means that a ship can be brought to life and powered up much earlier, reducing the time required to make her ready for sea trials. As submodules are assembled, they move down to the south end of the bay for stacking into one of the five major modules that make up a finished LHD. Each module is stacked and welded into place, and then its lines and connections are fused, just as a surgeon might graft arteries and tendons to rejoin a severed limb. Modules 1 (the bow) through 4 (the stern and well deck) are stacked together and joined into a single hull at the south edge of the assembly area. By this time, each module weighs several thousand tons. These huge chunks fit together with tolerances of a few millimeters or less. After the four hull modules are joined, Module 5, the island deckhouse is added. At over 500 tons, this last item is the largest structure ever lifted by a crane. At this point, the pile of rust-colored metal is beginning to look like a ship, but it is land-locked like a beached whale.

Now the ship can be connected to steam and power lines, and lighting and air-conditioning systems are turned on. This makes life more bearable for the workers on muggy Gulf Coast summer days. USS *Bataan* (LHD-5), already joined with all major modules in place, was being outfitted prior to the next step. This involves translating the completed hull sideways (at about 16 in./40.6 cm per minute) onto a floating drydock, moving the dock out into the channel of Mississippi Sound, and floating off the new ship. Once launched, the ship is towed to an outfitting berth on the south and east sides of the yard, where they prepare her for sea trials, commissioning, and delivery to the Navy.

Let's take a walk though the uncompleted *Bataan* to see how things are done. Wearing a hard hat, I joined Steve Davis, the General Ship Superintendent for *Wasp* (LHD-1), to tour the interior spaces. Each ship is assigned a Superintendent as the chief of construction until she is turned over to the Navy. Steve Davis has decades of shipbuilding experience on nuclear attack submarines, DDGs, and LHDs. After warnings about what not to touch, we entered the massive hull. While warm and smelling of burned metal, the interior of the LHD was surprisingly easy to move about in. It was smoky and dirty, but you could clearly see a warship emerging from the effort of hundreds of workers on board. Ingalls workers are clearly proud of their work, and Steve was anxious to show me how *Bataan* had been improved over his first LHD, *Wasp*. As we headed back outside, we stopped for a moment on the uncompleted hangar deck to talk with several of the outfitters, including Steve's son. Litton Ingalls is proud to be a family company, and it is not unusual to find two or three generations working at the Pascagoula shipyard.

Once a ship passes her builder's trials, she is ready for delivery to the Navy. Many sailors of the first crew, known in Navy tradition as "plank owners," actually join the ship during construction, to assist in the final fitting out and testing. This includes the final step in the manufacturing process, which they call "the Litton Miracle." Under the meticulous supervision of a lady named Annie Gese, the new warship is scrubbed spotless from stem to stern—even in corners and dark spots where inspectors would probably never look. Only then is the ship ready for commissioning in the fleet. As we headed back through the summer heat and humidity, Steve showed me partially assembled modules for USS *Bonhomme Richard* (LHD-6, named for John Paul Jones's Revolutionary War frigate) being stacked and made ready for mating as soon as *Bataan* was floated in 1996.

Litton Ingalls is a busy place, with over a dozen destroyers and LHDs in various stages of assembly and outfitting. Later, Steve and some of the senior Litton Ingalls executives expressed their hope that the next LHD, an as-yet-unnamed seventh ship, would be funded in the coming fiscal year. Less than a month later, they got their wish when the Congress approved LHD-7 as part of the FY-96 budget. This will guarantee the best possible price for the Navy, keep the work force stable, and keep suppliers healthy for future programs. In fact, when shipbuilding executives from the Far East and Europe want new ideas on how to build ships better, they come and look at how Litton Ingalls is doing things in the heart of Mississippi!

Even before a ship is delivered, the Navy has selected her first Captain. A good first skipper can make a ship "happy" or "lucky" and set the tone for every skipper and crew for years to come. As the first commanding officer of *Wasp*, the Navy chose Captain Len Picotte, who has become a Rear Admiral. Command of an LHD or one of the LHAs is particularly coveted in the Navy, since it is the largest surface vessel that can be commanded by a nonaviator. Because of the variety of missions that an LHD or LHA might draw, the Navy has decreed that if the captain is a surface line officer, the executive officer must be an aviator. This is reversed if the captain is an aviator, so the positions tend to switch off as officers move up and out. From the day she was laid down (May 30th, 1985), USS *Wasp* has been a lucky, happy ship. Unlike the LHAs, very few problems arose during design and construction. By the end of the Summer of 1987, she had been floated off (August 4th) and christened (September 19th). She passed her trials and was commissioned on July 29th, 1989. She entered into service with ARGs of the Atlantic fleet, and has been there ever since. In the fall of 1996, she goes into her first major overhaul and upgrade.

Let's go aboard the *Wasp* and get to know her a bit better. We'll enter through the landing craft well deck. As you move into position aft of the *Wasp*, a couple of things strike you almost immediately. How can anything so big move across the ocean? Then, as a helicopter comes in to land a few yards/meters above your head, you wonder how can anyone land on something so small. As the landing craft comes in to dock, you notice the slight downward tilt to the stern of the ship. This is because the stern gate has been lowered and the aft ballast tanks have been flooded down to provide the smoothly sloping artificial "beach" for the landing craft. If you're standing on the navigation bridge of an LCU, be sure to watch your head if you are over 6 ft/2 m tall. Lined with Douglas fir, the well deck is vast (322 ft/98.1 m long, 50 ft/15.25 m wide, and 28 ft/8.5 m high), but it seems crowded when a pair of LCUs or three LCACs are docked inside. Once the landing craft is beached and the bow ramp is lowered, you walk up a steep non-skid ramp, and you are on the vehicle deck.

Following Navy etiquette, we "request permission to come aboard" from the senior officer present.

Walking forward, you enter a stowage area for vehicles of the embarked MEU (SOC). On this deck and the one below are HMMWVs, 5-ton trucks, M198 155mm field howitzers, and trailers. Though the decks are stressed for armored vehicles as heavy as M1A1 Abrams tanks, AAV-7 amphibious tractors, and wheeled LAVs, you usually find these beasts over on the LSDs or LPDs of an ARG. On the "big deck" assault ship, planners prefer to keep only vehicles that can be lifted by the CH-53E Sea Stallion helicopter. Like a parking garage, the vehicle decks are linked by drive-up ramps. You can drive from the lower vehicle deck all the way up to the hangar and flight decks. Despite the vast stowage space, vehicles, cargo and equipment are packed together with only inches/centimeters of clearance. Even a ship as big as *Wasp* never has enough room for everything a MEU (SOC) commander wants. So the rule is to leave just enough room for a Marine to climb through a vehicle's window, door, or hatch, so that it can be driven out of its parking spot when a space develops. Shuffling vehicles and cargo around the stowage space of an amphibious ship is like that children's puzzle with movable tiles and one empty space. You have to move the tiles around incessantly to reach what's needed. The MEU (SOC) logistics (S-4) staff spends hours on their computers arranging load plans to maximize stowage. But with only 20,900 ft^2/1,941.7 m^2 of vehicle stowage space and 125,000 ft^3/ 3,539.3 m^3 of cargo space, you need the mind of an accountant with the imagination of an artist to figure it all out. A conveyer system on an overhead monorail with five hoists helps shift cargo pallets around the various bays. In addition, *Wasp* is equipped with fourteen electric two-ton forklifts, twenty-five three-ton diesel forklifts, two five-ton rough terrain forklifts, two pallet conveyers, five aircraft tow tractors, and four spotting dollies. There are also six six-ton cargo elevators to move things from the well deck and vehicle/cargo areas to the hangar and flight decks.

Walking up the vehicle ramp to the hangar deck, you emerge into a vast space which takes up almost a third of the *Wasp*'s length. Two full deck levels high, the hangar deck is the aircraft maintenance and stowage area. A typical air group includes a dozen or so CH-46 Sea Knights, four big CH-53E Sea Stallions, four AH-1W Cobras, and four UH-1N Iroquois. A half-dozen AV-8B Harrier II fighter/bombers are usually stowed up on the flight deck or "roof" as the crew members call it. This is because Harriers are designed to be weatherproof. While it is

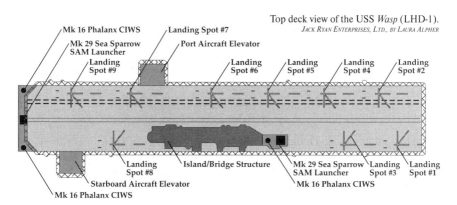

Top deck view of the USS *Wasp* (LHD-1).
JACK RYAN ENTERPRISES, LTD., BY LAURA ALPHER

Mk 16 Phalanx CIWS
Landing Spot #7
Mk 29 Sea Sparrow SAM Launcher
Port Aircraft Elevator
Landing Spot #9
Landing Spot #6
Landing Spot #5
Landing Spot #4
Landing Spot #2
Landing Spot #8
Island/Bridge Structure
Mk 29 Sea Sparrow SAM Launcher
Landing Spot #3
Landing Spot #1
Starboard Aircraft Elevator
Mk 16 Phalanx CIWS
Mk 16 Phalanx CIWS

theoretically possible for *Wasp* to operate up to forty-five CH-46E-sized aircraft, you usually find some of them up on deck, leaving some room to work down in the hangar. The deck and hangar are linked by two deck edge elevators, each capable of lifting up to 75,000 lb/34,090 kg. This is a change from the LHA, which had one elevator on the fantail. In addition to maintenance and stowage, the hangar deck is used by the embarked Marines for fitness and proficiency training (rappelling and other skills). It serves as a staging area for mission teams as they prepare for action. In the rafters are small office and control spaces for the air and maintenance departments, with windows for monitoring the activities below. Walking forward along the starboard side, you come to the flight deck ramp tunnel. This allows vehicles to drive up to the flight deck through the island structure without having to use the aircraft elevators. This usually is the way that the Marines march up to the flight deck to board helicopters. When the LHD was designed back in the early 1980s, the standard utility vehicle for the Marines was the old M151 Jeep. During the construction of the early LHDs in the 1980s, the Marines replaced their jeeps with HMMWVs, which turned out to be wider than the designers expected. Unfortunately, the dimensions of this access tunnel were already frozen, so HMMWVs must ride the elevators. It is a minor inconvenience; starting with LHD-2 they widened the tunnel. But the story underlines how long it takes to design and build new warships. Even though *Wasp* was based on the existing LHA design, it still took most of eight years to bring her into the fleet.

As you exit the island onto the flight deck, there is the feeling of leaving a huge cave and breaking into daylight and fresh air. Covered with a non-skid coating and dotted with aircraft tie-down points, the flight deck is the LHD's primary reason for existence. At 844 ft/257.25 m long and 107 ft/32.6 m wide, it defines the ship's largest dimensions. It is also the most dangerous place on the ship. You have probably seen film footage of flight operations on a big-decked supercarrier. It is a hot, noisy, hazardous place to work, filled with things that can kill you. Jets and helicopters loaded with fuel, weapons, and men race around the deck like crazed banshees. Well, the deck of the *Wasp* is all of that and much more. For one thing, it is smaller (about one third the size), and most of the weapons aboard the aircraft are

A CH-46E and CH-53E of HMM-264 sit folded on the port elevator of the USS *Wasp* (LHD-1) in the summer of 1995. The ship is equipped with two such elevators.
JOHN D. GRESHAM

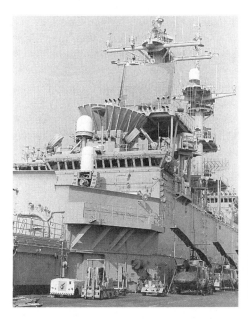

The massive island structure of the USS *Wasp* (LHD-1). Located on the starboard side of the main deck, it is packed with weapons, electronics, and other equipment vital to the operations of the ship.

JOHN D. GRESHAM

armed Marines, loaded with gear that can get loose and be sucked into a turbine engine. Though they make for some difficulties, those Marines are the reason why *Wasp* and her sisters were built.

The deck of the *Wasp* has nine takeoff and landing spots for helicopters. Each spot is numbered, running from starboard to port, front to rear. Thus, the starboard spot farthest forward is Spot 1, while the port spot farthest aft is Spot 9. Usually, Spots 1, 3, and 8 (along the starboard side) are parking areas for AH-1W Cobra and UH-1N Iroquois helicopters forward, and AV-8B Harrier IIs aft. This arrangement maximizes the use of the limited space in the hangar, and still leaves a large area for launch and recovery of aircraft up on the roof. As on the big supercarriers, the deck crews wear colored jerseys to designate their tasks. Red for fuel and ordnance, yellow for spotters, etc. These people operate in a world where noise is an enemy, and virtually all signals are by hand. They move and service $50-million aircraft with little more than gestures and nods for communication. When you consider that these sailors are about twenty years old (are you happy when a kid that age parks your car?), you can appreciate their burden of responsibility. Accidents do happen, and safety nets ring the perimeter of the flight deck. If a deck-hand should fall or be blown overboard (by wind or jet exhaust), he (hopefully) falls into a net before dropping sixty ft/twenty m or more to the sea. Around the perimeter are points for refueling, rearming, and servicing aircraft.

The deck level is the best place to observe *Wasp*'s weapons. Though the LHAs were a model for the LHDs, the armament of the *Wasp* shows how far 1990s technology has gone beyond the 1960s and 1970s. The LHA's armament provided a rudimentary defense against aircraft and limited capability against surface and shore targets. The designers of the LHDs dispensed with inshore bombardment, focusing instead on air and missile threats. The LHD design deletes the 5-in./127mm guns and the manned 20mm mounts. Instead, modern eight-cell RIM-7 Sea Sparrow launchers are fitted. The Sea Sparrow is a surface-launched version of the AIM-7 Sparrow air-to-air missile (AAM). Unlike the aerial version, Sea Sparrow has racked up an impressive record of reliability in three decades as a short range SAM ("point defense" the Navy calls it). The current version, the RIM-7M, has a range of around 10 nm/18.5 km, providing an inner layer of defense against incoming anti-ship missiles and aircraft. Like the airborne version of Sparrow, the RIM-7 utilizes semi-active radar guidance, which means that a radar on the ship "paints" the target, and the missile homes on the reflected microwave energy. The 450-lb/204-kg missile has a lethal 90-lb/40.8-kg warhead. Sea Sparrow is found on everything from aircraft carriers to frigates and supply ships, and has been widely exported to NATO and "friendly" countries. *Wasp* carries two eight-cell launchers (each with eight reloads) and a pair of Mk 91 illumination radars. One launcher sits at the front of the island structure, and the other is mounted on a sponson on the fantail.

In addition to the Sea Sparrow launchers, three Mk 16 Phalanx Close-In Weapons Systems (CIWS) are installed to deal with any missiles that "leak" through the area SAM defense of escorting destroyers and cruisers, or the point-defense systems. One unit is located at the front of the island structure, and the other two are mounted on either side of the Sea Sparrow launcher on the fantail sponson. Each CIWS is built around a 20mm General Electric Gatling gun like the M61 on the F-15 Eagle and F-16 Fighting Falcon. CIWS fires 3,000 rounds per minute in 200-round bursts, with tungsten penetrators designed to break up an incoming missile, or detonate its warhead. Each CIWS has a 1,550-round magazine, and carries its own search and track radars. It is a self-contained unit; once turned on, it automatically attacks any fast-moving target it identifies as hostile. It can hit targets up to 6,000 yards/5,488 meters away, but is most effective within about 1,625 yards/1,486 meters. When conducting flight operations, *Wasp* tends to keep her three CIWS turned off, in case they accidentally identify a "friendly" aircraft as "hostile." Electronic Identification Friend or Foe (IFF) systems still are not terribly reliable, and both sailors and aircrews take such things quite seriously. Several years ago, a RIM-7, fired accidentally by an American aircraft carrier, hit a Turkish destroyer on maneuvers, killing her captain and several crewmen.

While CIWS can defeat small anti-ship missiles like the French MM-38/ AM-39/MM-40 Exocet or the American A/RGM-84 Harpoon, it has trouble with large, fast sea-skimmers like the Russian SS-N-22 Sunburn with its 1,100-lb/500 kg warhead and Mach 2 speed (the subsonic Harpoon and Exocet have 250-to-500-lb/125-to-225-kg warheads). Even if CIWS detonates the incoming missile's warhead, it is moving so fast that missile fragments will "pepper" the ship. This is one reason why the structure of *Wasp* and all new U.S. Navy ships have been hardened with lightweight Kevlar armor panels. A new system called the RIM-116A Rolling Airframe Missile (RAM) will augment CIWS. This little missile combines the AIM-9 Sidewinder airframe with the FIM-92 Stinger seeker head. It can intercept targets out to 5 nm/9 km, far enough to avoid "fragging" the ship with high-velocity wreckage. RAM is fired by a twenty-four-round Ex-31 lightweight launcher. *Wasp* will get two Ex-31 RAM launchers when she comes in for her first major overhaul; new units will get them starting with *Bataan* (LHD-5). Eight M2 .50-cal. machine-gun mounts provide defense against small boats and frogmen. New units (and ships completing their first overhaul) will replace four machine guns with three 25mm Bushmaster cannon. An SLQ-25 Nixie torpedo-decoy system is mounted in the LHD's stern. Towed behind the ship, these acoustic/magnetic decoys (hopefully) decoy any incoming torpedo. Active "anti-torpedo torpedo" systems for installation on major warships may be ready in a few years.

The most noticeable difference between the LHAs and the LHDs is the smaller "island" structure on the newer vessels. The LHA island held all the control spaces for fighting and running the ship plus all of the planning and command spaces for the embarked Marines. This kept everything centrally located, but was very vulnerable to a single missile or bomb hit. Anti-ship-missile seeker heads usually "lock" onto the largest or hottest structures of a ship (the island with its boiler uptakes is perfect). Thus, on the LHD design, the massive island structure was cut down by two full decks and the Marine command spaces relocated below, deep within the ship. In addition to the weapons mounted on the island, most of the ship's sensors and communications antennas are mounted as high as possible. These include:

- **SPS-48E**–A 3-D search radar which provides air control and AAW battle management functions for the *Wasp*. This high-resolution radar has a reported range out to around 60 nm/110 km.

- **SPS-49 (V)5**–The best naval 2-D air-search radar of our time. Very reliable with a detection range of up to several hundred miles/kilometers, it is found on most major combatants in the U.S. Navy, as well as many foreign vessels.

- **SPS-64 (V)9**–This is primary a navigation radar for keeping formation and operating close to shore. It is a development of the classic LN-66 navigation radar in use for several decades.

- **SPS-67**–The SPS-67 is a general-purpose surface-search radar, designed to provide precise targeting data against surface targets.

- **Mk 23 Target Acquisition System (TAS)**–This is a small, fast-rotating radar for detecting sea-skimming or high-angle missile attacks. It feeds data into the SYS-2 (V)3 weapons-control system, which can automatically activate the RIM-7 Sea Sparrow or RIM-116 RAM systems.

- **SLQ-32 (V)3**–The SLQ-32 is a family of electronic-warfare systems which can be tailored to the protection requirements of a particular ship. The (V)3 version has a wide-band radar-warning receiver, a wide band radar jammer, and a bank of four Mk 137 Super Rapid Blooming Chaff (SRBOC) launchers. These six-barreled mortars throw up a cloud of chaff (metal-coated mylar strips) and infrared decoys to (hopefully) blind or confuse an incoming missile at the last minute.

These systems give *Wasp*'s commanding officer and battle staff great situational awareness of the battle space surrounding their ship and the ARG. Given the variety of threats that face an LHD, you can see why the Navy needs to defend this billion-dollar-plus asset.

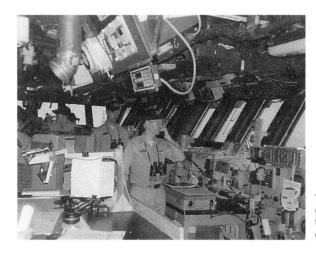

The bridge of the USS *Wasp* (LHD-1). From this position, the ship is maneuvered and operated. *JOHN D. GRESHAM*

By now you are probably getting warm out on the flight deck, so let's go inside. When you enter the island, a blast of cold air hits you immediately. The LHDs were designed to protect their crews against the possibility of chemical, biological, and nuclear warfare. A Collective Protection System (CPS) creates an environmental "citadel" inside the island and forward part of the ship. This sealed citadel provides clean, filtered air, allowing the crew to work in a shirt-sleeve environment. It also gives *Wasp* and her sisters the finest air conditioning in the fleet. *Wasp* was built with only five of her six planned chiller units, and she is almost too cold! The cool interior enables the crew and Marines to cope with the heat of tropical places; it also extends the life and reliability of the electronic equipment packed inside *Wasp*'s slab-sided hull. The CPS system does not extend to the hangar, cargo, vehicle, or well-deck areas. Thus, you have to get used to the CPS "zone" hatches and airlocks, each of which must be opened, closed, and dogged as you pass through.

The island structure is filled with steep ladders, and your leg muscles get a workout as you move around *Wasp*. To reach the bridge, you go up five levels and pass through a cipher lock and several more hatches. The bridge has exceptional visibility through green-tinted windows. Navigational instruments, map tables, and communications equipment are laid out neatly and logically. Spacious and comfortable, *Wasp*'s bridge is a model of functional design. Even the Captain's chair and day cabin are designed for comfort and ease of access. A wing bridge, protruding from the starboard side of the island, lets the bridge crew conn the ship during Underway Refueling and Provisioning (UNREP) and docking. In some seven years of operations, only one design problem affected the bridge: Some of the thick windows cracked during a 1994 winter operation in Norway, due to the intense difference between internal and external temperatures.

Exiting the bridge and heading aft, we find the "debark control" used by a Marine commander to monitor operation of amphibious vehicles. There also is a weather office which would be the envy of any large airport. Amphibious operations are extremely sensitive to weather conditions, and the Navy has invested heavily to make sure that the *Wasp* can keep an eye on what Mother Nature is up to. In fact, the crew regards the weather forecasters as a branch of the intelligence department. Heading up through several more ladders and cipher-locked doors, we enter Primary Flight, or "Pri-Fly," the ship's control tower for air operations, and home of the Air Boss. The Air Boss is a virtual god of the air and deck space around the ARG. Usually, the Air Boss is a commander (O-5) who has completed a squadron command tour. The Air Boss is assisted by various Landing Signals Officers (LSOs) who "wave aboard" the aircrews based aboard the *Wasp*. Big-deck aircraft carriers have a special platform at deck level for LSOs; but LHDs and other helicopter ships place their LSOs inside Pri-Fly. When aircraft are landing vertically, the best place to watch is above and to the side of the action. Each component of the Air Combat Element (based around a reinforced HMM) has one or two LSOs among its pilots, and one is always on duty whenever that kind of aircraft is flying.

Heading down from the island (as tough on the ankles as going up), we arrive in the *Wasp*'s main living and work areas. Down on the 02 Level (just below the flight deck) are the officers' berthing and mess areas, as well as most of the command and control spaces for the embarked Marines. The center of this activity is the officers' wardroom, which functions as restaurant, theater, town hall, and conference room at various times of the day. Four times a day, the *Wasp*'s mess specialists lay out meals (breakfast, lunch, dinner, and "mid-rats" at 11:00 P.M./2300 hours) for the Navy and Marine officers. Between meals, the wardroom is used for meetings, training, and final briefings prior to launch. Forward of the wardroom is the junior officer berthing area, composed of four- and six-man staterooms. Each officer has a comfortable bunk, stowage for personal gear, and a fold-down desk. A little personal space makes a six-or-seven-month cruise (normal for ARG ships these days) a lot more bearable. There is a certain etiquette in using the dormitory-style showers and head facilities. As a matter of basic hygiene, everyone wears shower slippers while bathing, to prevent the spread of foot infections which could devastate the marching ability of the embarked Marines! At the forward end of the 02 Level is the most popular area on the ship, *Wasp*'s fitness center. This is a beehive of activity around the clock for the sailors and Marines trying to to stay fit and work off some of the nervous tension and stress of shipboard life. You generally have to wait to get onto one of the machines or weight benches jammed into the space. Officers told me that this one room did more for crew morale than anything except food and the CNN satellite feed!

Heading aft past the wardroom takes you through "Officers Country," the berthing areas for senior Navy and Marine officers. Usually these are one- or two-man staterooms, with an attached head and shower. Don't envy these officers' comforts; very few get to spend much time in their racks. A department head or unit commander aboard an amphib often works a sixteen-to-twenty-hour day. You are lucky to get four to six hours of sleep (not always at night!). The Captain has his sea cabin on the 02 Level, and a day cabin on the bridge, but rarely gets to rest in either of them! Also located here is "Flag Country," berthing spaces for embarked admirals and their staff. The *Wasp* is big enough to accommodate such a staff without disrupting the ship's routine.

Continuing aft, you enter a series of darkened command and control spaces. As noted above, these were relocated from the island structure to protect their vital personnel and equipment. These spaces include:

- **Combat Information Center (CIC)**—The nerve center of the ship, with displays for all of the ship's sensors, as well as information acquired from data links and national sources (the DOD euphemism for "spy satellites"). Filled with consoles, terminals, and big-screen displays, this battle management center has separate zones for anti-sub, anti-air, and anti-surface warfare, communications, damage control, and other functions. Officers learn to assess fast-developing situations and act quickly. In World War II a good captain fought his ship from the bridge, but today's Burke or Vian would be found at a glowing console in a dimly lit CIC.

- **Landing Force Operations Center (LFOC)**—The LFOC is a mission control center for amphibious operations. Each embarked Marine unit has a console, with the MEU (SOC)'s console in the center at the rear with a clear view of the large-screen displays at the front of the compartment. Everything is tied into a computer, the Integrated Tactical Amphibious Warfare Data System (ITAWDS), linking the commander to embarked Marine units. At the rear of the LFOC is a conference area for the MEU (SOC) staff. Like the ship's captain, the embarked Marine commander usually fights his battle from here.

- **Flag Plot**—This is where the ARG commander and staff reside during operations. It is generally similar to CIC and the LFOC, and there are numerous repeaters for the various sensors and displays.

- **Ships Signals Exploitation Space (SSES)**—This small sealed space adjacent to the CIC is for secret stuff: "exploitation of enemy signals and electronic emissions." Equipped with data links to national and theater-level intelligence systems, the SSES can provide decision makers with up-to-date information on enemy intentions and activities. Only specially cleared intelligence and communications technicians are allowed inside.

- **Joint Intelligence Center**—The Joint Intelligence Center is a clearinghouse for information required by the ship, the ARG, and embarked Marine components. Analysts in the JIC can draw from vast databases of Defense Mapping Agency maps, satellite photography, and anything else the intelligence community provides. Even better, they can probably tell you what it means. The staff is a "rainbow" organization from every unit involved.

- **Tactical Logistics Group Center (TACLOG)**—Crammed with computers, phones, and people, TACLOG controls the logistics battle. Everything from the layout of vehicles in the stowage areas to the embarkation of troops by the ship's combat cargo staff is controlled from here.

- **Tactical Air Control Center (TACC)**—The air traffic control center for the ship and the ARG, the TACC monitors the airspace around the ARG, and generates the daily air tasking order (ATO).

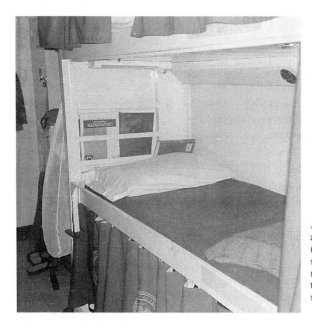

An enlisted berthing area aboard the USS *Wasp* (LHD-1). The bunks are stacked three-high, and are much more comfortable than those aboard nuclear submarines and older vessels.
JOHN D. GRESHAM

When an operation or exercise is underway, these spaces resemble a beehive without the buzz, on the job, around the clock, until it is finished.

One deck down (the 03 Level) is the LHD's medical department. One of the more chilling features of the original LHA was the provision for a large hospital facility (about 375 beds). It was almost doubled in size when the LHDs were being designed. In fact, when the *Wasp* is at home in Norfolk, Virginia, she is listed in the Virginia State disaster plan as the fourth-largest hospital in the state, with some 600 beds! Marines know how fast amphibious warfare can generate casualties when things go wrong. Except for the hospital ships, *Mercy* (T-AH-19) and *Comfort* (T-AH-20), these are the most capable medical facilities afloat. In addition to a large triage area, there are six operating theaters, eighteen post-operative/intensive-care beds, 6 isolation ward beds, and 36 primary-care beds. Using berthing space from disembarked Marines allows up to 536 additional bed cases. There are also oversized radiology and dental departments.

Below the Medical Department, on the 04 Level, are maintenance shops for mechanical equipment, electronics, and hydraulic systems. Further forward are accommodations for enlisted and non-commissioned (NCO) sailors and Marines—over two thousand berths, divided into many compartments. The chiefs and Marine NCOs live in "Goat Lockers" with about a dozen bunks (in two high racks) and recreational areas with tables and televisions. Enlisted personnel have racks stacked three-high, and you might find as many as sixty or seventy personnel in one such berthing space. While dense, the accommodations are much more comfortable than those we have seen previously on older vessels. Each sailor or Marine has an individual berth, and there is no "hot bunking" as aboard submarines. In addition to personal stowage, Marines also have armories for their weapons and combat equipment, so that they can rapidly assemble their gear in an emergency. These berths are located just forward of the Medical Department, and can become hospital beds if necessary.

Dining facilities for NCO and enlisted personnel look just like shore-based mess halls. The food is good and the service fast. It has to be when you consider that they serve almost twelve thousand meals every day. Sharing the same food and ship has a way of bonding everyone, no matter what their rank, as "shipmates." Admirals and generals walk the same passageways and share the same dangers with PFCs and chiefs. It makes for a unique shipboard society. I like it. It says good things about the Navy, the Marine Corps, and America. It says that when the work starts, we all work, and we all share.

The *Wasp* is a virtual city-at-sea, with all the needs of a city. One of the biggest is communications, both within the ship and to "the world." Communications systems include FM/HF/UHF/VHF radios, UHF/VHF/EHF satellite systems, video teleconferencing, and other command and control systems. For communications around the ship, there is a phone system, as well as the ever-present public-address system known as the 1MC. There are moves to bring the *Wasp* (and the Navy on the whole) into the computer network era as well. The *Wasp* is wired for a wide-area network (WAN) divided into departmental local-area networks (LANs). These in turn are being tied into the Navy's department-wide telecommunications system. Desktop and laptop computers are everywhere. You see young sailors in their bunks using them to tap out letters home, or officers creating briefing viewgraphs for the next landing exercise.

A ship-based cable television system broadcasts news and movies to every compartment. You see many small personal televisions (hooked to the ship's cable network), VCRs, and stereo systems used by crew and Marines for entertainment during the rare off-hours. A stabilized satellite television dish was recently fitted on the *Wasp*'s island structure. Officially, this allows intelligence specialists to monitor CNN and other twenty-four-hour news services, but it also brings the crew news and sports from home without the delay of videocassettes. Soon, it will be standard equipment on all Navy vessels. Other amenities for the crew include a well-stocked ship's store, a post office, and an efficient laundry service. All of these features make life more livable for over 2,500 people during *Wasp*'s six- or seven-month cruises.

A ship is nothing but a cold hulk unless it can generate power. We'll finish our tour of the *Wasp* in the heart of the ship—engineering and propulsion. You have to go into the very bowels of the ship, below the vehicle and cargo decks, to enter the "land of the snipes," the nickname for boiler and engineering technicians. Rather than the gas turbines or marine diesels that drive most modern warships, LHDs continue the tradition of oil-fired steam plants. The *Wasp* is powered by a pair of 2,600-PSI/41.7-kg-per-cm^2 Combustion Engineering boilers, which generate steam for the two Westinghouse turbines, for a total of 70,000 horsepower to the twin shafts. This translates to a cruising speed of around 22 kt/40.25 kph, and a maximum speed of approximately 24 kt/43.9 kph. While it may not quite match the 30+ kt/55 kph of a supercarrier or destroyer, it is adequate for the job. With a full load of fuel, steaming at approximately 20 kt/36.6 kph, the *Wasp* has an unrefueled range of approximately 9,500 nm/17,600 km, which means that it can transit to most potential trouble spots with a bare minimum of support shipping.

The *Wasp*'s vast electrical requirements are met by a series of motor generators supplying different types of power (220 V and 110 V AC, 12 V and 15 V DC, etc.). The freshwater distillation plant produces enough water for every member of the crew to take a "Hollywood" shower every day. Distilled water is quite soft and pure, without the chlorine taste prevalent in city tap water. The "snipes" of the Engineering Division also manage *Wasp*'s fuel and fluid systems, including hydraulics, jet fuel, and diesel for the vehicles of the embarked Marines. They play a key role in damage control effort, since without power, *Wasp* would quickly succumb to damage from missiles, bombs, torpedoes, or even accidental fire. Warships are collections of combustible, flammable, and explosive stuff; all of these demand intense vigilance. Damage control is something of an obsession with Navy captains and crews. Our experience in the Persian Gulf and that of the British in the Falklands in 1982 emphasized the survival value of damage control. As noted in *Submarine*, the Navy has worked hard to deploy improved fire fighting systems like Aqueous Film Forming Foam (AFFF) fire extinguishers and improved emergency breathing apparatus. Everywhere on *Wasp* you see Day-Glo orange containers with emergency breathing masks for survival in the smoke of a fire.

Not just a packing crate for Marines and their equipment, *Wasp* is a platform capable of many different missions, from amphibious raids and assaults, to sea control (escorting convoys and protecting sea lanes). It is perhaps for this reason that the *Wasp* (LHD-1) and her sister ships, *Essex* (LHD-2), *Kearsarge* (LHD-3), and *Boxer* (LHD-4), have become the most sought-after ships in the Navy. When the next three LHDs, *Bataan* (LHD-5), *Bonhomme Richard* (LHD-6), and the still unnamed seventh unit of the class, join the fleet in a few years, it will give all twelve ARGs a big-deck aviation ship. The final three ships have significant improvements over the earlier LHDs. The Ex-31 RAM launchers and 25mm Bushmaster cannon mounts will be built in from the start, along with smaller superstructures, more aviation fuel capacity, and improved communications, damage control, and medical capabilities. There will also be accommodations for female personnel, under the "Women at Sea" program (see the LPD-17 below for more on this). These features will be retrofitted to earlier units during their first major overhauls. The *Wasp* and her sister ships represent the core of America's forced-entry capability, and will be so for decades to come.

USS *Whidby Island* (LSD-41)

At almost $1.25 billion dollars each, *Wasp*-class LHDs are hardly the most economical solution for every amphibious task. Sometimes, you need a ship that does just one or two things well. So the Landing Ship Dock (LSD) was created. The LSD is a transport and service platform for landing craft. At first, they were simple ships with well decks and minimal stowage or troop capacity. They could "flood down" to launch landing craft. Later, LSDs evolved into general-purpose vessels, with long-term accommodations for embarked troops and equipment, and limited helicopter capability. The design of the *Anchorage* (LSD-36) class, constructed in the 1960s and 1970s, emphasized carrying large numbers of landing craft. These five ships served effectively in ARGs for almost three decades. But they are at the end of their service lives. The *Whidbey Island* (LSD-41) class will replace them.

The USS *Whidbey Island* (LSD-41) leaves Cadiz, Spain, on February 16th, 1996, headed home from her 1995/96 Mediterranean cruise.
JOHN D. GRESHAM

The *Whidbey Island* class supplements the capabilities of the big-deck aviation ships of an ARG. In the event of a need to "split" an ARG, the LSD always accompanies the LHD, LHA, or LPD. This lets the ARG commander retain a forced-entry capability, due to the numerous landing craft the two ships carry. While the LSDs lack the command and control capabilities of the LHDs and LHAs, and the cargo capacity of the LPDs, they serve a vital role as amphibious delivery systems. Let's get to know *Whidbey Island* a bit better.

In the early 1980s, planners at Naval Sea Systems Command (NAVSEA) began to think about the mix of ships they wanted for the ARGs of the 1990s and beyond. Even before the decision to build *Wasp*-class LHDs, they knew that standoff from the enemy shore would dominate future amphibious ship design. While the old *Anchorage*-class LSDs could carry and operate the new air cushioned landing craft, it was clear that more LCACs would be needed in an ARG to replace the slower, more vulnerable LCUs. NAVSEA set about designing a new ship, known as the LSD-41, and selecting a contractor. The first three ships went to Lockheed Shipbuilding in Seattle, Washington. *Whidbey Island* was laid down on August 4th, 1981, launched on June 10th, 1983, and commissioned two years later on February 9th, 1985, with further units at one-year intervals. When Lockheed decided to leave the ship construction industry in the 1980s, the rest of the class was awarded to Avondale Industries of New Orleans, Louisiana. Avondale, an old Navy contractor, built the *Knox*-class (FF-1052) ASW frigates in the 1960s and 1970s. Set on the banks of the Mississippi, the yard uses more conventional technology than Litton Ingalls. Avondale's old-style slipways and serial assembly methods may suffer in head-to-head price comparisons with foreign competitors, but they do build quality ships.

The *Whidbey Island*-class ships are relatively conventional, being evolutionary follow-ons to the LSD-36 class, with small but significant improvements. Only 609 ft/185.8 m long and 84 ft/25.6 m in beam, they are much smaller ships than the *Wasp*. Displacement is just 17,745 tons fully loaded. The have a shallow draft of 19.5 ft/6 m versus 26 2/3 ft/8.1 m for the LHDs. *Whidbey Island* is powered by medium-speed marine diesels, rather than steam turbines. The four SEMT-Pielstick engines deliver a combined total of 41,600 hp to twin shafts, for a top speed of 22 kt/40.25 kph. At an economical 20 kt/36.5 kph, they can cruise for 8,000 nm/

USS *Whidbey Island* (LSD–41)

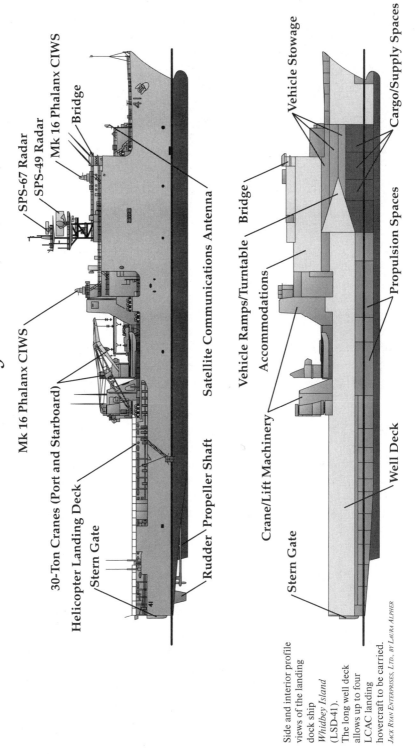

Mk 16 Phalanx CIWS

SPS-67 Radar

SPS-49 Radar

Mk 16 Phalanx CIWS

Bridge

30-Ton Cranes (Port and Starboard)

Helicopter Landing Deck

Stern Gate

Satellite Communications Antenna

Rudder Propeller Shaft

Vehicle Stowage

Vehicle Ramps/Turntable   Bridge

Accommodations

Cargo/Supply Spaces

Propulsion Spaces

Crane/Lift Machinery

Stern Gate

Well Deck

Side and interior profile
views of the landing
dock ship
Whidbey Island
(LSD-41).
The long well deck
allows up to four
LCAC landing
hovercraft to be carried.

Jack Ryan Enterprises, Ltd., by Laura Alpher

Looking forward into the cavernous well deck of the USS *Whidbey Island* (LSD-41). This deck can accommodate up to four LCACs or three LCUs, or be used to store vehicles.

JOHN D. GRESHAM

14,816 km without refueling—an excellent match for the LHDs and LHAs. A relatively small crew of 334 officers and enlisted sailors reduces operating costs.

The most notable features of the *Whidbey Island* class are the large deckhouse forward for stowage and accommodations, and long well deck, topped by a flight deck with a pair of landing spots for helicopters up to the size of a CH-53E Sea Stallion. Since the class lacks hangar or support facilities, no helicopters are based aboard while on cruise, and LSD-41s can only refuel helicopters based on other ships. The well deck has room for up to four LCACs, three LCUs, or ten LCM-8s should there be a need to utilize those older craft. The well deck resembles the one on *Wasp*, using ballast tanks to lower the stern and flood the deck so landing craft can arrive or depart. Measuring 440 ft/141.1 m in length, 50 ft/15.2 m in width, and 27 ft./8.2 m. in height, it is the largest well deck on any amphibious ship. Landing craft park end-to-end, as on *Wasp*, and can be loaded by driving vehicle through one landing craft to get to another.

Despite their minimalist design, the LSD-41s are quite capable at handling landing craft and off-loading cargo. They carry two electric two-ton forklifts, two pallet jacks, two five-ton rough terrain forklifts, an eight-ton cargo elevator, and three large cranes with (fifteen, twenty, and sixty ton capacity). A special turntable in the ramp between the well and helicopter decks speeds vehicle movement and handling. With 13,500 ft^2 of vehicle space and 5,100 ft^3 of cargo space, they are smaller than the LHDs and LHAs, but still capable of carrying a useful payload. Berthing space for up to 454 Marines is similar to what we saw on *Wasp*.

The LSD-41s lack many of the features found on the *Wasp* class. These include:

- **Command and Control Facilities**—The LSD-41s have only a CIC and a Tactical Landing Support Group space. There are no provisions for a flag staff, and no flag plot.

- **Medical Facilities**—The *Whidbey Island* class has only a single operating theater and eight beds (one intensive care, two isolation, five primary care), with no real overflow capability. It depends on the large-deck amphibs for medical support.

USS *Harpers Ferry* (LSD-49)

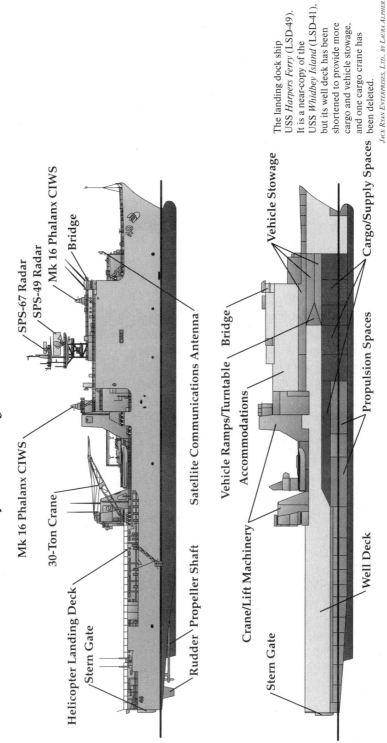

Mk 16 Phalanx CIWS

SPS-67 Radar

SPS-49 Radar

Mk 16 Phalanx CIWS

Bridge

Satellite Communications Antenna

30-Ton Crane

Helicopter Landing Deck

Stern Gate

Rudder Propeller Shaft

Vehicle Stowage

Vehicle Ramps/Turntable Bridge

Accommodations

Cargo/Supply Spaces

Crane/Lift Machinery

Propulsion Spaces

Stern Gate

Well Deck

The landing dock ship USS *Harpers Ferry* (LSD-49). It is a near-copy of the USS *Whidbey Island* (LSD-41), but its well deck has been shortened to provide more cargo and vehicle stowage, and one cargo crane has been deleted.

JACK RYAN ENTERPRISES, LTD., BY LAURA ALPHER

- **Sensors**—SPS-49 air search, SPS-64 (V)9 navigation, and SPS-67 sur-face-search radars are carried. There are no fire-control systems of any kind.

- **Defensive Armament** – The LSD-41s are armed with only a pair of Mk 16 20mm CIWS, two Mk 67 25mm Bushmaster cannon mounts, and two mounts for M2 .50-cal. machine guns. The basic SLQ-32 (V)1 electronic warfare suite has only a radar-warning receiver and four Mk 137 SRBOC/decoy launchers. They also have an SLQ-25 Nixie system. No radar jammer is fitted. These ships require combatant escort to survive in a hostile environment.

Though the Whidbey Island ships seem austere compared to *Wasp*-class LHDs, they do have features that make them valuable amphibs, including:

- **Structures/Protection**—The LSD-41s have the same structural protec-tion as the *Wasp* class, including hardening of the deckhouse against fragment damage from a near miss.

- **Environmental Protection**—The *Whidbey Island* class has the same kind of nuclear/chemical/biological CPS system as the LHDs, and thus the same levels of air-conditioned comfort.

How would all this work in combat? Consider the following example. In most cases, the ARG staff will load up the LSD-41 with heavy vehicles like M1A1 Abrams tanks and wheeled LAVs. This provides an armored punch for the early waves of a Marine assault or raid. Once *Whidbey Island*'s own load of equipment and cargo is off-loaded, the landing craft help other ships to unload vehicles and cargo, thus speeding the flow of combat power to the beach. This secondary role of landing craft base is what makes the LSDs so valuable to an ARG commander.

A total of eight LSD-41s were built. These include three Lockheed-built units; *Whidbey Island* (LSD-41), *Germantown* (LSD-42), and *Fort McHenry* (LSD-43), plus five Avondale-produced ships: *Gunston Hall* (LSD-44), *Comstock* (LSD-45), *Tortuga* (LSD-46), *Rushmore* (LSD-47), and *Ashland* (LSD-48). Four additional units are being built to a modified configuration that has an interesting origin. You see, the new amphibious ships, when combined with over-the-horizon delivery systems like the CH-53E Sea Stallion and LCAC, can actually put troops, vehicles, and cargo onto a beach faster than Navy beachmasters can handle it. There is a physical limit to how fast you can move stuff over a beach, and the beach control parties that serve as the ARGs "traffic cops" have hit that limit. The LCACs turned out to be faster at doing their jobs than expected. This gave NAVSEA an opportunity to modify the last four ships of the LSD-41 class. Since the new LHDs could carry up to three LCACs, and the older LPD-4 class assault ships could carry two, this meant that an ARG only required two more to reach the desired level of seven such craft. So, the last four units of the *Whidbey Island* class, redesignated the *Harpers Ferry* (LSD-49) class, were redesigned with a shortened well deck (only 184 ft/56 m long). The remaining space would be used to enlarge the vehicle and cargo footprints of the new ships, as the table below shows:

Class	Troops	Cargo^2 (Ft^2)	Cargo^3 (Ft^3)	LCACs	Helicopters
LSD-41	454	13,500	5,100	4	0
LSD-49	454	13,500	50,700	2	0
LSD-36	302	8,400	1,400	3	0

As you can see, the cargo^2/vehicle space in LSD-49 has been expanded by 15% over LSD-41, and the cargo^3 space by a whopping 994%. This makes the LSD-49s very valuable amphibious ships. Any CO of a forward-deployed unit will tell you that they never have enough stowage space for "stuff," and the trade-off on these ships makes them an outstanding value for the money. All four—*Harpers Ferry* (LSD-49), *Carter Hall* (LSD-50), *Oak Hill* (LSD-51), and *Pearl Harbor* (LSD-52—named for the facility, not the battle!)—are built by Avondale in New Orleans. The first two are already in service, and the other two are scheduled for completion by early 1997. One LSD-41/49 will be assigned to each of the Navy's 12 ARGs. Right now, 12 ARGs only provide about 2.5 MEBs of lift, as opposed to the 3 that the Marine Corps considers necessary to meet mission requirements. Additional LSDs are unlikely though, since the Navy is committed to construct new LPD-17-class assault ships to replace aging *Austin*-class LPDs.

USS *Shreveport* (LPD-12)

The Landing Assault Ship USS *Shreveport* (LPD-12) is a living legacy of the 1960s-era shipbuilding program that has been the backbone of the amphibious force for three decades. While old by warship standards (she was commissioned in 1970), cramped, and antiquated compared to contemporary designs, she still has many years of service ahead. Part of the eleven-ship *Austin* class (LPD-4 to -15), *Shreveport* may serve for another ten to fifteen years. The LPD is the "swing" ship—a virtual "utility infielder" among the three ships that usually comprise an ARG. While the LHDs/LHAs and LSDs work together as the "big" decks of the ARG, the LPD is a general-purpose workhorse, taking on missions that used to be assigned to the LSTs and LKAs. When an ARG splits to undertake more than one mission at a time, the

The USS *Shreveport* (LPD-12) leaves Morehead City, N.C., on August 29th, 1995, on her way to the Mediterranean. She is fully loaded for "split ARG" operations, and is headed to Bulgaria for an exercise.
JOHN D. GRESHAM

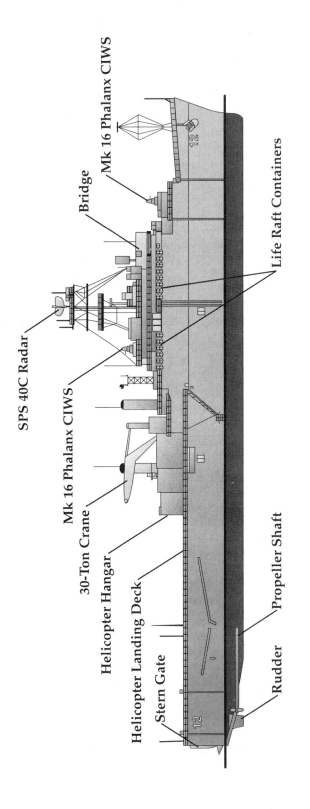

SPS 40C Radar

Bridge

Mk 16 Phalanx CIWS

Mk 16 Phalanx CIWS

30-Ton Crane

Helicopter Hangar

Helicopter Landing Deck

Stern Gate

Life Raft Containers

Propeller Shaft

Rudder

A side view of the multipurpose amphibious ship USS *Shreveport* (LPD-12).
JACK RYAN ENTERPRISES, LTD., BY LAURA ALPHER

LPD is frequently on her own. LPDs tend to pick up the stray "cats and dogs" of the embarked MEU (SOC), such as amphibious tractors, Force Recon teams, and the SEAL team. They act as a floating Forward Fuel and Arming Point (FFARP) for helicopters, and a base for the AH-1W Cobra attack helicopters and the embarked Pioneer UAV unit. That's a lot to ask of an old ship like *Shreveport* (LPD-12), but she does her best in a world where she is little loved, but heavily used.

The original LPDs of the *Raleigh* class (LPD-1) were designed in the late 1950s to transport a large load of amphibious troops and supplies, at the expense of off-load capability. LPDs have relatively small well decks compared to the LHAs, LHDs, and LSDs, as well as smaller aviation facilities, with only a single helicopter landing pad. They are nevertheless one of the three types of amphibious ships that will survive (along with the big deck LHDs/LHAs and the LSD) into the 21st century. There even are plans to build a new class of twelve (the LPD-17s), though the LPD-4s will stay around for almost a decade before these new ships enter service. Following the three *Raleigh*-class ships, a further class of LPDs was constructed in the late 1960s. These became the *Austin* class (LPD-4), in service around the world today.

The *Shreveport* and her sister ships look a lot like the older LSD-36-class dock-ships, except that they have a larger superstructure, as well as a shorter main deck/helicopter platform and well deck. She is some 570 ft/173.7 m long, with a beam of 84 ft/25.6 m, and a nominal draft (with the ballast tanks dry) of 23 ft/7 m. Full displacement is 16,905 tons. The twelve ships of the class were constructed in three separate shipyards. USS *Austin* (LPD-4), USS *Ogden* (LPD-5), and USS *Duluth* (LPD-6) were built at the government-owned New York Naval Shipyard, some of the last U.S. warships built there. Ingalls built USS *Cleveland* (LPD-7) and USS *Dubuque* (LPD-8) at Ingalls in Pascagoula, Mississippi. USS *Denver* (LPD-9), USS *Juneau* (LPD-10), USS *Coronado* (LPD-11), USS *Shreveport* (LPD-12), USS *Nashville* (LPD-13), USS *Trenton* (LPD-14), and USS *Ponce* (LPD-15), were all built by Lockheed Shipbuilding in Seattle. *Coronado* (LPD-11) was converted into a command ship.

Shreveport (LPD-12) was laid down in Seattle, Washington, on December 27th, 1965, launched on October 25th, 1966, and commissioned on December 12th, 1970. She is powered by two 2,600-PSI Babcock and Wilcox boilers feeding a pair of De Laval steam turbines for a total of 24,000 hp to the twin shafts. Maximum speed is 21 kt/38.4 kph, though the efficiency of the powerplant allows it to cruise at 20 kt/36.6 kph. The steam plant is old and cranky by comparison to newer Navy steam, diesel, and gas turbine ships. Nevertheless, her dedicated "snipes" keep her going. *Shreveport* is one of nine ships in the class with extra bridge and berthing space, so it can act as a squadron flagship in "split ARG" operations.

When you walk around *Shreveport*, you find it generally similar to other Navy warships: gray paint, the overhead crowded with piping, conduits, and wiring runs, and hatches that need to be opened and closed by hand. But *Shreveport* is different from the ships we have visited so far. While some systems have been updated, there is a 1960s "feel" to the structure you see. Austin-class (LPD-4) ships were designed for a crew of drafted conscripts instead of volunteer professionals. The ship's systems had minimal automation (which required costly analog electronics) and maximum utilization of manpower, which was comparatively cheap (and more reliable!) in those days. Warship designers knew that a larger crew increases the ability of a ship to take damage and survive. Damage control is labor-intensive; and until recently, packing lots of men into a small hull was a good thing. You see this in *Shreveport* and her sister ships.

Let's go to particulars. Down in the crew and passenger (one of the Navy terms for "Marine") accommodation areas, you find the bunks are smaller and a bit shorter, and personal stowage space is more limited, than on *Wasp* or *Whidbey Island*. You find almost no recreational or fitness facilities. And *Shreveport* lacks the environmental-control systems found on every new warship today. In fact, her air-conditioning is even more cranky than her power plant, which can be tough on the crew and embarked Marines. During the MEU (SOC) workup in the summer of 1995, most of *Shreveport*'s air-conditioning system went out during a major heat wave. Even though the ARG was at sea, temperatures in the Marine berthing areas quickly rose to over 90° F/32° C with high humidity. Little could be done other than to push cold fluids to the men, and to shift some smaller units over to spare berthing on *Wasp* and *Whidbey Island*. Everyone took it in stride, but such problems sometimes occur in older vessels.

Passenger comfort is not why warships are built; and despite her advancing years, *Shreveport* is well equipped to operate not only as an ARG flagship, if necessary, but as an independent amphibious unit. *Shreveport*'s systems include:

- **Command and Control Capabilities—**In addition to accommodations as a flagship, *Shreveport* has full command and control facilities, although smaller and more limited than those aboard an LHD or LHD. These include a CIC, LFOC, SSES, and data links and communications gear.
- **Troop Capacity—**Along with her crew of 402 (plus a flag staff of 90 if carried), *Shreveport* can carry up to 840 Marines.
- **Vehicle/Cargo Capacity—**While she was designed before automated cargo handling, the *Shreveport* has 14,000 ft^2/1,301 m^2 of vehicle space, as well as 51,100 ft^3/1,447 m^3 for cargo. This is far more than *Whidbey Island* (LSD-41), allowing a great deal of autonomy if the ship must operate alone.
- **Transport/Off-load Capability—***Shreveport*'s robust aviation and transport facilities also enable her to to operate independently if required. These include a helicopter pad with two landing spots, as well as a hangar and air traffic control. The well deck can berth and support a LCAC or LCU, or up to four LCM-8s.
- **Cargo Handling Capacity—***Shreveport*'s cargo handling gear includes ten two-ton forklifts, a pair of three-ton rough terrain forklifts, three pallet conveyers, an eight-ton weapons and cargo elevator, and six cargo monorails like those aboard *Wasp*. There is also a thirty-ton deck crane for general-purpose lifting.

Shreveport can hold up her end of the amphibious task, either as part of an ARG, or all by herself, should that be required. *Shreveport*'s armament is typical of her generation. Back in the 1960s the Navy did not expect that amphibious ships would have to defend themselves; that was the job of aircraft carriers, surface escorts, and submarines. Times have changed since then, though, and *Shreveport* has been fitted for basic self-defense. In addition to an SPS-10F surface-search and SPS-40C air-search radar, she carries the SLQ-32 (V1) ESM package, which can detect an incoming missile and attempt to confuse it with chaff or decoys from four Mk 137 SRBOC launchers. Two of the original four twin 3-in/76-mm gun mounts have been removed, and replaced with a pair of 20mm Phalanx CIWS mounts. There is none of the splinter armor that you find aboard *Wasp* or *Whidbey Island*. This means that she could suffer severe fragmentation damage from a sea-skimming cruise missile even if the CIWS detonates the warhead before impact.

As the *Shreveport* and her sisters enter the twilight of their careers, you might expect the Navy to ease up a bit and try to stretch out their remaining service life. But the LPD-4s will stay at the forefront of amphibious operations until the new LPD-17-class assault ships arrive in the early part of the 21st century. The plan is to stretch the life of the class from the normal thirty years to roughly forty to fifty years! This will demand improvements to environmental systems, some communications, a fiber-optical data network, and perhaps even the Cooperative Engagement System designed into the LPD-17. These will be difficult to fund in the current budget environment. But the LPD-4s are a national asset, and you can expect General Krulak to fight like a "big dog" to ensure these venerable ships stay ready to land Marines.

Landing Craft

Ever since Stone Age men built the first raft to raid the neighbors downstream, small boats have been essential to amphibious operations. Captains of amphibs do not like to bring their large and sometimes vulnerable vessels within range of enemy artillery as they close a hostile shore. After the retirement of the last LST-1179-class ships, the option of running an ocean-going amphib up onto a beach (and getting her off again) will be gone forever. Given the dangers from mines, missiles, and guns, this is probably no great loss to our capabilities.

The amphibious equivalent of a delivery truck is the landing craft. As noted earlier, the development of landing craft during World War II was one of the key technologies that made amphibious warfare possible. Today, the Navy's landing craft range from the high-tech LCAC (Landing Craft, Air Cushioned) to conventional Landing Craft, Utility (LCU) and Landing Craft, Medium (LCM). While older craft are on their way out, they still provide amphibious planners with a range of delivery options. This is critical as the Navy and Marine Corps wait for long-delayed systems like the AAAV and MV-22B Osprey to enter service in the early 21st Century. The older landing craft provide vital support to Maritime Prepositioning Force (MPF) units for contingency and follow-on forces. Let's take a look at these delivery vans. Other than the Marines themselves, nothing is closer to the tip of the amphibious spear.

Landing Craft, Air Cushioned (LCAC)

When you first see one on its concrete pad at Little Creek, Virginia, it looks like a pile of Leggo blocks on a flattened inner tube. It is hard to believe that such an

Landing Craft, Air Cushioned (LCACs) of Amphibious Craft Unit Four (ACU-4) operate during a 26th MEU (SOC) exercise in Tunisia in 1995.
OFFICIAL U.S. MARINE CORPS PHOTO

odd machine changed the face of amphibious warfare. When they first appeared in the late 1930s, landing craft were never called "revolutionary" or "world shaking." But the Navy's introduction of the Landing Craft, Air Cushioned (LCAC) in the 1980s produced the biggest change in amphibious doctrine since the helicopter thirty years earlier. Pretty impressive for something that looks like a prop from a low-budget science fiction movie. Let's look LCAC over, and see for ourselves.

Amphibious planners always want to carry more payload, farther and faster. They dream of assault craft that don't need pleasant stretches of gently graded beach for landing zones. Conventional landing craft are limited to landing under optimal tidal and beach conditions—which means they have access to only 17% of the world's coastline. Traditional flat-bottomed assault boats severely restrict a planner's options. What was needed was new technology that did not require pushing a boxy hull through the water. The requirement was for a magic carpet, to whisk a seventy-ton battle tank across the water to the beach, and even inland.

The solution they found was a surface-effect vehicle: the hovercraft. A hover-craft floats on a cushion of air contained by a rubber skirt. Like a puck in an air hockey game, it barely touches the surface, but "floats" on the boundary interface. Riding a virtually frictionless layer of air, it needs relatively little thrust to move and maneuver. Hovercraft have great agility and speed, and they can carry a good payload with efficiency and economy. They are also relatively immune to rough weather and high seas. And they transition easily from water to ground, allowing the same craft to transport payloads some distance inland. Civilian hovercraft serve as high-speed ferryboats across the English Channel, and between Hong Kong and Macao in the Far East.

The Soviet Union, with its poor road network and vast marshlands, led the world in developing and deploying military hovercraft. During the Cold War, it built several types of amphibious assault hovercraft for the Northern, Baltic, and Black Sea fleets. Their planned targets were rocky coasts where conventional landing craft have little or no utility. But with hovercraft able to cross something like 70% of the worlds coastlines (versus 17% for conventional landing craft), they became a natural choice for Soviet Naval Infantry. Large troop- and vehicle-carrying hovercraft, known by NATO reporting names like *Aist* ("Stork"), *Lebed* ("Swan"), and *Pomornik* ("Skua"), could reach speeds up to 70 kt/128 kph, carrying heavy tanks, artillery, and troops. Technical intelligence reports made Western military forces sit up and take notice.

Early Western hovercraft were smaller, like the British-designed SR.N5 (called the PACV-series, when built by Bell for U.S. service), carrying an infantry squad or platoon. Field trials included combat deployments to Vietnam and Malaysia, with mixed results. The plus side was their speed and agility across rivers, swamps, and bays. The downside was vulnerability, especially their rubber skirts and propulsion systems. Despite this, Great Britain and Iran (under the last Shah) purchased many patrol hovercraft. Several factors kept hovercraft from entering Navy service as quickly—mainly money. The war in Vietnam was a huge financial drain in the 1960s and 1970s. The Navy and Marine Corps only began developing an amphibious hovercraft in the 1970s.

In late 1976 the Navy formalized a requirement and opened the competition for a Landing Craft, Air Cushioned (LCAC). Two contractors, Aerojet-General and Bell Aerospace (now Bell-Textron Land-Marine Systems in New Orleans, Louisiana), designed and built prototypes in the hope of winning a production contract for a planned fleet of over one hundred LCACs. The requirement included specifications for payload (up to 150,000 lb/68,182 kg), speed (greater than 50 kt/91 kph), and range (up to 200 nm/365 km at cruising speed). The Aerojet-General prototype was called JEFF-A; the Bell entry was JEFF-B. They looked similar when placed side by side. The competition was fierce, with both designs showing advantages and faults. In the end, Bell's JEFF-B design won, entering production as the Navy's new LCAC. JEFF-B's shorter length (87 ft/26.5 m versus 100 ft/30.5 m for the JEFF-A) and lower displacement (160 tons versus 162.5 tons) were decisive factors. In 1982, the Navy issued the first production contract for three LCACs. First delivery came in 1984, followed by ship compatibility trials. Lockheed Shipbuilding (later acquired by Avondale Shipbuilding) was certified as a second-source contractor, but Bell-Textron has built the majority of the craft.

By the late 1980s, several dozen LCACs were in service with the Navy, aboard a dozen amphibious ships in the Pacific and Atlantic fleets. Seventeen LCACs served on six LSDs during Desert Shield and Desert Storm, providing much of the lift during those operations. Though they did not conduct any assault landings, the amphibious forces offshore tied down over seven Iraqi divisions in coastal defenses around Kuwait City. The LCACs maintained a 100% availability rate throughout nine months of operations in the Persian Gulf, giving ARG commanders great confidence in their reliability. Since that time, the fleet has shifted the bulk of landing craft duty to the LCACs. In humanitarian and peacekeeping operations in Bangladesh, Haiti, and Somalia, and regular operations in ARGs, the LCACs again proved their worth. The total force of 91 LCACs was nearly complete by early 1996. More were planned, but the Navy's drawdown cut the original target of 107 units. The force of 91 LCACs is a national treasure which is being used hard.

To understand the LCACs, you need to visit one of two bases constructed to service them. I visited the LCAC facility at the Naval Amphibious Base at Little

Creek, near Norfolk, Virginia. This is the home of Assault Craft Unit (ACU) 4, the core unit for Atlantic Fleet-based LCACs. A similar facility services ACU 5 (the Pacific Fleet unit) at Camp Pendleton, California. ACU 4 operates roughly forty LCACs, providing detachments of hovercraft to Atlantic Fleet amphibious ships. The size of these detachments varies according to ship type. The following table summarizes the LCAC capacity of various ships:

Class	LCAC Capacity
USS *Wasp* (LHD-1)	3
USS *Tarawa* (LHA-1)	1
USS *Whidbey Island* (LSD-41)	4
USS *Harpers Ferry* (LSD-49)	2
USS *Anchorage* (LSD-36)	3
USS *Austin* (LPD-4)	1
LPD-17 (Projected)	2

Given the mix of ships within an ARG, a MEU (SOC) commander might have between six and nine LCACs in his well decks. That is a lot of capability to project Marines and firepower in just a few small packages. The ARG commander must manage this handful of LCACs carefully.

As you walk up to an LCAC on the ramp at Little Creek, the first thing you notice is that it looks much more like an aircraft or spacecraft than a warship. Much of the design for the LCAC was based on aircraft structures and technology to reduce weight and maximize payload. The LCAC is basically a platform with lift fans underneath, and twin deckhouses and engines along the sides. There are ramps at both ends, and a large rubber skirt running around the sides. Most of the structure is aluminum alloy, with some ceramic splinter armor. LCAC has to be able to survive hits when it works inshore. The threats range from artillery to anti-tank guided missiles. The four Avco-Lycoming TF-40B gas-turbine power plants provide a total of 12,444 shp/11,800 kw, and are mounted in pairs. Two engines drive the four 5.25-ft/1.6-m-diameter lift fans. The other pair drive the two 11.75-ft/3.6-m-diameter propulsion fans. Steering is done with variable-pitch propellers, aerodynamic rudders, and a pair of rotatable bow thruster nozzles. With a nominal load of fuel and a sixty-ton payload, LCAC can sustain up to 50 kt/91 kph in seastate 2 (a light chop) for a range of up to 328 nm/607 km. By cutting the payload, longer ranges can be obtained.

As you walk up the bow ramp, you enter a large (67-by-27-ft/20.4-by-8.2 m) cargo stowage area. Cargo tie-down points stud the decking, and there is a decided "crown" (or hump) to the deck to drain off any seawater. A nominal load of 119,980 lb/54,421 kg can be spread over 1,809.5 ft^2/168.1 m^2 of space. If necessary, this can be raised to an overload of 149,978 lb/68,027 kg as long as the seastate is moderate (the pounding of the waves in a high seastate can cause structural damage). Along with the deck cargo, there is room in the deckhouses for twenty-three passengers. Passenger accommodations are decidedly austere and very noisy when the LCAC is underway.

On the starboard side is the control cab, where the crew of five is located. This includes the LCAC commander, pilot, engineer, and navigator. U.S. Navy landing craft are commanded by a chief petty officer instead of a commissioned officer. This tends to make life aboard the landing craft a bit more relaxed and earthy than what you find aboard large amphibs; but don't think the enlisted crews of landing craft are lax about their responsibilities. On the contrary, they are highly professional, and over the last five decades, have won their share of Medals of Honor and Navy Crosses. Accommodations on the LCACs are spartan, with few of the "homey" amenities that we would find in the LCUs. Crews live on-board the ships where they are based, since LCACs lack galley and berthing facilities.

The control cab is laid out like an aircraft cockpit, which makes sense when you consider that an LCAC is more an aircraft than a surface craft. In fact, LCAC missions are listed on the daily ARG/MEU (SOC) air tasking order, to avoid interference with flight operations by helicopters and V/STOL aircraft. The control stations for the navigator, engineer, and pilot are laid out left to right. In addition to the throttle controls for the four TF-40B gas-turbine engines, there is a helm control station with instruments to assist in steering and navigation. These include a modified LN-66 navigation radar (to detect surface targets and land masses); an inertial system, known as the Attitude Heading and Reference Unit (AHRU); and a speedometer known as the High-Speed Velocity Log (HSVL). Like the Doppler sensing systems used on helicopters, described in *Armored Cav*, these sensors determine position, heading, and speed. A GPS receiver feeds into both the AHRU and HSVL systems, which makes pinpoint, split-second accurate landing possible for the first time. Now, all of this data is worthless if you cannot share it over a secure and robust communications system. The LCACs are fitted with a variety of VHF, UHF/VHF, HF, and FM transceivers, ranging from Motorola "Handy-Talkies" to fully encrypted digital radio systems.

The LCAC's role makes good communications a mission-critical feature. The LCAC is much faster than any previous landing craft. Speeds of up to 50+ kt/91+ kph are common, depending on load and seastate. This capability means that the big amphibious ships that operate LCACs no longer need to stand a few thousand yards/meters off of an enemy coastline, vulnerable to enemy fire. In fact, LCAC-equipped ships can stay up to 50 nm/91 km offshore and still be able to put a wave of loaded LCACs onto a beach every three hours. This three-hour cycle time is the normal turnaround used by Navy and Marine planners in landing operations. It assumes an hour each way for transit time, plus a half hour on each end for loading and unloading. This is what "standoff" really means, and LCAC is the first of three new systems (LCAC, the MV-22B, and the AAAV) that makes standoff amphibious assault possible.

You may wonder why so many navigational systems are necessary. If you have ever tried to navigate a boat 50 nm/91 km offshore, you would understand! As you approach a coastline, the reference points you use to determine your course and position are slow to appear, and even easier to miss. Now add in fog, rain, spray, darkness, currents, and uncharted rocks. Getting lost at sea is easy! History is replete with stories of amphibious landings which hit the wrong beach, even when the right one was in sight from the amphibious ships a few thousand yards away. Now, just imagine what kinds of errors are possible from 50 nm/91 km out!

The GPS receiver, with positional accuracy of a few yards/meters and timing accuracy within milliseconds, is the most valuable navigational system for keeping

LCAC on course and on time. But a new system is coming on-line to assist that. Known as the Amphibious Assault Direction System (AN/KSQ-1), it ties every ship, aircraft, and landing craft in an ARG/MEU (SOC) into a common network, feeding positional data from each unit's onboard GPS system. This lets the LFOC and CIC monitor real-time positional, heading, and velocity information on every friendly unit in the area. This system should eliminate many of the coordination problems inherent to amphibious operations.

Riding aboard an LCAC is different from any other boating experience you will ever have. First, the entire LCAC is buttoned up and the bow and stern ramps raised. When the turbine engines start, the noise is tremendous, and safety rules prohibit any exposed personnel on deck during transit. Even inside the deckhouses, earplugs and/or hearing protection is a necessity to make the turbine whine endurable. To back out of the well deck of a ship like *Wasp* or *Whidbey Island*, the pilot reverses the forward maneuvering thrustors to ease out. One advantage of the LCAC over conventional landing craft like the LCU or LCM is that the mother ship's well deck does not have to be "flooded down." Because of their ability to "climb" over obstacles up to 4 ft/1.2 m high, the LCAC can easily cross the lowered stern gate of an LHD, LHA, LSD, or LPD, simplifying operations for the ship's crew. This also reduces the seawater spray thrown up by the LCACs. This salt spray gets into nooks and crannies in the well deck overheads, causing corrosion that requires a lot of labor to repair. In fact, NAVSEA has plans for future dockships with "dry" well decks specifically designed for LCAC-type landing craft. Meanwhile, the Navy is experimenting with new-corrosion control techniques, including flame-sprayed coatings to prevent rust.

Once clear of the well deck, the pilot usually takes the craft to a holding/assembly area where it waits for any other LCACs being launched. If necessary in a "hot" area, the LCAC(s) pick up an escort of AH-1W Cobra attack helicopters. Now the pilot turns the LCAC to its desired heading, and takes off. The acceleration is smooth and rapid, and you have the feeling of riding on a magic carpet, or perhaps a really fast vacuum cleaner! While there is a fair amount of vibration, it is not the pounding that you feel in a conventional landing craft on a rough sea. The lift air flowing under the skirt tends to smooth out the wave action, making transits under all but the worst conditions quite tolerable. Speeds of 40 to 50 kt/73.2 to 91.4 kph can easily be maintained except for handing a maximum (sixty-ton-plus) load in heavy seas. For the pilot, the LCAC is easy to handle, though it tends to sideslip in a hard turn. This is because there is no keel or rudder to "bite" into the water to hold it steady. The LCAC is actually "flying" above the water, and the sensation is not unlike riding in a low-flying helicopter. The LCAC is quite maneuverable at all speed ranges. And it is stable and easy to handle, even at slow speeds in confined areas like a well deck or narrow rivers or swamps.

During transit, the navigator constantly passes course corrections and speed recommendations to the pilot, so that they will hit the target area accurately and on time. This notion that a landing craft can transit 50 nm/91 km or more and arrive on time at a pre-planned point is still a source of wonder to old amphibious warfare veterans. In fact, as noted earlier, the ability of beachmasters of the Navy's beach control teams to receive troops, vehicles, and cargo has not kept up with the ability of ships to off-load them, even from over the horizon. Even the introduction of computerized bar-code tracking linked to satellite communication systems has not solved the traffic jams that develop on a busy beach. This is one reason why LCACs don't always stop at the surf-line to dump their cargo. The LCAC's capability to transit from water to land, and continue inland for a distance, is still being explored. For example, with a pre-surveyed GPS navigational point, an LCAC might unload an artillery battery several thousand yards/meters inland, far away from the maddening traffic jam of the beach. Such concepts are being integrated into the doctrine of Marine amphibious units right now.

As you approach the shoreline, the beach comes up fast, and there is the feeling of an impending crash into a oncoming wall. Then the pilot begins to retard the throttles a bit and decides where to transit onto the beach. In fact, when you actually "hit" dry land, the feeling is like going up the ramp of a parking garage. The pilot then follows the instructions of the beach control party on where to stop and unload. The lift fans are killed, the skirt deflates, and the LCAC is ready to disembark its cargo. Once the bow and/or stern ramps are lowered, vehicles and troops can off-load in just a minute or two. For palletized cargo or containers, it takes a bit longer, as a forklift or palletized lifter vehicle is needed to unload the cargo deck. Unloading completed, the crew buttons up, fires up the engines, and heads back to the mother ship for another load. In the case of a LHD or LSD where two or more LCAC may be vying for space in a well deck, the craft are parked nose to tail. Then, with the bow and stern ramps lowered, vehicles drive through one LCACs to reach the other one.

While the LCAC has done quite well in its first decade of service, don't think that hauling cargo, vehicles, and Marines is all that the Navy wants to do with it. Concepts to expand the options for LCAC include increased personnel capacity, using a cargo deck passenger module. LCAC is now limited to just 23 passengers in the deckhouse spaces, but the module can carry up to 180 personnel (plus the 23 in the deckhouse) per trip. Configured for medical evacuation, the same module might carry up to 50 litter cases per trip, as well as 23 walking wounded in the deckhouse. This is important to the Marines, given the "golden hour" of combat trauma cases. Survival rates for wounded personnel are directly related to how quickly they reach medical facilities aboard the LHD/LHD or LPD. The Navy has ordered a number of these modules, and they should be coming into the force soon.

Another use for LCACs is in mine warfare. The Navy has funded demonstrations of LCACs equipped to both lay and sweep underwater mines, as well as a rocket propelled system that throws an explosive mine-clearing charge over a beach landing zone from offshore. There have also been studies of using the LCAC as a gunboat to support landing operations. Though the LCACs are unarmed (mounts for three machine guns are normally not used), there are concepts for mounting 20mm and 25mm cannons. The Marines have demonstrated the ability to fire vehicle mounted weapons such as the LAV's 25mm Bushmaster cannon and the 120mm gun on a M-1A1 tank from landing craft.

With only ninety-one LCACs either delivered or under contract, it is likely that the Navy and Marines will jealously guard them for their primary mission as ship-to-shore delivery systems. In this role it is not, of course, ideal (like all designs, it is a set of engineering compromises). For one thing, it is more vulnerable than conventional landing craft to enemy fire, but has the speed and maneuverability to avoid many threats. And the LCAC cannot handle extreme seastates as well as a conventional landing craft like the LCU or LCM, but it can land cargo under a wider variety of coastal conditions. Still, don't get the idea that LCACs are not tough. One unit, LCAC-42 (landing craft have only pennant numbers, not names), has survived two major incidents, and is still in service. It hit a protruding coral head sideways during one exercise, and struck a large navigation buoy on another, but got off with only minor damage and is still hauling for the Pacific Fleet. In over ten years of LCAC service, the U.S. Navy has yet to lose even one in operations. Plan on seeing LCACs around for a long time to come. A SLEP (Service Life Extension Program) will extend the planned twenty-year service life of the LCAC fleet to a full thirty years. Next-generation landing craft will be air-cushioned. Scaled-down designs for LCM-sized LCACs are being considered as general-purpose deliver platforms for the ARGs of the mid-21st century. Not bad for a giant air hockey puck.

Landing Craft, Utility (LCU)

It might surprise you that in an era of satellite navigation and computerized logistics, a large percentage of landing craft used by the Navy and Marines are virtually identical to World War II types. Many such craft will continue to serve well into the 21st century. Currently, the largest of these is the Landing Craft, Utility (LCU). In fact, the LCU is the largest Navy vessel that is not commanded by an officer. The LCU is a *ship*, with full crew accommodations (galley, berthing, heads, etc.) for its crew of ten (fourteen in wartime). It has enough range (up to 1,200 nm/2,195 km at economical speeds) to transit the Mediterranean or Baltic Seas in even the worst weather. LCUs are the heavy haulers among landing craft, in the twilight of their years, but still doing a vital job. Let's have a look.

Like other conventional landing craft, the LCU design dates back to the 1940s. The idea behind the LCU was simple. Take the largest possible cargo/vehicle

A Landing Craft, Utility (LCU) of Assault Craft Unit Two (ACU-2) leaves Cadiz Harbor on February 16th, 1996, to mate up with the USS *Whidbey Island* (LSD-41) for the homeward leg of its Mediterranean 1995/96 cruise. *JOHN D. GRESHAM*

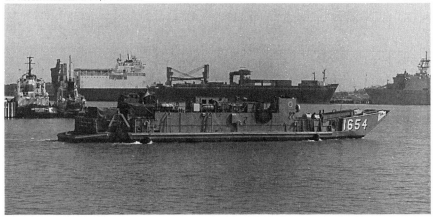

load possible, deliver it to and from a hostile shore, and then return to a mother ship—usually one of the first-generation LSDs. The LCU can carry up to 180 tons of vehicles, troops, and cargo at speeds approaching 12 kt/22 kph in virtually any seast-ate or weather, and deliver them to a "hot" shoreline. It is a big, brutish sort of craft, with none of the LCAC's futuristic look. In fact, the beast looks like it could seri-ously hurt a bigger vessel by ramming (this is no joke; it probably could!). These classic landing craft, loved by their crews and prized by the ARG and MEU (SOC) commanders, are still finding new ways to serve.

Like the LCAC, the LCU is a "double-ended" design, with ramps at both ends allowing vehicles to load by driving through one LCU to get to the next one. They are constructed of heavy steel, welded back in the days where the quality control test was a long swing with a sledgehammer! The LCU may be one of the most bullet-resistant craft in the Navy, which explains why they are frequently used as gunboats and escorts for rubber boats and AAV-7s. LCUs were built by many contractors, such as Defoe Shipbuilders of Wisconsin, General Ship & Engine Works of Boston, Gunderson Brothers of Oregon, Moss Point Marine of Mississippi, and Southern Shipbuilders of Louisiana. Their construction was simple, requiring no special skills or equipment. Though the original LCUs date back to 1951, the class currently in service, the LCU-1610s, were built between 1959 and 1985. During all that time, the design was essentially unchanged, except for one experimental unit constructed of aluminum.

The LCU is essentially a floating steel box or barge, with a deckhouse to star-board, fore and aft loading ramps, and some side plating to keep passengers in and water out. Powered by four GM/Detroit Diesel engines (each delivering 300 hp), they are some of the most powerful ships per ton of displacement in the Navy. They are even used as tugs when actual tugboats are not available to push barges and lighters around. When you climb up the bow ramp of a LCU, you are immediately struck by how functional everything is. The chief petty officers who run the LCUs do so in a no-nonsense fashion, without pretensions to polishing the brass or keep-ing the paint clean. But I defy you to find a line out of place, corrosion forming, or a hatch left undogged. This is the Navy of the old chiefs, where you find little of the high technology or political correctness that permeate the big ships of the "real" Navy. Aside from a portable GPS receiver in the pilothouse and a small homegrown cable TV/VCR network down in the crew berthing spaces, everything on the LCUs of the 1990s would be familiar to your grandfather, if he was a sailor in the 1940s. The steel deck has tie-down stanchions to keep heavy gear and cargo from shifting in heavy seas; and since the cargo deck is open to the elements, the crew quickly hands you a life preserver. There is a winch-driven anchor system to drag the LCU off of the beach if the tide goes out while it is beached.

The 121-by-25 ft/36.9-by-7.6-m cargo deck takes up most of the LCU's 134.75-ft/41.1-m length. The cargo deck can handle up to 1,850 ft^2/171.9 m^2 of vehicles, troops, and cargo, up to a weight limit of 180 tons! Given that the LCU can deliver this load in almost any seastate, you can see why the Marines like to have LCUs hauling their heavy gear like 70-ton M1A1 Abrams tanks and large palletized-loading-system (PLS) trucks. In a seastate where an LCAC would be unable to haul a single M1A1, an LCU can carry two of the armored monsters, with space and capacity to spare.

The LCUs' long range means that they can be used as utility transports in closed waters (like the Baltic and Adriatic), returning to base to haul fresh food, spare parts, and that vital commodity, mail. LCU crews take working inshore quite seriously, and frequently mount machine guns, grenade launchers, and other weapons. They have even fired 25mm and 120mm cannons of embarked LAVs and M1A1s, which is awesome firepower. The LCU crews see themselves on the cutting edge of the recently reborn art of riverine warfare, and they practice it often in exercises.

As noted earlier, LCUs are warships, with their own berthing, galley, and head facilities. The galley, aft of the pilothouse in the starboard deckhouse, can whip up a full meal. In fact, when they are in the well decks of their mother ships, they require only power, water, and sewage hookups (some also ask for access to the ship's cable TV system) to live independently from the ship's company. They buy their own food from the mother ship's supply system, and even have their own communications call signs for message traffic from higher commands. The living facilities are located belowdecks, along with the engine rooms (there are two, separated to improve survivability), machine shop, and other necessities. You might call the living conditions spartan, but LCU crews like them just fine. In fact, life in an LCU is reminiscent of life aboard a submarine, with many of the same benefits and drawbacks. As with a submarine, the only private space is the captain's cabin, though the commander of an LCU is only a chief petty officer! Don't say "only" a chief, though, because these men know their stuff! There is a saying in the Navy that if you want someone to think, ask an officer. But if you want it done, ask a chief...nicely!

For all of their age, the LCUs are a pleasure to ride. One of the joys of preparing this book was a late summer ride out to the USS *Wasp* (LHD-1) on the bridge (above the pilothouse) of an LCU. Stable as a rock as we headed into the huge well deck, we could not help feeling that we had rediscovered something wonderful about the world. The LCUs ride well, even in a heavy or following sea, and can handle almost any climate from the heat and dust of North Africa to the ice and cold of Norway. They also fit well aboard amphibious ships, as the following chart shows:

Amphibious Ship LCU Capacity

Class	LCU Capacity
USS *Wasp* (LHD-1)	2
USS *Tarawa* (LHA-1)	4
USS *Whidbey Island* (LSD-41)	3
USS *Harpers Ferry* (LSD-49)	1
USS *Anchorage* (LSD-36)	1/3*
USS *Austin* (LPD-4)	1
LPD-17 (Projected)	1

* LCU Capacity with Mezzanine Deck Removed

As you can see, amphibious ships trade about two LCACs for each LCU. Given the LCU's compatibility with older ships like the LHAs and LSD-36s (for which they were designed), it's a shell game to mix and match ships and landing craft to obtain the ideal combination of landing craft for a particular mission. For example, when Captain C.C. Buchanan (Commander of Amphibious Squadron Four, PHIBRON-4) was configuring his force for the 1995/96 cruise of PHIBRON-4 and its embarked Marine unit, the 26th MEU (SOC), he decided on the following mix. Aboard USS *Wasp* (LHD-1, his flagship), he embarked three LCACs from ACU-4 at Little Creek, Virginia. He then ordered up one LCU each for USS *Whidbey Island* (LSD-41) and USS *Shreveport* (LPD-12) from ACU-2 (the Atlantic Fleet LCU unit: ACU-1 services the Pacific Fleet), also at Little Creek. This mix made optimum use of available well deck space, and provided maximum lift capacity for the coming Mediterranean cruise. It was a prudent decision. Sailors and Marines are conservative, and they believe in the reliability of the big steel LCUs. In fact, the LCUs are scheduled to get the fancy new AN/KSQ-1 Amphibious Assault Direction System, which says something about their longevity in the eyes of Naval planners. "Rusty but trusty," the LCUs fill a vital role in the amphibious Navy.

Landing Craft, Medium (LCM)

The last landing craft we will look at is by far the eldest: the venerable Landing Craft Medium, Mark 8. The LCM-8 is the last direct link with the kind of landing craft you see in old war movies storming the beaches of Normandy or Iwo Jima. The basic design of this long-serving utility craft dates back to a British vessel of the early 1940s. Back then, the requirement was to haul a thirty-ton tank or equivalent load from an offshore transport. Other than increasing the payload capacity to accommodate a modern main battle tank, not much has changed.

The basic LCM-8 is a metal box, with a retractable bow ramp and a pair of 165 hp marine diesels. Most of the LCM-8s are made of high-tensile steel, though some units were welded aluminum to reduce weight for stowage aboard LKA-113-class assault cargo ships. Aft is a small pilothouse. And that is about it. There are armament or berthing facilities for the crew of five (they live aboard their mother ship). The cargo area is open to the elements. The LCM-8 can make about 10 kt/18 kph for a range of about 190 nm/347 km with a sixty ton cargo load or perhaps 125 Marines. An LCM-8 can carry every piece of ground equipment in a MAGTF, except the M1A1 Abrams tank. The LCM-8s roll a fair amount, and can ride decidedly rough in heavy seas. Nevertheless, they are quite seaworthy, despite the pounding that they deliver to their passengers and cargo.

Currently, though the capability does still exist, an ARG carrying an MEU (SOC) would almost never carry LCM-8s. Where you find the LCM-8 is in the three maritime preposition squadrons. There they function as cargo carriers for vehicles and equipment. They act as tugs for barges, and transport personnel between ships. Many allied forces, including Britain's Royal Navy, use the LCM, and will continue to for some time. After a half century of service, however, the LCM's retirement from the U.S. Navy is finally at hand. Within the next ten to fifteen years, the last LCMs will leave U.S. service, becoming a fond memory to the sailors that crewed them. They have served in wars from the Pacific to the South Atlantic, with distinction.

What will replace them? By about 2010, the Navy will need a landing craft with a cargo capacity in the thirty-five-to-fifty-ton range. A logical successor might be a downsized LCAC. In addition to carrying cargo, a gunboat version able to escort

LCACs or AAAVs would be very useful. The problem, of course, is money. There simply is no budget for anything but paper studies, and no program office has been chartered to solve the problem. Given the fiscal limitations of the next decade or so, you might see LCMs serving well into the first quarter of the 21st century. They are simple. They work. That alone may keep them around for some time to come.

The Maritime Prepositioning Force (MPF)

During the past two or three decades, the U.S. has managed to abandon or get thrown out of most overseas bases for its forward deployed forces. It was our own fault really. Sometimes we backed the wrong dictators (Marcos in the Philippines or Noriega in Panama). Sometimes we just got our butts kicked out, as happened in France, Vietnam, and Libya. And sometimes nature takes a hand, as when the volcano Mount Pinatabo erupted, wrecking Clark Field and hastening our exit from the Philippines. As a result, the U.S. Navy is currently limited to a handful of overseas bases, usually on old colonial possessions or territories of our best allies. These include Guam, Diego Garcia, the Azores, and Okinawa. Unfortunately, such bases are separated by thousands of miles/kilometers from the continental U.S. and from the most likely potential flash points.

This caused serious difficulties in the late 1970s when the U.S. had virtually no bases in Southwest Asia to confront the Islamic revolution in Iran or the Soviet invasion of Afghanistan. The only U.S. base in the Indian Ocean, Diego Garcia (leased from Great Britain), is almost 2,000 nm/3,700 km from the Straits of Hormuz, at the head of the Persian Gulf. This situation was compounded by drastic cuts in the Navy's budget, slashing the power-projection capability it had possessed just five years earlier at the end of the Vietnam War. The drawdown of U.S. military by the Carter Administration probably encouraged the actions of the Soviets and Iranians in 1979. Then-Secretary of Defense Harold Brown authorized a study in 1979 to find ways to reverse the downward slide of forward-based U.S. forces around the world. Several alternatives were considered, including:

- Construct a vast new fleet of amphibious ships, roughly doubling Navy lift capacity.
- Build additional strategic airlift aircraft (C-5s, C-141s, etc.) to rush units of regiment and brigade size to crisis areas.
- Find new ways to forward base units and equipment for rapid deployment to a crisis.

The third alternative won out: prepositioning stocks of military equipment close to potential trouble spots, allowing troops to fly in and form up their units on the spot. Called Prepositioning Of Materiel Configured in Unit Sets (POMCUS), it was a key element of NATO strategy during the Cold War. It was much cheaper than maintaining full-time units on the inter-German border, and allowed ground forces to be based mainly in the continental U.S., saving vast sums of money. The Marines already had prepositioned stocks in Norway, stored in large caves in the Oslo area. POMCUS sites are also used in Korea to deter aggression by a belligerent neighbor. The problem in Southwest Asia in 1979 was that the U.S. had no allies willing to allow basing of equipment on their territory. Some way had to be found to base enough equipment for a Marine brigade (about 18,500 personnel) without upsetting the neighbors.

The answer was a pair of commercial shipping technologies that came of age in the 1970s. The first, containerized cargo handling, allowed long-term packing and storage of equipment and supplies, with computerized tracking to provide rapid access to the contents of any particular container. The other technology was the Roll-On, Roll-Off (Ro-Ro) ship, which allowed vehicles to drive on or off a ship without special handling equipment or personnel. All that was required was a jetty or wharf where the ship could drop its ramp. The vehicles did the rest themselves. Ro-Ro ships were fairly common by the late 1970s, and it was quite possible to package a complete Marine brigade on a group of such ships. You could have ships sit in an island lagoon or just steam offshore from the crisis area. All they would need was a port facility to off-load, and an airfield to fly in personnel and aircraft. The ships would carry enough supplies (water, fuel, food, ammunition, etc.) to support a Marine brigade long enough for follow-on forces and supplies to arrive from the United States.

By 1980, an interim force of seven leased merchant Ro-Ro ships (enough for a reduced 11,000-man Marine brigade) was stationed at Diego Garcia in the Indian Ocean. This was only a temporary stopgap, so in 1981, the Maritime Prepositioning Force (MPF) was established as a permanent unit. MPF leased thirteen converted Ro-Ro ships, forming three Maritime Preposition Squadrons (MPSRONs). Each MPSRON can equip, supply, and support an 18,500-man brigade-sized MAGTF for up to thirty days. With three such units on permanent station, at least one would be within seven days steaming from anywhere in the world they might be needed. As history turned out, this has been the case.

The thirteen ships procured under the 1981 MPF program are technically of three different types, though they fall into two classes for size and capacity comparison. The first five were Norwegian-owned Maersk-type Ro-Ro vessels. Three American-built Waterman-class Ro-Ro ships were converted, with five additional units of the AmSea/Braintree class purpose-built by General Dynamics, Quincy Shipbuilding Division. Conversion of the original eight ships involved splitting them, adding a large cargo section in the middle, a helicopter platform aft, and heavy lift cranes forward. Their general characteristics are summarized in the table below.

MPF Ship Class Characteristics

Class	Waterman	Maersk	AmSea
Loaded Displacement (LT)	51,612	44,088	46,111
Length	821' 1/2"	755' 5"	673' 2"
Beam	105' 8"	90'	105' 6"
Draft	33' 6"	32' 10"	32' 1"
Propulsion	Steam Turbine	Marine Diesel	Marine Diesel
Speed (Kt)	17.7	16.4	17.7
Range (Nm)	11,107	10,802	11,107
Cargo ^{2} (Ft ^{2})	608,839	607,975	608,740

MPF Ship Class Characteristics (Continued)			
Class	Waterman	Maersk	AmSea
20-Ft Cargo Container Capacity	2,311	2,020	2,311
Bulk Fuel Capacity (Gal)	6,419,192	6,643,350	6,542,760
Bulk Water Capacity (Gal)	374,808	424,620	395,976
Water Production Capacity (Gal)	100,000	125,000	100,000

The ships were built to commercial standards, with comfortable accommodations for their small crews. This is important, because they may be deployed for months at remote sites around the world. Each ship has several vehicle/cargo decks, where everything from main battle tanks to cargo containers are stored. These can be rolled off the stern ramp onto a pier, or lifted out by deck cranes. Each MPF ship has a large stowage capacity for fuel and water, and equipment to distill up to 100,000 gallons/377,358 liters of freshwater per day. Finally, each MPSRON has an afloat commodore (usually a senior captain) and staff as the Navy command element.

The MPF conversions took several years to complete, and the ships a while longer to outfit and equip. Nevertheless, by 1986, they were ready for service. All thirteen were then leased back to the Navy to form three MPSRONs. To support the MPF program, a maintenance facility was established at Blount Island near Jacksonville, Florida. Every thirty months, each ship rotates through Blount Island for a few weeks. All of its equipment and supplies are off-loaded. Then everything is inspected and replaced as required; equipment and vehicles are cleaned and modified to the latest USMC standards. In this way, twelve out of thirteen MPF ships are always on station with their MPSRONs.

Each squadron is based a few days steaming time from its primary Area of Responsibility (AOR, the diplomatic euphemism for "trouble spot"). Their organizational structure looks like this:

The maritime prepositioning ship *PFC James Anderson, Jr.,* sits alongside at Blount Island near Jacksonville, Fla., prior to her return to Maritime Prepositioning Squadron Two (MPSRON-2) at Diego Garcia Atoll in the Indian Ocean.

JOHN D. GRESHAM

MPF Ship/Squadron Organization

MPSRON-1 (Afloat, Mediterranean Sea)	MPSRON-2 (Diego Garcia, Indian Ocean)	MPSRON-3 (Agana Harbor, Guam)
PFC. Obregon (Waterman)*	*CPL. Hauge* (Maersk)*	*1st Lt. Lummus* (AmSea)*
2nd Lt. Bobo (AmSea)**	*Pvt. Phillips* (Maersk)**	*Sgt. Button* (AmSea)**
Sgt. Kocak (Waterman)	*1st. Lt. Bonnyman* (Maersk)	*1st Lt. Lopez* (AmSea)
Maj. Pless (Waterman)	*PFC. Baugh* (Maersk)	*PFC. Williams* (AmSea)
	PFC. Anderson (Maersk) (Maersk)	

* – MPSRON Flagship ** – MPSRON Alternative Flagship

The Waterman- and AmSea-class ships have roughly the same stowage footprint, while the Maersk-class ships have somewhat less (mostly in the area of containerized cargo). Thus, MPSRON-2 has the five Maersk-class Ro-Ro vessels, while MPSRON-1 and -3 each have four of the other types. All vehicles are combat-loaded, fueled, and armed, ready to drive down the stern ramp, directly into battle if necessary.

Just how much stuff does a MPSRON carry? Well, a lot! The following matrix measures the typical loadout for all three MPSRONs. It should be noted that MAGTF equipment and supplies are evenly distributed across the ships of an MPSRON, so that loss or damage of one ship will not cripple the entire force:

Marine Brigade MAGTAF Personnel/Equipment Matrix

MAGTAF/Naval Personnel*	MAGTAF Equipment	Aircraft**
Command Element—883	M1A1 Abrams Tanks—30	F/A-18C/D Hornet—36
Ground Combat Element—6,414	LAV/AAV—25/109	AV-8B Harrier II—20
Aviation Combat Element—7,005	Towed 155mm Howitzers—30	EA-6B Prowler—5
Combat Service Support—3,039	HMMWVs—631 (129 Armed, 72 with TOW Missiles)	KC-130 Hercules—12
Naval Cargo Handling and Port Group—306	Hawk/Stinger SAM Launchers—8/45	CH-53E Super Stallion—24
Naval Beach Group—763	ROWPUs—41	CH-46 Sea Knight—12
Naval Security Group—124	Trucks (5 Tons or Larger)/MHE—489/121	AH-1W Super Cobra—18
		UH-1N Iroquois—9
Total—18,534 Personnel		**Total—124 Aircraft**

*—Deployed Via AMC/CRAF/Charter Aircraft   ** – Self Deployed

In addition to the equipment stowed on board, there are stocks of rations (*lots* of MREs!), clothing and individual equipment, fuel and lubricants, construction materials, ammunition, medical and dental supplies, and repair parts. All you need to add is personnel and aircraft. These are flown into a friendly airfield, then "marry up" with the shipborne equipment and supplies. More on this later.

Assume that a crisis has broken out somewhere in the AOR of an MPSRON, and the national command authorities decide to insert a Marine Brigade MAGTF to stabilize the situation. If a friendly host nation exists (the preferred option), then the MPSRON begins to steam for a port or anchorage where it can unload. If there is no friendly host nation, the next step is a "kick-in-the-door" operation by one of the MEU (SOC)/ARG teams, perhaps with the help of an Army unit like the alert brigade of the 82nd Airborne Division at Fort Bragg, North Carolina. However they are secured, the keys to a successful MPF operation are a ten-thousand-foot/three-thousand-meter runway and a port facility or calm stretch of beach.

Approximately ninety hours prior to the start of unloading, a Navy team flies out to the MPF ships to help prepare the vehicles and unloading equipment. This includes installing batteries in vehicles and preparing cranes and lighterage. At the same time, ground troops and air units prepare for deployment. The 18,500 Marines deploy on Air Force C-5/17/141 transports, Civil Reserve Air Fleet (CRAF) airliners, and commercial charter aircraft. The tactical aircraft self deploy with the help of Air Force tanker aircraft, while helicopters are partly disassembled for shipment on C-5/17/141 transports. All told, it takes about 250 airlift sorties to bring the entire force in, with several dozen additional daily sorties to support the operation once it gets going.

Just prior to the fly-in, the ships begin unloading. If a port is available, then the vehicles exit off the stern ramps, and their crews take possession and drive them to assembly areas (or right into combat if the situation is really urgent). Cargo containers are then off-loaded onto trailers or the docks, and the operation is completed. This scenario has been tested in exercises and real-world deployments, and refined down to a science. With a decent port facility, every vehicle can off-load in just eighteen hours, and all the cargo in three days. Following this, the ship remains in harbor only if local fuel and water supplies are not available, or to re-embark everything at the end of the operation.

Things get a little tougher if no port facility is available. To deal with this contingency, called an unload "in-stream," each MPSRON carries landing craft (LCM-8s) and lighterage (floating causeways and barges) to move vehicles and cargo ashore. The breakdown of each MPSRON's equipment is shown in the table below:

MPSRON Lighterage/Causeway Equipment

MPSRON-1 (Afloat, Mediterranean Sea)	MPSRON-2 (Diego Garcia, Indian Ocean)	MPSRON-3 (Agana Harbor, Guam)
8 LCM-8	10 LCM-8	8 LCM-8
20 SLWT Powered Causeways	20 SLWT Powered Causeways	20 SLWT Powered Causeways
24 Non-Powered Causeways	24 Non-Powered Causeways	24 Non-Powered Causeways
4 LARC V	4 LARC V	4 LARC V

As you might imagine, unloading in-stream is slower than in a port facility. The LCM-8s move heavy vehicles and equipment like tanks and artillery, while the causeways move the rest of the MAGTF's supplies. Also, the stern ramp can be used to launch amphibious tractors like the AAV-7 or AAAV so that they can land under their own power. Under these conditions, it takes about three days to get the vehicles ashore, and at least two more days to unload cargo. Each MPF ship is equipped with supply conduits for water and fuel. These floating pipes allow the ships to stand up to four thousand yards/meters offshore and supply the needs of the MAGTF. This scenario is difficult and dangerous, as it forces the MPF ships to come close to shore and stay there for the better part of a week. Nevertheless, it may be the only option that puts a sizable entry force into a crisis area.

Since their inception, MPSRONs have been some of the busiest units in the Navy. In the 1990s, MPSRON-2 (based at Diego Garcia) has made three Persian Gulf deployments in response to Iraqi aggression. In 1990, MPSRON-2 delivered the first heavy units and equipment (the 7th MEB and 3rd MAW from California) during Operation Desert Shield. It also provided the first sustained logistical support for Army units flown into Saudi Arabia with almost no supplies or ammunition other than what they carried on their backs. These units drew from the stocks on the MPF ships, holding the line until follow-on forces and supplies began to arrive in late August 1990. MPSRON-2 deployed to Kuwait in 1994 and 1995, in response to threatening moves by Iraqi forces near Basra. These last two deployments took place less than ten months apart, demonstrating the value of a forward mobile base force like MPF. In addition, individual MPF ships deployed to support relief and peacekeeping operations in the Balkans and Somalia. By any standard of success, the MPF has vindicated those who originated the concept some fifteen years ago. At a minimal cost, the U.S. has reversed the downward spiral of crisis-response capability, without requiring permission from foreign governments to conduct operations. It has been quite a bargain.

As we move into the 21st century, the future of maritime prepositioning has never looked better. The Navy/Marine Corps MPF program is going strong and continues to be well funded by Congress. Meanwhile, both the U.S. Army and U.S. Air Force have bought their own fleets of prepositioning ships and are beginning to station them around the world. (Some of the Army MPF forces will share space at Diego Garcia and Guam with their Navy counterparts.) The Army vessels are larger and have a deeper draft than the Navy MPF ships, but they were built from the keel up for the job (they are Army-owned rather than leased, like the Navy ships), and have better vehicle handling facilities. Given the joint nature of military operations these Army and Navy days, it is likely these units will work together in future contingencies.

The future of the Navy MPF program is an open issue. Halfway through their projected thirty-year service life, the leased ships are in good shape, though it is time to consider eventual replacements. NAVSEA has proposed building a dozen new MPF ships to replace the existing force around 2015. Another option the Marines are evaluating is called Project Seabase, which would dispense with ships and build a huge floating base, which could move into a crisis area. With the stowage capacity of an entire MPSRON, it could operate and maintain all of the aircraft and helicopters assigned to the MAGTF. This Mobile Base concept originated with Admiral Bill Owens (the retired Vice Chairman of the Joint Chiefs of Staff). He envisioned a series of linked platforms, like those used for oil drilling, that could handle aircraft up to the size of a C-130 Hercules, or even a C-17 Globemaster III. The base's propulsion system would move it at about eight knots, and it would anchor between twenty-five and fifty nm offshore. In this way, the deployed MAGTF would have no need for an airfield and port complex to unload. LCACs, V-22s, and other delivery systems would move units ashore, eliminating the need for an MEU (SOC) to make a forcible entry. The problem with this concept is that it would be terribly expensive, probably costing more than a nuclear-powered aircraft carrier. A second generation of MPF ships will probably be the most economical way to sustain our forward-based equipment stocks. Whatever solution is chosen, there can be little doubt that this successful program will continue into the next century. MPF has provided exceptional value to the American taxpayer, and has been a major force in keeping the peace.

The Future: The LPD-17

This chapter has examined ships that represent an amphibious modernization program conceived over twenty years ago. This program was designed to replace the fleet of amphibious shipping constructed during the 1960s and early 1970s at the height of the Cold War. Despite all of the shipbuilding that we have described thus far, there is still a huge shortfall. This is the cargo footprint currently carried by forty-one ships of the LST-1179, LKA-113, LSD-36, and LPD-4 classes. These ships, whose average age ranged from twenty-three to twenty-six in 1995, are quickly reaching the end of their service lives. The Navy's answer is a new class of twelve ships, called the LPD-17 class, to replace all forty-one ships that will retire over the next ten years or so.

The LPD-17s will reflect everything the American shipbuilding industry has learned over the last three decades. These twelve ships (the amphibious "ships-of-the-line" that General Krulak discussed in Chapter 2) will form the inshore leg of 21st century ARGs. They will have to be extraordinarily versatile to replace the fleet of older vessels that are going to the scrapyard. One measure of this is to compare the tonnage and manning of the old ships versus the new. The twelve LPD-17s, with a total crew of 5,200 and displacing just 300,000 tons, will replace forty-one ships with over 13,000 crew and displacing 525,000 tons. Quite a lot to expect from a ship which has not even had its final weapons suite decided. Let's look further.

The Navy views the design of LPD-17 in several ways. Recall the five footprints that I described earlier. The chart below summarizes these in comparison with the LPD-4 class they will replace:

LPD-17 vs. LPD-4

Class	Troops	Cargo^2 (Ft^2)	Cargo^3 (Ft^3)	LCAC	Helicopters
LPD-17	720	25,200	29,500	2	Up to 6
LPD-4	788	11,800	38,300	1	Up to 4

As the chart clearly shows, the LPD-17 class will have a significant edge in most of the key footprints that are of concern to the Navy and Marine Corps. Cargo^2 is vastly improved, along with facilities for landing craft and aircraft. While there is a significant decrease in cargo^3 space, this has been compensated for in the design of the *Whidbey Island/Harpers Ferry*-class (LSD-41/49) landing dockships.

The slight decrease in troop berthing has also been dealt with in the design of other amphibious ships. The 720 Marine berthing spaces provided will be among the most comfortable and spacious of any vessel ever built for the Navy.

About the year 2005, these new ships will take their place in the ARGs, becoming the standard transport for the seven MEU (SOC) units. The table below shows two notional ARG configurations we might expect to see:

Proposed 21st Century ARG Ship Mixes

ARG Option	Aviation Dockship	Landing Dockship	Personnel Ship	Total ARG LCACs	Total ARG Aircraft
LHA	LHA-1	LSD-41	LPD-17	7	48
LHD	LHD-1	LSD-49	LPD-17	7	51

As you can see, the commanders of an ARG and a MEU (SOC) will get roughly fifty aircraft spots and seven LCACs to support their operations. It should be noted that these two mixes represent minimum ARG capabilities. Other combinations are possible. The LPD-17 will be the inshore ship in the ARG, required to go further into harm's way than either the LHA/LHDs or the LSDs. In fact, the LPD-17s will regularly operate about 25 nm/45.7 km offshore, while other ships in the ARG (the LHAs/LHDs and LSDs) remain 50 to 200 nm/91.4 to 365.8 km out, since their maximum standoff from the target areas is determined by the speed of the LCAC (over 40 kt), and the new MV-22B (over 200 kt). The LPD-17's 25-nm/45.7-km standoff is dictated by the transit speed of the new AAAV. LPD-17 will be the primary platform for the AAAV, while providing facilities for other elements of the ARG and the MEU (SOC). For example, the LPD-17 will be the platform for the MEU (SOC) force of AH-1W Cobra attack helicopters as well as the embarked unit of UAVs. The LPD-17s will also be the lone wolf during "split ARG" operations; functioning as a mini-MEU (SOC), which General Krulak described in Chapter 2. The LPD-17 will become the utility infielder for the 'Gator Navy. Thus, if one of the ships in the ARG is going to get hit by an enemy attack, the LPD-17 will likely be the target.

The LPD-17 has therefore been designed to be the most defensible and survivable amphibious ship ever built. Structurally, the LPD-17 is going to be the world's toughest warship per ton. Even the long-ignored threat of mine warfare has been anticipated. In the LPD-17 design, NAVSEA has devoted over two hundred tons of structural stiffening to reduce damage from hull "whipping," when an underwater mine goes off nearby. Like *Wasp*-class LHDs, the LPD-17 will have a chemical/biological overpressure protection system, improved fire-zone protection, blast-resistant bulkheads, and fragmentation armor topside. The lessons in stealthy shaping gained from the *Arleigh Burke* (DDG-51)-class destroyers have been applied to the LPD-17. When you study a drawing of the LPD-17, you see that the angles and curves resemble those on the DDG-51, and even the Lockheed F-117A Night Hawk stealth fighter. This is no coincidence; the principles of radar diffraction discussed in *Fighter Wing* apply to ships just as well as aircraft. Blankets and coatings of radar-absorbing material will be incorporated into the LPD-17, along with reduced acoustic and infrared signatures. NAVSEA claims that the LPD-17 will have only 1/100th the radar signature of the *Whidbey Island/Harpers Ferry*-class (LSD-41/49) landing dock ships.

LPD–17 (Notional Configuration)

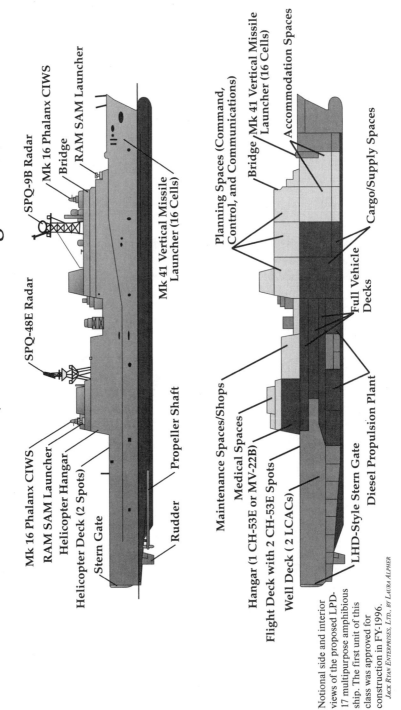

Mk 16 Phalanx CIWS
RAM SAM Launcher
Helicopter Hangar
Helicopter Deck (2 Spots)
Stern Gate
Rudder
Propeller Shaft

SPQ-48E Radar
SPQ-9B Radar
Mk 16 Phalanx CIWS
Bridge
RAM SAM Launcher

Mk 41 Vertical Missile
Launcher (16 Cells)

Planning Spaces (Command,
Control, and Communications)
Bridge
Mk 41 Vertical Missile
Launcher (16 Cells)
Accommodation Spaces

Maintenance Spaces/Shops
Medical Spaces
Hangar (1 CH-53E or MV-22B)
Flight Deck with 2 CH-53E Spots
Well Deck (2 LCACs)
LHD-Style Stern Gate
Diesel Propulsion Plant
Full Vehicle
Decks
Cargo/Supply Spaces

Notional side and interior
views of the proposed LPD-
17 multipurpose amphibious
ship. The first unit of this
class was approved for
construction in FY-1996.

JACK RYAN ENTERPRISES, LTD., BY LAURA ALPHER

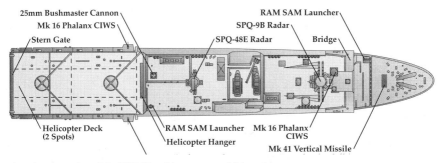

A notional top view of the LPD-17 multipurpose amphibious ship. *Jack Ryan Enterprises, Ltd., by Laura Alpher*

Another issue is active defensive measures. While the armament package of the LPD-17 class is still under study, likely weapons systems have been identified. Up forward, room is allocated for a sixteen-cell Mk 41 vertical launch system (VLS), like those on *Spruance*-class (DD-963) destroyers, *Ticonderoga*-class (CG-47) cruisers, and *Arleigh Burke*-class (DDG-51) destroyers. While this would theoretically allow the LPD-17s to fire RIM-66 Standard SAMs and BGM-109 Tomahawk cruise missiles, the primary weapons system being considered for the VLS launcher is the new Enhanced Sea Sparrow Missile (ESSM). Packaged into four-round launch canisters (for a total of forty-eight ESSM missiles), these will provide the LPD-17 with better anti-air and anti-missile defense than the existing RIM-7 Sea Sparrow. The LPD-17 will probably carry a pair of Ex-31 RAM launchers (each with twenty-one ready missiles) as well as a pair of 20mm CIWS for last-ditch defense against incoming "leaker" missiles and aircraft. Finally, there will probably be a pair of Mk 38 25mm Bushmaster cannon mounts, and mounts for four M2 .50 caliber machine guns to deal with small craft and swimmers (such as enemy frogmen). The LPD-17s will be the most heavily armed amphibious ships built since World War II. Backing up all this firepower will be a new "Cooperative Engagement Capability" (CEC). When the CEC system is retrofitted to all the ships in the fleet (aircraft carriers, escorts, amphibs, support ships, etc.), it will automatically coordinate the employment of every AAW weapon in a group of ships, right down to the level of point -defense systems like Sea Sparrow and RAM. Backing up the "shooting" defensive systems will be an AN/SLQ-32 (V3) electronic warfare system tied to six Mk 137 SRBOC decoy launchers and an active radar jammer. The LPD-17 will also carry four AN/SLQ-49 "Rubber Duck" decoy launchers, which release an inflatable radar decoy which mimics a ship's radar cross section. With an appropriate escort (such as a DDG-51), the LPD-17 will be very hard to hit and kill.

Program officials at NAVSEA like to call it the "25[3] ship," because its displacement is around 25,000 tons, and its cargo[2] and cargo[3] both run around 25,000 ft. The crew and embarked Marines will total around 1,200. The ship will hold a sizable fraction of the ARG's total vehicles, equipment, and supplies. Key features will include:

- 1,190 permanent berthing spaces.
- A fiber-optic computer network using the new super-fast asynchronous transfer mode (ATM) protocol. This replaces tons of copper wire.

- A full Landing Force Operations Center, so that the LPD-17 can conduct independent "split ARG" operations.
- Over 25,000 cubic feet of cargo stowage space.
- Three full-sized vehicle decks with over 25,000 square feet of vehicle storage space.
- A well deck with room for a pair of LCACs.
- A VTOL flight deck with up to four landing spots.
- A helicopter hangar with room for two CH-46s, or a single CH-53E or MV-22B

As mentioned earlier, the LPD-17 will be among the most comfortable warships ever built. This is important when you consider that cruises of over six months are typical of ARG operations. It also will be the first warship ever designed from the keel up, with facilities for female crew members. This reflects the Navy's "Women at Sea" initiative, and is the biggest cultural change for the Navy since President Truman integrated the armed forces in the late 1940s. Crews of ships like the amphibs will be between 10% and 25% female. As existing ships enter their major overhauls, they receive a package of upgrades generically known in the fleet as "Fem Mods." Once they are completed, up to 25% of the crew accommodations can be assigned to women, without disrupting normal ship operations.

As stated earlier, the LPD-17 will be the first U.S. Navy ship with the Women at Sea features designed in from the start. The LPD-17's habitability improvements include:

- Berthing in organizational units. For example, the berthing for an entire Marine platoon, including armory and recreation areas, will be together.
- Berthing spaces for same-sex personnel with attached heads. This will include petty officer/senior non-commissioned officer berthing compartments with only six bunks, and enlisted berthing compartments with just forty-two bunks.
- Unisex heads for use by all crew member regardless of their sex. At the time of this writing, no urinals are planned aboard the LPD-17s, though options are still under study. Shower facilities will be segregated.
- Medical facilities with heads and examination facilities suitable for both men and women.

While designers and engineers have worked hard to make the LPD-17s good for the people who will be on board, that is not the only customer the Navy has to satisfy. There's the American taxpayer. Ships need to be affordable. Remember, cost overruns are why there are only five LHAs instead of nine. For this reason, LPD-17 program officials are positively vicious about cost containment. On a "per ton" basis, the LPD-17s will be exceptionally economical to build. Current budget projections have the lead ship of the class costing $974 million in FY-96 dollars; and later ships in the class are expected to cost between 15% and 20% less. The planned production rate will build the entire class of twelve in just eight fiscal years. With the contract going to a single contractor, this should go a long way towards keeping costs under control.

Two teams are competing for the contract. One team has Litton-Ingalls (builder of the LHDs and DDG-51s) and Tenneco-Newport News Shipbuilding (nuclear-powered aircraft carriers and submarines), with Hughes GM as the systems integrator. The other team combines General Dynamics-Bath Ironworks (they build DDG-51s) and Avondale (the construction yard for the LSD-41s/49s), with Loral as systems integrator. The competition is already fierce, and given the probable rewards, will become even hotter. Total value of the contract will probably exceed $10 billion. The winning team will be selected in the summer of 1997, with the first unit funded in FY-96, for delivery to the fleet in 2002. After several years of testing, LHD-17 will enter service with an ARG around 2004. Subsequent ships will be procured, two a year, until all twelve have been built.

At the same time that the Navy is looking at the cost of buying the LPD-17s, it is closely examining the costs of operating them. Not all of the costs are financial. One hidden cost is environmental pollution. No warship is worth having if it attracts protests every time it goes out to sea. For this reason—and for other, more altruistic ones—the Navy has put major effort into reducing the amount of pollution and waste ships generate. Current plans have the LPD-17s being powered by medium-speed marine diesels, which are very efficient to operate. But diesels generate pollutants that can damage the ozone layer, so there will be systems to reduce the emission of the LPD-17's power plants. The LPD-17 will also be fitted with several features to reduce adverse environmental impacts. These will include:

- Environmental control systems (air-conditioning, refrigeration, etc.) completely free of CFCs that can harm the ozone layer.
- Oil-pollution control systems, including an oil/wastewater separator, and no oil drains into the bilges.
- A hazardous-materials storage locker, which will allow storage of sixty days worth of such materials. There will be a compactor for the containers.
- To reduce the volume of solid waste, there will be a food-waste grinder/pulper. A plastic grinder will be installed, with provisions for storage and recycling of plastic containers.
- A series of "Black Water" (sewage) and "Gray Water" (shower runoff, dishwashing water, etc.) storage tanks, allowing the storage of up to twelve hours worth of such waste, so that sewage dumps can be made in deep water offshore, rather than close inshore.

Many of these systems will eventually be retrofitted into older ships like the *Wasp* and the *Whidbey Island*. But the LPD-17 will be the first designed from scratch to reflect these new values. Maybe you're thinking that concern for "environmental correctness" outweighed combat capability in the LPD-17's design. Nothing could be further from the truth. In fact, the LPD-17 program manager, Captain Maurice Gauthier, would tell you the Navy is simply coming to the realization that we cannot have a fleet that protects our society and nation while it strangles the planet. Remember that the LPD-17s will probably retire around 2050, long after many of you reading this have passed away! Navy/Marine planners have to think a half century or more into the future.

A Guided Tour of the 26th MEU (SOC)

It had been a rough week for Air Force Captain Scott O'Grady. On June 2nd, 1995, while flying Basher 52, a F-16C Fighting Falcon fighter of the 555th Fighter Squadron (FS) of the 31st Fighter Wing out of Aviano Air Force Base, Italy, his aircraft had been hit by an SA-6 Gainful missile from a Bosnian Serb SAM battery. O'Grady had ejected from the dying aircraft as it fell into the cloud base below, denying his wingman any knowledge of whether he had survived or not. Over the next six days, the young Air Force officer had done a textbook job of escape and evasion, while hoping to reach a friendly aircraft on his rescue radio. Then, the night before, another F-16 from Aviano had finally found him and had stayed overhead until just a short time earlier. After authenticating his true identity, the pilot had contacted NATO Allied Forces Southern Region, and told O'Grady to hang tough and there would be someone to get him out soon.

The morning of June 8th, 1995, dawned cool and foggy as O'Grady began his sixth day on the ground in Bosnia-Herzegovina. The first indication of action came around 6:00 A.M. local time, when a pair of two-seat Marine F/A-18Ds roared over him, fixing his position and setting up top cover for what was about to begin. About this time, the young flyer was probably beginning to wonder just who was coming to get him out. Would it be one of the big MH-53J Pave Low helicopters from the USAF's Special Operations Group escorted by huge AC-130 Combat Talon gunships? Or would it be a team of Army Rangers, flown in by MH-60K Blackhawks, escorted by AH-60 attack helicopters? Then, the answer came. Through the wet morning fog at around 6:40 A.M. came the familiar "whomp-whomp" sound of twin-bladed helicopters, Marine AH-1W Cobras. Like their menacing namesakes, they surveyed the area around O'Grady, looking for any threat to the rest of the rescue force that was approaching. Overhead, a flight of AV-8B Harrier II attack jets joined the F-18s in covering the operation. Then, after contacting O'Grady on his rapidly dying radio and marking his position with a smoke grenade, they called for the rescue force.

On the ground O'Grady heard the dull roar of helicopters. *Big* helicopters. Through the wispy ground fog came a pair of Marine CH-53E Super Stallion assault helicopters, loaded with Marines and Navy corpsmen from the 24th Marine Expeditionary Unit–Special Operations Capable MEU (SOC). As the first CH-53 flared in for a landing, the 3rd Battalion/8th Marine Regiment's mortar platoon dashed off to set up a security perimeter for the rescue force along with their battalion commander, Lieutenant Colonel Chris Gunther. Then, as the second Super Stallion came into land, O'Grady made his move. Wearing a Day-Glo orange "beanie" cap, and clutching his radio and 9mm pistol, he dashed for the second helicopter, and was pulled aboard by the crew chief, Sergeant Scott Pfister. Several minutes later, after retrieving the mortar platoon, Lieutenant Colonel Gunther ordered the two helicopters to lift off and head home.

Inside the second helicopter, O'Grady was being taken care of by more Marines, including the 24th MEU (SOC)'s commander, Colonel Martin Berndt, and his senior NCO, Sergeant Major Angel Castro, Jr. After O'Grady was given some water, part of an MRE, and Colonel Berndt's Gortex parka, he settled in for the ride home. But even this task was to prove an adventure for the young pilot and his rescuers. As the CH-53s and AH-1Ws passed near a small town, anti-aircraft and small-arms fire erupted, hitting both transport helicopters. And then three man-portable SA-7 Grail SAMs were fired from below, requiring evasive maneuvering by the four choppers. It didn't take long after that for the airborne task force to break clear of the danger and go "feet wet" over the Adriatic, headed for home aboard the USS *Kearsarge* (LHD-3). Twenty minutes later, all were safely aboard, and another page had been written into Marine history.

What had saved the young Air Force captain was not a special operations force in the traditional sense. From our experience of movies and television, we tend to think of such forces as supermen, rescuing hostages and "taking down" terrorists in their lairs. Built around clandestine units like the Army's Delta Force and the Navy's SEa-Air-Land (SEAL) teams, these units maintain a low profile and tend to keep out of the public view. The MEU (SOC)s are different. While quite capable, they are not special operations forces per se. On the contrary, they are regular Marine units, drawn from around the Corps, which are given special training to make them capable across a limited but important range of conventional and special operations missions. This distinction is the reason why the MEU (SOC)s have remained independent at a time when most American special operations forces are under the unified command of the U.S. Special Operations Command (USSO-COM), located at MacDill AFB near Tampa, Florida.

The story of the MEU (SOC)s is the story of how the Marine Corps has enhanced its ability to make "kick-in-the-door" forced entries (i.e. invasions and raids) into enemy territory. This is a lot to ask from just seven battalion-sized units, of which only three or four are deployed on cruise at any particular time. General Krulak likes to call the Marines "the risk force"; the MEU (SOC)s are the diamond-tipped point of that force's spear. As it exists today, the MEU (SOC) can be seen as the evolutionary result of over two millennia of experience in amphibious warfare. More immediately and practically, it is one of the most compact, responsive, and capable military units in the world today, with multiple means of delivering its weapons and personnel onto or even behind an enemy coastline. The MEU (SOC)'s special training gives it versatility across a finite but significant range of possible special missions, including raids, rescues, and security operations. And its ability to plug itself into a variety of "joint" military operations makes it a valuable addition to any military force. Thus, regional CinCs covet a MEU (SOC) whenever they can get one. Finally, because it can be, and usually is, forward-deployed into possible trouble spots along with its own aviation and logistical components, it is fast, mobile, and self-contained. It needs nothing to get the ball rolling on an operation and keep it going for up to fifteen days without external support. So let us look at this unique family of units, and get to know their organizations, missions, and history.

Beginnings: The Road to the MEU (SOC)

The beginnings of the MEU (SOC) concept date to just after the end of the Second World War. As early as the late 1940s, the need for forces based close to potential trouble spots was already posing a problem for the U.S. and its Cold War allies. One result was the Marine afloat battalion, which became something of a standard unit in the decades ahead. These were created by using some of the vast amphibious shipping tonnage that had been built up in World War II and a few battalions of the dwindling Marine Corps of the time. Quickly, they began to prove their worth. Each one was a typical Marine Air-Ground Task Force (MAGTF), with ground, air, and logistical units (or components as they are called), matched together into a fighting team. This turned out to be an excellent idea. In the Taiwan Straits (1957), Lebanon (1958), Cuba (1961 and 1962), and the Dominican Republic (1965), the forward-deployed Marine units aboard U.S. Navy ships were to make themselves felt. Even during the height of the Vietnam War, amphibious ready groups (ARGs) with their Marine units aboard prowled the oceans and seas of the world, protecting American interests.

Following the end of the Vietnam war and the rough years of the 1970s, things began to become a bit more regular within the Marine afloat battalions. Redesignated as Marine Amphibious Units (MAUs), they now had a formal headquarters unit, which would then fill out its component parts from regular Marine units from around the Corps. Previously, the units just were thrown together for the duration of their cruise afloat. This move to a formal headquarters structure was more than just cosmetic; it meant that the Corps had begun to consider the MAUs one of their premier MAGTF organizations. Now they would become fully integrated MAGTFs under the command of a full Colonel (O-6), capable of a wider variety of tasks and missions. In fact, with the drawdown of Navy and Marine forces in the late 1970s during the Carter Administration, the compact MAUs aboard their ARGs quickly became the only U.S. military units that could begin to rapidly respond to a crisis around the world.

The coming of the Reagan Administration in 1981 brought the MAUs the opportunity to prove themselves in combat. Initially, the results were decidedly mixed. On the plus side was Operation Urgent Fury, the invasion of Grenada in October 1982. The 22nd MAU provided much of the combat muscle for that operation. Things unfortunately did not go so well for the 24th MAU. Two days before Urgent Fury hit the beaches of Grenada, while it was on "peacekeeping" duty in the war-torn city of Beirut, Lebanon, an Iranian driving a truck bomb wiped out much of the 24th's ground component. Over two hundred Marines were killed in the explosion, which occurred early on a Sunday morning. It today remains one of the worst disasters in U.S. military history, and had a variety of effects on the Marines and their MAUs.

The Beirut disaster and problems in other operations began to show that the Marines had some problems in their combat doctrine. Much like their sister services, who had seen such difficulties following the Vietnam War, the Corps was beginning to experience some serious shortcomings in its ability to carry out even traditional missions like amphibious invasions and raids. Grenada, while successful, had been costly and poorly coordinated. Luckily, the solution to these shortcomings came in the form of a new senior Marine Corps leader, General Alfred M. Gray, who would eventually become the 29th Commandant in 1987. While he was the commander of Fleet Marine Force, Atlantic (FMFLANT), General Gray began his campaign to promote "warfighting" as the primary task of the Marine Corps in the 1980s and 1990s. Much like visionary thinkers in the other services, Gray helped promote the idea that combat was the core ability of the Marines (this earned him the nickname "the warfighter"). What made his effort unusual was that he felt it was not enough just to know how to shoot and blow things up. He urged Marines of all ranks, officer and enlisted, to apply intellectual power as a force multiplier for the Marine ethos. This began to pay immediate benefits. He also promoted the use of the word "expeditionary" to describe the inherent characteristics of Marine units of all sizes. In particular, he pushed renaming the MAUs as MEUs (the E obviously standing for "expeditionary"), to reflect the kinds of missions he wanted the Marine Corps to be ready for.

Along with these intellectual developments, General Gray began to think about the kinds of units that the Marines had formed over the years, and just what kinds of missions each was capable of. One particular kind of mission which had come to be vital in the 1980s was special operations. The failed Iranian hostage rescue mission in 1980 had forced all of the services to look at their capabilities in this area. Out of this came a 1983 study that examined what the Marine Corps would need to become a credible player in future low-intensity ("short-of-war") conflicts. Unlike the other services, there was no drive within the USMC to create new and separate special operations units. Instead, it was decided that regular units within the Corps would be given special training prior to a deployment. This would make them "special operations capable" (SOC) across a fixed range of missions and tasks.

In 1984, Marine Corps Headquarters ordered FMFLANT (at the time commanded by General Gray) to put together a program to create a special-operations-capable Marine unit, and deploy it on an ARG for an overseas cruise of some six months duration. General Gray and then-Colonel James Myatt (who eventually rose to the rank of Major General and commanded the 1st Marine Division during Desert Storm), came up with a list of special missions and equipment that they wanted to put into the unit assembled. Along the way, Gray and Myatt made several key decisions. These included:

- FMFLANT would modify one of the MEUs to produce a battalion-sized SOC-capable MAGTF that could carry out the special missions that they had in mind.
- They made plans to establish a training and certification program to make sure that every unit would go through a standardized curriculum.
- The actual unit, to be called a MEU (SOC), would be given an extra infusion of equipment and personnel to support its expanded mission.

The units for the first MEU (SOC) were taken from a regular MEU, the 26th, preparing for a deployment to the Mediterranean. Personally selected to command the first MEU (SOC) deployment, Colonel Myatt took the 26th out for a six-month cruise in 1986.

Now it should be said that this first MEU (SOC) cruise did not shake the world. The 26th did support the Navy carrier groups that were operating against

Libya at the time, and these actions were generally successful. But more importantly, that first MEU (SOC) deployment brought home valuable lessons that were immediately applied to the next cruise, and all the others that followed. Even though actual combat eluded them for the next few years, MEU (SOC)s were nevertheless very active. First combat by a MEU (SOC) occurred on April 18th, 1989, when a strike force of Marines from the 22nd MEU (SOC), was tasked to take part in Operation Preying Mantis. Preying Mantis was a rapid response to the mining, several days earlier, of the USS *Samuel B. Roberts* (FFG-58) by Iranian forces in the Persian Gulf. The operation was designed to take out several Iranian oil platforms that were being used as targeting bases for attacks against tankers coming down the Gulf. The 22nd, along with several surface-action groups (SAGs) of U.S. warships, was tasked to capture and then demolish the platforms, while aircraft from Carrier Air Wing Ten (CVW-10) embarked on USS *Enterprise* (CVN-65) provided cover against Iranian aircraft and ships. The results were startling. By the end of the day, the oil platforms had been destroyed, and most of the Iranian Navy had either been sunk or disabled. The MEU (SOC) lost one AH-1 Cobra attack helicopter with both of its crew, but it was an impressive combat debut for the new unit, and it went almost unnoticed by the world.

Just fourteen months later, the MEU (SOC)s hit their stride, thanks to a rash of crisis situations that erupted in the summer and fall of 1990. The trouble started when a civil war in Liberia escalated. Initially, the 26th MEU (SOC) was sent to handle any possible evacuation of U.S. nationals and embassy personnel. It had been planned to relieve the 26th with the 22nd MEU (SOC), but the eruption of hostilities in the Persian Gulf in August meant that both units had to stay out to cover both problems. Eventually, the 22nd handled the evacuation, and the 26th went on to support operations in the Mediterranean. At the same time, the 13th MEU (SOC) from the West Coast was rapidly moving into position in the Persian Gulf, supporting maritime embargo operations and acting as a floating reserve for the 1st MEF in Saudi Arabia. Then, in December of 1990, with the eruption of the civil war in Somalia, heliborne Marines from the amphibious group in the Persian Gulf conducted an evacuation from the American embassy in Mogadishu.

The period following Desert Storm has been a busy one for the MEU (SOC)s. In Somalia, Haiti, and now Bosnia, they have led the way for American efforts and forces. In the case of our pullout from Mogadishu, they have even covered our withdrawal from a dangerous and risky situation. Given their level of activity over the past ten years, it is amazing that it took the O'Grady rescue to bring them any sort of public notice. Despite the lack of public credit, the MEU (SOC) deployments are going along like clockwork. Originally, the MEU (SOC) effort was limited to just one such unit on each coast, but no more. Because of the desires of the regional CinCs to have at least one of them available for any crisis that might arise, all the MEUs are being given SOC certification before they are deployed on cruise. The O'Grady rescue just highlights the many desirable qualities of these unique units, and provides a jumping off point for our own explorations of them. So follow us, and we'll show you how they work, and are put together.

The MEU (SOC) Concept

From early raids on British forts during the Revolutionary War to the embassy evacuations and rescues of today, when you have trouble that needs to be taken care of good and fast, you call the Marines. Every branch of the military has special operations forces, and these sometimes overlap. So how does a small and underfunded service like the Marine Corps justify such a capability, both from a financial as well as institutional point of view? The Marines' answer: a hybrid, dual-purpose special operations/amphibious unit, the MEU (SOC). To repeat, the MEU (SOC) is based upon the concept that given special training and equipment, regular units can be made capable of accomplishing both their normal duties and extraordinary missions. This notion runs contrary to the "snake-eater" tradition of the many special operations units around the world. For most of these, including the British Special Air Service (SAS), the U.S. Army Delta Force, and the German GSG-9, selection is limited to the physical and mental elite of a particular service. These are highly specialized units, lavishly expensive to create and operate, with a strong focus on hostage rescue and counter-insurgency warfare. Consequently, the leaders of their nations tend to view special operations units with the same kind of restraint and reserve they might have toward nuclear weapons. You only use them when you really need to, and when you do, you open yourself up to an extreme level of political risk. This is the reason why you see such limited use of special operations units, and why so many of them tend to hang around inside their compounds, practicing and waiting.

As a matter of fact, several of the most significant and remarkable special operations missions in history did *not* involve actual, purpose-built special operations forces. During World War II, for example, the famous bombing mission on Tokyo by then-Lieutenant Colonel James H. Doolittle and his raiders was accomplished by personnel and aircraft drawn from regular Army Air Corps bomber units. Thanks to several months of special training, as well as special modifications to their B-25 Mitchell bombers, they made history on April 18th, 1942, as the first force to bomb Japan during the war. Also in World War II, a regular British unit, given special training and equipment, was responsible for the less well known but equally valorous action at Pegasus Bridge on D-Day. On the night of June 5th/6th, 1944, a specially trained gliderborne unit, Company "D," drawn from the Oxfordshire and Buckinghamshire Light Infantry (the "Ox and Bucks") of the 6th Airborne Division, conducted a *coup de main* on a pair of vital bridges over the Orne River and Caen Canal. The tiny force, led by the charismatic Major John Howard, took the bridges and held them until relieved by British commandos coming inland on D-Day. Finally, there was the Entebbe Raid. A Palestinian terrorist group held a number of hostages from a hijacked French Airbus in a terminal at the Entebbe Airport in Uganda. The raid was designed to release and retrieve them. As soon as the crisis started, the Israeli Defense Force formed an ad hoc rescue force out of various regular paratroop units. On July 4th, 1976, after a long flight on a picked force of C-130 Hercules transports, the rescuers assaulted the terminal and freed the hostages with minimal losses—in the process killing most of the terrorists. Again, a clearly defined goal, supported by extremely strong leadership, led to success in a special operation

The official emblem of the 26th MEU (SOC).
JACK RYAN ENTERPRISES, LTD., BY LAURA ALPHER

by "pick up" units with special training. Such units, given the time and training, can achieve wonders. And because they are drawn from regular units, they are cheaper to run and less expensive to risk.

The MEU (SOC) joins the responsiveness and professionalism of a task-specific special operations force with the costs and success records of specially trained, ad hoc special operations units. Composed of regular units from around the Marine Corps, the MEU (SOC) is a MAGTF based around a reinforced rifle battalion, with the special training and equipment that makes it capable of a limited number of special operations missions. One of the interesting characteristics of MEU (SOC)s is that they are not composed of the same units every time they go out on a cruise. Since they are formed from battalion landing teams (BLTs), medium Marine helicopter squadrons (HMMs), MEU service support groups (MSSGs), and ARGs, the various components can be mixed and matched as required. And since their special operations capability is layered on top of their existing conventional amphibious/heliborne capabilities, the MEU (SOC)s are actually quite a bargain for the taxpayers. Finally, and this may be the greatest benefit of all, they can be forward-deployed and based aboard their own ARG, requiring little or no approval from foreign governments or allies for their use. Given the frustrations that such foreign interference has caused in the past, this probably provides the American national command authorities all the justification required to continue operating and maintaining the seven MEU (SOC)s.

History and Structure: The 26th MEU (SOC)

Though the 26th was the first MEU (SOC) to go out on cruise in 1985—it was known then as a MAU (SOC)—the luck of the draw has not been kind to it...if luck means getting involved like other MEU (SOC)s in something flashy. Nevertheless, in the years since it first took the concept out for its first test, the 26th has done yeo-

Colonel James Battaglini on August 29th,
1995, as he prepared to lead the deployment of
the 26th MEU (SOC) to the Mediterranean.
JOHN D. GRESHAM

man work. Over the last decade, the 26th has supported evacuation operations from Liberia and been stationed off the coast of Somalia.

As the 26th was headed into its training and workup cycle in the winter of 1994/95, it acquired a new commanding officer (CO) to act as its brain, father, and caretaker. The new CO, Colonel James R. Battaglini, is an imposing figure; his mere presence in a room, on a deck, or in a landing zone (LZ) is enough to tell you that the boss has arrived and is in charge. A tall, lean man with a hard look in his eyes, Colonel Battaglini loves his Marines more than almost anything else in his life. A native of Washington, D.C., a graduate of Mount Saint Mary's College, and a holder of two master's degrees (in management and national security studies), he can talk about the merits of satellite communications systems at one moment, and tell you his opinions of nonlethal weapons doctrine in the next. As he rose up the chain in the Corps, he commanded virtually every kind of Marine unit from a reconnaissance platoon to the 1st Battalion of the 8th Marine Regiment (1/8), during its 1991/92 deployment with the 22nd MEU (SOC). Along the way, he picked up a Bronze Star for valor in combat during Desert Storm. Backing up Colonel Battaglini is his senior enlisted advisor, Sergeant Major W. R. Creech, himself a veteran of over twenty years in the Corps.

What Battaglini and Creech were putting together for a six-month Mediterranean cruise in 1995 was a team built of many interlocking components. Like all other deployed Marine forces, the MEU (SOC) is built along the classic MAGTF structure. And like all expeditionary units deployed by the Corps, it has ground, aviation, and logistical components. The essential parts of this structure are:

- **Command Element (CE)**—This is a company-sized (28 officers, 186 enlisted) unit which provides the leadership, command, control, and communications for the entire MEU (SOC). The 26th MEU (SOC) CE is based at Camp Lejeune, North Carolina.

26th Marine Expeditionary Unit – Special Operations Capable MEU (SOC)

Command Element (CE – 26th MEU (SOC))

Ground Combat Element (GCE – BLT 2/6)

Air Combat Element (ACE – HMM-264)

Combat Service Support Element (CSSE – MSSG-26)

The organization of the 26th MEU (SOC). The three components (ground, air, and support) are standard to all Marine MAGTFs. *Jack Ryan Enterprises, Ltd., by Laura Alpher*

- **Ground Combat Element (GCE)**—The GCE is a reinforced battalion landing team (BLT—54 officers, 1,178 enlisted), designed to fit in the limited space aboard a three-ship ARG. For the 1995/96 26th MEU (SOC) deployment, the GCE was built around the 2nd BLT of the 6th Marine Regiment (2/6). This unit is part of the 2nd Marine Division at Camp Lejeune, North Carolina.

- **Aviation Combat Element (ACE)**—A MEU (SOC) ACE is composed of a reinforced medium Marine helicopter squadron (HMM—55 officers, 263 enlisted), made up of a mix of CH-46 Sea Knights, CH-53E Super Stallions, AH-1W Cobras, UH-1N Iroquois, and AV-8B Harrier IIs. In addition, the MEU (SOC) can be reinforced with a land-based force of KC-130 Hercules airborne tankers. The 26th MEU (SOC)'s ACE is built around HMM-264, which is based at the Marine Corps Air Station at New River, North Carolina (adjacent to Camp Lejeune). The Harrier detachment is drawn from Marine Corps Attack Squadron 231 (VMA-231) at MACS Cherry Point, North Carolina.

- **Combat Service Support Element (CSSE)**—The CSSE is a company-sized unit (13 officers, 234 enlisted), composed of a series of eight platoons covering areas such as supply, engineering, transport, maintenance, and medical services. The 26th MEU (SOC)'s CSSE is MEU Service Support Group 26 (MSSG-26), which is also based at Camp Lejeune.

These four elements, the CE, GCE, ACE, and CSSE, make up an MEU (SOC) MAGTF like the 26th. In addition, each MEU (SOC) commander gets to tailor the unit's structure to match the planned mission and his own operational style.

Frequently, these modifications and additions are based upon suggestions and lessons learned from the MEU (SOC)s ahead of them in the rotation schedule. For example, while the 26th was working up in the summer of 1995, they were taking to heart the lessons from the 22nd MEU (SOC), which had just returned, and Marty Berndt's 24th MEU (SOC), which was then on cruise out in the Mediterranean. Out of the 24th's experiences (such as the O'Grady rescue) came the suggestion that the 26th's ACE be enlarged with additional CH-53E Super Stallion and AH-1W Cobra helicopters, to support possible evacuation operations in Bosnia-Herzegovina. Now let us take a look at each of the 26th MEU (SOC)'s individual complements and see what gives this little unit such a big bite.

Leadership: The 26th MEU (SOC) Headquarters

The Command Element of the 26th MEU (SOC), headed by Colonel Battaglini, is composed of a traditional U.S. military executive staff structure. The Executive Officer, or XO, Lieutenant Colonel Fletcher "Fletch" W. Ferguson, Jr., coordinates and supervises the executive staff. In the 26th MEU (SOC), he is also the Commanding Officer of Troops aboard the ARG flagship, and is the Officer in Charge of the Forward Command Element. The MEU (SOC) Sergeant Major, Sergeant Major William Creech, performs those duties which generally pertain to discipline, welfare, conduct, morale, and leadership of the enlisted personnel. The rest breaks down like this:

- **S-1—Adjutant:** This is the personnel and administrative section, and is headed by Captain Daniel McDyre.
- **S-2—Intelligence:** Headed by Major Phil Gentile, the MEU (SOC) intelligence section has staff responsibility for matters pertaining to weather, enemy, and terrain within the MEU (SOC)'s area of operations. It determines the intelligence requirements and directs the effort for collection of information. It then processes information into intelligence and disseminates it to those who need to know. It is augmented with the following detachments:

    —An Interrogator Translator Team detachment that provides enhanced human intelligence support through the interrogation, debriefing, and screening of those personnel with intelligence value.

    —A Force Imagery Interpreter Unit detachment that provides limited imagery interpretation support.

    —A Counterintelligence Team detachment that provides counterintelligence support.

    —A Topographic Platoon detachment that provides limited cartography and terrain-model-building capability.

    —A Radio Battalion detachment that provides an enhanced capability for signal intelligence collection, analysis, and electronic warfare. A radio reconnaissance team capability is included for advanced tactical employment during selected operations.

- **S-3—Training and Operations:** When augmented with the attachments joining the MEU (SOC), this is the largest section in the Command Element. Headed by Lieutenant Colonel Steve Lauer, the S-3 has the responsibility for matters pertaining to organization, training, and tactical operations. The S-3 operates the Landing Force Operations Center (LFOC) aboard the ARG flagship. Upon activation of the MEU, it is augmented with the following attachments:

  —A Force Reconnaissance Company detachment.

  —An Air and Naval Gunfire Liaison Company (ANGLICO) detachment consisting of two supporting arms liaison teams and a firepower control team.

  —A detachment from the Marine Air Control Group with a Marine Air Support Squadron detachment and the Low Altitude Air Defense (LAAD) Battery detachment that provides low-level, close-in air defense. The LAAD battery is composed of two of the new Avenger SAM vehicles. Composed of an HMMWV chassis with eight Stinger SAMs and a .50-caliber machine gun, it is a potent point-defense asset. In addition, there are three manpack Stinger teams, each of which is transported by an HMMWV.

- **S-4—Logistics:** This is the section responsible for all logistics matters and the combat service support functions of supply, maintenance, embarkation, medical/dental care, passenger and freight transportation, landing support, material handling, food services, and financial management. The S-4, headed by Major Dennis Arinello, operates the Tactical Logistics Center (TACLOG) aboard the ARG flagship.

- **S-6—Communications:** This section plans, coordinates, and operates the communications and automated data-processing systems for the MEU (SOC). Headed by Captain James Dillon, the S-6 supervises cryptographic operations, operates the Landing Force Communications center, provides radio operators for the LFOC, and publishes and disseminates the Communications Electronics Operating Instruction(s) for the MEU (SOC). It includes a Communications Battalion detachment that provides command and control communications for execution of all operations.

As might be imagined, the MEU (SOC) staff is a fairly "lean and mean" type of organization. Thus, the jobs listed above have to be accomplished by highly motivated officers and enlisted personnel. For example, the MEU (SOC) S-4 officer, Major Dennis Arinello, operates the entire logistic effort of the MAGTF with a staff of about a dozen Marines, armed only with a battery of computers, phones, and a seemingly endless supply of coffee. In addition to the formal branch structure, there are several special staff officers who perform a variety of duties not specifically assigned to one of the "S" sections. These include the staff judge advocate, disbursing officer, and chaplain. The MEU (SOC) commander also has a small unit known as the Maritime Special Purpose Force (MSPF). The MSPF is task-organized from MEU (SOC) assets to provide a special-operations-capable

26th MEU (SOC) Combat Equipment Breakdown

M1A1 Abrams Main Battle Tanks

CH-46E Sea Knight Transport Helicopters

HMMWVs (Armed with TOW, Mk 19 40mm Grenade Launchers, .50-cal. MGs, and 81mm Mortars)

Light Armored Vehicles (4 LAV-25/2 LAV-AT)

CH-53E Super Stallion Heavy Transport Helicopters (Normally 4)

M198 Towed 155mm Howitzers

AAV-7/LVTP-7 Armored Amphibious Tractors

AH-1W Cobra Attack Helicopters (Normally 4)

AV-8B Harrier II Attack Jets

UH-1N "Huey" Command and Control Helicopters

The combat equipment breakdown of the 26th MEU (SOC).   JACK RYAN ENTERPRISES, LTD., BY LAURA ALPHER

force that can be quickly tailored to accomplish a specific mission and employed either as a complement to conventional naval operations or in the execution of a directed maritime special mission operation. Command of the MSPF remains under the control of the MEU (SOC) commander.

The Ground Component: BLT 2/6 Marines

The GCE of the 26th MEU (SOC) is the heavy combat element of the MAGTF. Composed of a reinforced BLT, it is designed to provide Colonel Battaglini and his CE with the necessary personnel and equipment for anything from making a forced-entry amphibious assault, to attempting a non-combatant evacuation of an embassy or other facility. The 26th MEU (SOC)'s GCE is made up of BLT 2/6 out of Camp Lejeune. It is a proud unit, with a long history of service to the nation. The 2/6's combat record dates back to Belleau Wood in World War I, and includes service in such varied actions as Shanghai, Tarawa, Iwo Jima, and Beirut. The 2/6 is currently commanded by Lieutenant Colonel John R. Allen, a Naval Academy graduate of the Class of 1976, who also carries around a pair of master's degrees in government and strategic intelligence studies. A native of Fort Belvoir, Virginia, he has the distinction of having served on the very first MEU (SOC) deployment by the 26th back in 1985. It almost killed him. Injured in a CH-46 crash during the unit's workup, he recovered and has stayed with the Corps. Colonel Allen is a cerebral sort of Marine who is always considering new ways to use the force that the Corps has given him to command. Whether it's exploring the use of AAV-7A1 amphibious tractors as gunboats for riverine warfare, or figuring out new means of deception to help cover his reconnaissance elements in the field, John Allen is thinking all the time. His senior enlisted advisor is Sergeant Major James Rogers, who looks after the enlisted personnel of the BLT.

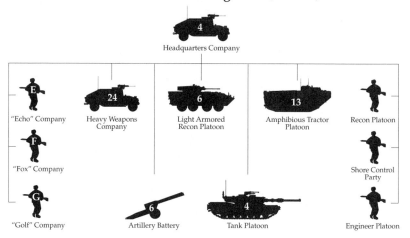

Battalion Landing Team, 2nd Battalion/6th Marine Regiment (BLT 2/6)

Headquarters Company

"Echo" Company

Heavy Weapons Company

Light Armored Recon Platoon

Amphibious Tractor Platoon

Recon Platoon

"Fox" Company

Shore Control Party

"Golf" Company

Artillery Battery

Tank Platoon

Engineer Platoon

An organizational chart showing the breakdown of Battalion Landing Team (BLT) 2/6 (2nd Battalion/6th Marines). *JACK RYAN ENTERPRISES, LTD., BY LAURA ALPHER*

BLT 2/6 is a standard Marine rifle battalion, with some extras added to help conduct landing operations. To understand this, we need to have a look at the building blocks of the BLT. As mentioned earlier, the basic Marines combat unit is the four-man fire team. The team leader (usually a corporal equipped with an M16A2 with a clip-on M203 grenade launcher), is assigned two riflemen (also with M16A2s) and an automatic rifleman (with a M249 Squad Automatic Weapon—SAW). By matching up three fire teams with a sergeant, you get a squad. Three squads, with a second lieutenant and platoon (staff) sergeant in command, make up a platoon. From here things get a bit more complicated. By combining three infantry platoons with a heavy weapons platoon (M240G machine guns, M224 60mm mortars, and Mk 153 SRAWs), you get a Marine rifle company, under the command of a captain and his first sergeant. These company and platoon-sized units are the basic elements of the BLT, and are combined as follows:

- **BLT Headquarters and Headquarters & Service Company**—The BLT headquarters is structured much like that of the MEU (SOC), with "S"-style staff branches to support administration, operations, intelligence, etc.

- **Rifle Companies (3)**—There are three of the companies, each with about 150 personnel. Each company is designated by a phonetic letter name, based upon its order within the regiment that it is assigned to. For example, the 1/6 BLT is assigned Alpha, Bravo, and Charlie as their infantry companies. Thus, the 2nd BLT's three infantry companies are designated Echo, Fox, and Golf.

- **Heavy Weapons Company (1)**—This is the very lethal big brother of the weapons platoons organic to the infantry companies. This company is composed of three platoons: the 81mm Mortar Platoon (eight M252 81mm mortars), the Heavy Weapons Platoon (eight TOW II launchers, six Mk 19 40mm automatic grenade launchers, and six M2 .50-caliber machine guns on armored HMMWVs), and the Anti-Armor Platoon (eight M-47 Dragon launchers). The Heavy Weapons Platoon is organized in three Combined Anti-Armor Teams (CAAT) which combine the high mobility of the HMMWV with the effects and capabilities of the three weapons. In a pinch, these weapons can be mounted on the BLT's six Fast Attack Vehicles (FAVs—black-painted "dune buggies") which can be carried internally and inserted by the MEU's heavy lift helicopters.

- **Artillery Battery (1)**—This is a battery of six M198 towed howitzers, with 5-ton trucks as their prime movers. Additional trucks provide support and ammunition carriage for the battery.

- **Light Armored Reconnaissance Platoon (1)**—Also known as Task Force Mosby (after the famous Confederate guerrilla fighter) when combined with armed HMMWVs from the heavy weapons company in the 2/6, this is a mixed platoon of wheeled Light Armored Vehicles (LAVs), which are used as an armored reconnaissance and trip-wire force. It usually includes four LAV-25s (with 25mm Bushmaster cannons), two LAV-ATs (with TOW anti-tank missiles), and a LAV-R recovery vehicle.

- **Assault Amphibian (AAV) Platoon (1)**—The AAV platoon consists of thirteen AAV-7 amphibious assault tractors, along with one each of the recovery and command versions of the AAV-7.

- **Surface Rubber Boat Raid and Cliff Assault Company (1)**—In addition to the LCACs, LCUs, and AAVs, the BLT can use twenty F470 Zodiac combat rubber raiding craft to support landings, riverine operations, or special insertions into enemy territory.

- **Tank Platoon (1)**—For the first time in many years, a heavy tank platoon has been included in a MEU (SOC). Composed of four M1A1 Abrams main battle tanks and an M88A1 tank recovery vehicle, it is designed to provide a heavy direct-fire punch for the BLT.

- **Combat Engineer (Sapper) Platoon (1)**—To support breaching of obstacles, building of causeways, emplacements, and bunkers, as well as other civil engineering functions, BLT 2/6 has been assigned a small but capable combat engineering platoon. Equipped with a bulldozer and other equipment, the sappers provide a variety of engineering and construction services for Lieutenant Colonel Allen and his Marines.

- **Reconnaissance Platoon (1)**—In addition to the other Force Reconnaissance units that are part of the 26th MEU (SOC), the BLT has its own reconnaissance platoon, which can be combined with the scout/sniper platoon from the BLT headquarters. This concentration of reconnaissance assets is no coincidence, but a planned inclusion. It

Lieutenant Colonel John Allen, the commander of BLT 2/6 during 1995/96, with the author (right). Allen was a junior officer on the first MEU (SOC) cruise back in 1985, and is currently assigned as General Krulak's aide at the Pentagon.
JOHN D. GRESHAM

is based upon the hard-won knowledge that you can never have too many eyes on the battlefield.

- **Shore Fire Control Party (1)**—This is a small, but vital unit which controls the critical task of planning and executing naval gunfire support. Composed of Marines and sailors, this platoon is capable of rapidly calling for fire, even during the confusing first moments of an amphibious assault

All of these assets make BLT 2/6 a highly mobile and compact striking force, with the ability to do a variety of exciting things. For example, by combining the tanks, LAVs, and AAVs, you can rapidly assemble a reinforced armored infantry task force, which can be used for everything from assault operations to rapid response in peacekeeping operations. At the same time, the Marines of BLT 2/6 can be delivered any number of ways to its targets. They can ride on helicopters, LCACs, LCUs, AAVs, or rubber raider boats to get ashore. Maybe more importantly, the ARG and MEU (SOC) have the necessary lift to bring all of the combat power of the BLT ashore nearly simultaneously. This means that BLT 2/6 can hit in many different ways, all at the same time. If there is a weakness, it is that the BLT is mostly composed of "leg" infantry; it is short on vehicles to help it move about the battlefield. It is also only a single battalion. So Colonel Allen and his staff must carefully pick their fights, maneuvering for position and opportunity to get the most of their limited resources.

Flying Leathernecks: HMM-264

The 26th MEU (SOC)'s ACE is a composite organization, much like the 366th Wing that we visited in *Fighter Wing*. But unlike that unit, the primary mission of this ACE is the transport and support of operations by the 26th's GCE, BLT 2/6. The

LEFT: The official emblem of Marine Medium Helicopter Squadron 264 (HMM-264), the "Black Knights."
JACK RYAN ENTERPRISES, LTD., BY LAURA ALPHER

BELOW: The organization and equipment of HMM-264. The Harrier detachment is drawn from Marine Attack Squadron 231 (VMA-231) at MCAS Cherry Point, N.C.
JACK RYAN ENTERPRISES, LTD., BY LAURA ALPHER

Marine Medium Helicopter Squadron (Reinforced) 264 (HMM-264)

26th's ACE is commanded by Lieutenant Colonel David T. "Peso" Kerrick, who also commands HMM-264, the core unit in the air component. A native of Elizabeth, Kentucky, and a 1976 Naval Academy graduate, he has spent his career in the Corps as a CH-46 *Sea* Knight pilot, moving up to command of the squadron in 1995. He is supported by Sergeant Major Ronald Trombley, who looks out for the well-being of his enlisted Marines.

The HMM-264 "Black Knights" are one of the original helicopter transport units in the Marine Corps, with service in the Dominican Republic, Beirut, Liberia, and northern Iraq. As configured when it left on its 1995/96 deployment in September of 1995, HMM-264 and its attachments looked like this:

- **CH-46E Sea Knights (12)**—The core of HMM-264 is its dozen CH-46E medium lift helicopters. Though decidedly elderly and quite limited in both carrying capacity and range, the Sea Knight is still the prime mover for the 26th MEU (SOC), and will be until the introduction of the MV-22B Osprey in the early 21st century.

- **CH-53E Super Stallions (8)**—The heavy haulers of HMM-264 are eight of the big CH-53E Super Stallion transport helicopters. Normally an MEU (SOC) ACE only has four CH-53s. But following the suggestions of the 24th MEU (SOC), the number was doubled for this deployment, given the distances and payloads that might need to be handled in the event of an evacuation of U.S./UN personnel in Bosnia-Herzegovina.

- **AH-1W Cobras (8)**—Like their CH-53E brethren, HMM-264's normal complement of AH-1W Cobras was doubled from four to eight. This decision was also based upon experience from the 24th MEU (SOC) in Bosnia-Herzegovina, where the need for additional firepower and helicopter escorts became clear in 1995. Initially, it had been hoped that the upgraded version of the AH-1W with the new Night Targeting System (NTS) would be available to take on the cruise. Unfortunately, spares shortages meant that the first deployment of the NTS-equipped Cobras would have to wait until 1996.

- **UH-1N Iroquois (3)**—To help provide command and control for the 26th MEU (SOC) and its various components, the HMM-264 is equipped with three UH-1N Iroquois helicopters. Better known as "Hueys," they are assigned to command elements of the MEU (SOC) components to provide transport around the units AOR. One of the Hueys is equipped with a Nite Eagle laser targeting system. Originally designed for use on the now-defunct U.S. Army Aquila UAV, Nite Eagle is composed of an FLIR and laser-designation system mounted in a turret under the nose of a UH-1N. It was highly successful during Desert Storm. There are three of these systems around the Marine Corps, and each deployed MEU (SOC) usually gets one to provide designation for the AGM-114 Hellfire missiles fired from the AH-1W Cobras.

- **AV-8B Harrier II (6)**—In addition to the helicopters, HMM-264 contains a small six-aircraft detachment of AV-8B Harrier II attack jets. Drawn from VMA-231 (known as the "Ace of Spades" squadron), these are older-model Harriers that were produced in the early 1980s. Like the NTS-Cobras, the 26th MEU (SOC) ACE just missed being the first unit to deploy the new AV-8B Plus Harrier II operationally. The relative newness of the birds in squadron service meant that the first radar Harrier deployment would have to wait until 1996. However, the AV-8Bs that VMA-231 deployed with the 26th MEU (SOC) in 1995 were quite capable, able to employ 25mm GAU-12 cannon, iron and cluster bombs, 2.75-in./70mm rockets, as well as AIM-9 Sidewinder and AGM-65 Maverick missiles.

- **KC-130 Hercules (2)**—While an aircraft the size of a KC-130F Hercules aerial tanker obviously cannot take off or land aboard the

Wasp (LHD-1), the 26th MEU (SOC) ACE retains the services of two such aircraft, operated from nearby land bases if available. Both the CH-53E and AV-8B have in-flight refueling probes, and can take fuel from the KC-130. These are deployed from Marine Air Group 14 (MAG-14) at MCAS Cherry Point, North Carolina, at the request of the MEU (SOC) commander. MAG-14 operates two Marine Refueling and Transport squadrons of KC-130Fs, VMGR-252 and VMGR(T)-253. Should there be a requirement for airborne fuel (the receiving aircrews call it "Texaco"), MAG-14 can rapidly dispatch a pair of the big four-engined tankers to support the MEU (SOC).

- **Helicopter Expeditionary Refueling System (HERS)**—This is a system that allows the MEU (SOC) ACE to deploy a mobile refueling system ashore. Either flown or landed ashore, it is used to establish a Forward Arming and Refueling Point (FARP) for the ACE's helicopters to refuel, without having to return to the ships of the ARG.

When you put all of these pieces together, the 26th's ACE becomes a highly potent and capable composite air unit. Capable of delivering personnel and firepower on a moment's notice, HMM-264 represents a vital capability to Colonel Battaglini and his staff. If the ACE has a weakness, it is to be found in the CH-46Es. These elderly birds are headed into their fourth decade of service, and are only capable of hauling eight to twelve loaded Marines, depending on the range to the target. When the MV-22B Osprey arrives, it will be able to carry up to twenty-four combat-loaded troops out to many times the range of the Sea Knight, with much greater speed and survivability. But until the first of these new birds arrives in 2001, the old Bullfrogs will have to do.

Beans and Bullets:
MEU Service Support Group -26 (MSSG-26)

No military unit anywhere does anything without a steady supply of food, fuel, water, ammunition, and all the other things that keep them going. The Marine Corps recognizes this, and has given each MAGTF its own Combat Service Support Element (CSSE) to keep it supplied and ready for action. The logistical component of the 26th MEU (SOC) is the MSSG-26, and is commanded by Lieutenant Colonel Donald K. Cooper of Greensboro, Maryland (he is a 1971 graduate of Wake Forest), and his senior enlisted advisor, First Sergeant Ralph Drake. It is composed of approximately 275 personnel in eight platoon-sized units that break down like this:

- **Headquarters Platoon**—Like the other components of the MEU (SOC), MSSG-26 has a headquarters unit, with the appropriate "S"-coded sections.
- **Communications Platoon**—Because of their heavy inventory-control requirements, the MSSG-26 communications platoon has somewhat more robust communications and computer capability than other such units in the MEU (SOC).

- **Landing Support Platoon**—The Landing Support Platoon might best be described as the receiving and inventory-control agency for the MEU (SOC) either on the beach landing site (BLS) or in a helicopter landing zone (HLZ). Utilizing a computerized bar-code system, they scan every item coming ashore, and then monitor and track its position until it leaves the BLS or HLZ.

- **Engineer Support Platoon**—This platoon provides deliberate engineering support. The platoon can produce potable water with its Reverse Osmosis Water Purification Units (ROWPU), provide mobile electric power via an assortment of generators, store and distribute bulk fuel, and provide material-handling support with its own forklifts and bulldozer.

- **Supply Platoon**—The Supply Platoon is just that, the central repository and distribution point for most of the supply line items used by the MEU (SOC). This includes repair parts, packaged fuel, rations, clothing, etc., and is controlled through a computerized asset-tracking system.

- **Motor Transport Platoon**—The job of the Motor Transport Platoon is to distribute what the Supply Platoon issues to the units of the MEU (SOC) as well as transporting troops. To this end, they are equipped with 5-ton trucks, HMMWVs, fuel and water tankers, and several Logistics Vehicle Systems (LVSs).

- **Maintenance Platoon**—This unit provides maintenance services and support for all of the other units within the MEU (SOC), except for elements of the ACE. To this end, they repair and maintain everything from wheeled and tracked vehicles, howitzers, individual and crew-served weapons, and other powered equipment, to computers and other communications/electronics equipment.

- **Medical Platoon**—Though most of the medical services for the MEU (SOC) are provided by Navy doctors and corpsmen aboard ship, the MSSG has a small medical platoon staffed with Navy personnel to provide field support. It consists of one doctor and twenty Navy corpsmen, who provide a forward-aid-station function for the MEU (SOC), where they can resuscitate and stabilize the patients and then evacuate them back to the ships of the ARG for more definitive care.

Through the efforts of the MSSG and the resources of the ARG offshore, the MEU (SOC) is designed to sustain an operation of up to fifteen days duration. Beyond that, follow-on logistical support such as that of a MPSRON would be required to sustain further operations. Nevertheless, this makes the MEU (SOC) a highly capable and independent unit for short-duration operations. In fact, because of their relatively robust logistics capability compared to the units of other services, they are quite capable. For example, during a joint exercise during the summer of 1995, the 26th MEU (SOC) provided food (MREs) and water to a battalion of the 82nd Airborne Division that was taking part in an assault on an airfield. This is why folks like Lieutenant General Tony Zinni (the commander of I MEF) like to call units like the 26th MEU (SOC) "self-licking ice cream cones!"

Captain C. C. "Skip" Buchanan, the commander of Amphibious Squadron Four (PHIBRON 4), wearing his familiar blue coveralls in the wardroom of the USS *Wasp* (LHD-1), his flagship.

JOHN D. GRESHAM

Getting There: PHIBRON 4

In Chapter 6, we looked at the ships that make up the ARGs that carry MEU (SOC)s around the oceans of the world. The three ships of the ARG that carry the 26th MEU (SOC) around these days are assigned to Amphibious Squadron Four (PHIBRON 4), which is based out of the amphibious base at Little Creek, Virginia. Home-ported there are the USS *Whidbey Island* (LSD-41) and USS *Shreveport* (LPD-12). Because of her large size, the other ship of PHIBRON 4, USS *Wasp* (LHD-1), is based over at the main naval base at Norfolk, Virginia, next to the wharf where the supercarriers are berthed. PHIBRON 4 is commanded by Captain C. C. "Skip" Buchanan. Skip Buchanan, another Naval Academy graduate (Class of 1967), is a sunny, well-rounded man. You can usually recognize him from his faded blue jumpsuit coveralls. He prefers to wear these whenever possible, rather than the more customary khaki uniforms. Short and stout, he is a quiet man who keeps to himself, and who likes to watch and listen before he speaks his mind. But when he does, people listen! The 1995/96 cruise of PHIBRON 4 was going to be his last, as he planned to retire in 1997. It was to be the end of a long and productive career, and he planned to end it in style with a highly successful cruise with the 26th MEU (SOC). The Navy ARG is commanded by an officer of the same rank as the MEU (SOC) it carries. This means that Skip Buchanan and Jim Battaglini work as a team, sharing both the powers and responsibilities that their jobs demand. Both commanders try to make things work as well as possible for the roughly five thousand personnel under their joint commands.

Today PHIBRON 4 is made up of just three ships, where just a few years ago there might have been five. The retirement of the LSTs and the great size and capacity of the *Wasp* (LHD-1) and *Whidbey Island* (LSD-41) have made this smaller ARG possible. Composed of only three amphibious ships (an LHD or LHA, an LSD, and an LPD), this is the kind of force that you will see as the Navy goes into the 21st century. Interestingly, if you add up the total of the various ship footprints, you will actually find that the three ships of today's ARGs provide an embarked Marine unit with a great deal more space than the earlier five-ship group.

In addition to the embarked MEU (SOC), there are some small units and equipment that are specific to the ARG. These include:

- **HC-8**—While technically not part of the 26th MEU (SOC)'s ACE, there is a two-aircraft detachment of UH-46D Sea Knight vertical-replenishment (VERTREP) helicopters from Navy Helicopter Combat Support Squadron Eight (HC-8—the "Dragon Whales") out of NAS Norfolk, Virginia. Normally controlled by the ARG commander, these Navy Bullfrogs are used to support utility, replenishment, and "Angel" search and rescue duties around the ships.

- **VC-6, Detachment H**—The ARG controls a detachment of U.S. Navy Pioneer UAVs. The five UAVs with their supporting equipment and personnel belong to Detachment H, fleet Composite Squadron Six (VC-6). While the Marines have two UAV companies operating their force of Pioneers, the Navy continues to have the two detachments which used to be based aboard the retired battleships *Missouri* (BB-63) and *Wisconsin* (BB-64). One of these is based aboard the *Shreveport* (LPD-12) and provides the ARG/MEU (SOC) force with their own limited reconnaissance capability.

- **SEAL Team**—Organic to the ARG is a detachment from one of the U.S. Navy's Sea-Air-Land (SEAL) special-warfare teams. Their job is to provide physical security for the ships of the ARG when visiting foreign ports, as well as giving the ARG/MEU (SOC) team an additional reconnaissance and covert-action asset.

- **Assault Craft Unit Two (ACU-2)**—To provide ship-to-shore transport for the MEU (SOC), Captain Buchanan controls two LCUs from ACU-2 at the amphibious base located at Little Creek, Virginia. One is based aboard the USS *Whidbey Island* (LSD-41), while the other one operates from the USS *Shreveport* (LPD-12).

- **Assault Craft Unit Four (ACU-4)**—In addition to the LCUs, the ARG commander also controls three LCACs from ACU-4, also headquartered at the amphibious base in Little Creek, Virginia. These are based aboard USS *Wasp* (LHD-1).

- **Beach Control Party**—These are the famous "Beachmasters" who run the receiving end of an amphibious landing. These parties, along with their Marine counterparts in the CSSE, have the job of getting personnel, vehicles, equipment, and supplies off the landing craft and into action.

Since you don't fit a unit like the 26th on to just one ship, even one like USS *Wasp* (LHD-1), you have to break it into pieces and load it carefully aboard the various ships of the ARG. This is the job of the combat cargo shops on each of the ships, as well as the MEU (SOC)'s S-4 section led by Major Arinello. These organizations work to pack everything and everyone aboard as tightly as possible, while still accessing it when the time for its use arrives.

Captain Ray Duffy, the commanding officer of USS *Wasp* (LHD-1), on the bridge of his ship.

The official emblem of the USS *Wasp* (LHD-1).

Let's look at how the 26th MEU (SOC) was loaded into the ships of PHI-BRON 4 when they embarked at Camp Lejeune and Moorehead City, North Carolina, in late August of 1995:

- **USS *Wasp* (LHD-1)**—Commanded by Captain Raymond Duffy (Villanova, Class of 1970), the *Wasp* carries almost half of the personnel and equipment of the 26th, as well as the vast majority of the aircraft and support equipment. The breakdown looks something like this:

  —ARG—The entire PHIBRON 4 staff and their supporting equipment is carried aboard the *Wasp*. In addition, the two HC-8 helicopters are based aboard, as well as the three embarked LCACs from ACU-4.

The official emblem of the USS *Whidbey Island* (LSD-41).   *U.S. NAVY*

The official emblem of the USS *Shreveport* (LPD-12).   *U.S. NAVY*

—**CE**—Almost the entire 26th MEU (SOC) headquarters and its supporting units are based aboard the *Wasp*.

—**GCE**—the *Wasp* carries roughly half of the embarked BLT, including the Battle Staff, Headquarters and Service Company, one rifle company, the heavy weapons company, the artillery battery, and Task Force Mosby, which consists of the Light Armored Vehicle detachment (four LAV 25s, two TOW, one logistics variant) and the sixteen hardbacked HMMWVs (eight with TOW II launchers and eight equipped with either Mk 19 40mm grenade launchers or .50-cal. machine guns).

—**ACE**—The entire ACE is carried aboard the *Wasp*.

—**CSSE**—A small detachment of MSSG 26 is loaded aboard the *Wasp*.

• **USS *Whidbey Island* (LSD-41)**—Commanded by Commander T. E. McKnight (VMI, Class of 1978), the *Whidbey Island* carries the following 26th MEU (SOC) assets:

—**ARG**—The *Whidbey Island* operates one of the ACU-2 LCUs.

—**GCE**—One rifle company, the AAV Platoon, and the tank platoon are embarked on the LSD.

—**CSSE**—Small detachments of various MSSG platoons also reside aboard the *Whidbey Island*.

Because of her rather limited vehicle space (cargo²) compared to the *Wasp* and *Shreveport*, the *Whidbey Island* (LSD-41) is loaded with extremely high density cargo, such as armored vehicles and their supporting infantry.

• **USS *Shreveport* (LPD-12)**—*Shreveport* is commanded by Captain John M. Carter (University of Missouri, Class of 1969). Despite her advanced age, this old lady of the ARG represents the ability for Captain Buchanan to split the ARG into several parts and conduct operations separately. In 1995, she carried the following:

–ARG–PHIBRON 4 had several important components embarked on the *Shreveport*. In addition to the VC-6 UAV detachment, one of the HC-8 UH-46Ds is based aboard, as well, a LCU from ACU-2. She also carries the ARG SEAL team.

–GCE–*Shreveport* carries the BLT Reconnaissance Platoon, and one rifle company, which serves as the surface rubber boat raid and cliff-assault force.

–ACE–While none of the ACE aircraft are normally carried aboard, the *Shreveport* is set up to provide a mobile FARP for the helicopters of the HMM-264.

–CSSE–*Shreveport* has small detachments of various 26th MSSG platoons loaded on board, including the headquarters platoon.

While the loadouts of the ships that we have shown you above is representative, the staffs of PHIBRON 4 and the 26th MEU (SOC) are always trying to improve things. All the time, they are modifying the mix of units and equipment on each ship, based upon the requirements of a particular mission.

Cats and Dogs: Attached Units

Though more self-contained than most military units, the 26th MEU (SOC) and PHIBRON 4 are far too small and exposed to operate without some sort of escort and supporting forces. While the number of units that might be involved in the operations of a MEU (SOC) is almost limitless, a number of the units commonly associated with the 26th while they are out on cruise include the following:

- **Carrier Battle Group**—One of the interesting side effects of the drawdown of U.S. military forces over the last few years has been the wholesale retirement of whole classes of surface escorts. The same drawdown has reduced the number of aircraft carrier battle groups (CVBGs) to eleven. The result of all this has been the decision to "team" CVBGs and ARGs/MEU (SOC)s, so they might share the available escorts and work more closely together. This meant that for their 1995/96 cruise, PHIBRON 4 and the 26th MEU (SOC) was teamed with the battle group of the supercarrier USS *America* (CV-66). In addition to *America*, the battle group is made up of Destroyer Squadron 14 (DESRON 14), as well as several submarines and support ships. The group included the nuclear cruiser *South Carolina* (CGN-37), the Aegis cruisers *Normandy* (CG-60) and *Monterey* (CG-61), the guided-missile destroyer *Scott* (DDG-995), the guided-missile frigates *DeWert* (FFG-45) and *Boone* (FFG-28), the nuclear attack submarines *Hampton* (SSN-767) and *Oklahoma City* (SSN-723), the ammunition ship *Butte* (AE-27), and the fleet oiler *Monongahela* (AO-178).

As for the carrier *America*, her embarked Carrier Air Wing One (CVW-1) is composed of fourteen F-14As (VF-102, the Diamondbacks), thirty-six F/A-18Cs (VFA-82, the Marauders; VFA-86, the Sidewinders; and VMFA-251, the Thunderbolts), four EA-6B Prowlers (VMAQ-3), four E-2C Hawkeyes (VAW-123, the Screwtops), eight S-3B Vikings (VS-32, the Maulers), eight SH-60F and four

HH-60H Seahawks (HS-11, the Dragon Slayers), and a pair of ES-3A Shadow surveillance aircraft. By teaming a CVBG/CVW with an ARG/MEU (SOC), the American national command authorities have a force of immense power, flexibility, and balance. Also, when the CVBG and ARG operate separately, the ARG takes several of the battle group's ships to act as escorts. During the 1995/96 cruise, the *Normandy* (CG-60) and *Scott* (DDG-995) were usually teamed with PHIBRON 4 to provide naval gunfire and SAM support.

- **MPSRONs**—One of the key missions for the MEU (SOC) is to open up a beachhead for follow-on forces to exploit. Current U.S. plans have these forces centered around one of the three Navy/Marine MPSRONs. These ships can be used in a variety of ways. For beginners, the MPSRON can provide additional logistical support to extend the life of a MEU (SOC) operation beyond the fifteen days of supplies carried by the ARG. In addition, the MPF follow-on brigade can be used to expand a forced entry by a MEU (SOC) into an enemy territory. Finally, because they can be rapidly deployed to help out in a crisis, the MPF ships represent an excellent means of supporting humanitarian and/or peacekeeping operations.

- **Airborne Units**—A really exciting inter-service partnership has developed in the past few years: Long-range airborne assault is combined with an amphibious assault by a MEU (SOC). Now, this is not exactly a new idea: The Allies did it several times in World War II. What's new is that the airborne unit, from the 82nd Airborne Division at Fort Bragg/Pope AFB, North Carolina, can fly the unit directly to the target area, anywhere in the world. Thanks to in-flight refueling, strategic airlift aircraft like C-141s, C-5s, and C-17s can make the trip non-stop. The way it works: Airborne troopers from the 82nd drop onto something like an airport or other transportation facility, so that other follow-on units can be flown in. While they are doing this, the MEU (SOC) takes a nearby port or beach and drives inland to link up with the airborne unit. Once this is done, the airborne unit draws its support and sustenance from the MEU (SOC) until follow-on units join the operation. This particular scenario is practiced regularly by the 82nd Airborne in joint maneuvers with the Navy, Air Force, and Marines.

- **Land-Based Air Support**—There are provisions for the MEU (SOC) to take advantage of land-based air support in the form of a detachment of KC-130 airborne tankers, should an air base be close enough to be useful. This has been the case in the Adriatic, during Operation Joint Endeavor in Bosnia-Herzegovina. In addition to tankers, land-based air support could likely include units such as the following:

  —**Marine Fighter Support**—Along with tanker support, the Corps can deploy squadrons of two-seat F/A-18D Hornet all-weather strike fighters to support MEU (SOC) operations. Equipped with a Night Hawk laser targeting pod for LGBs, as well as AIM-120 AMRAAM and AGM-65 Maverick missiles, the F/A-18Ds are highly capable strike fighters.

—Tanker Support—In addition to the tanker support already mentioned, there are other tanker assets that can make the life of the MEU (SOC) ACE easier. The AV-8Bs with their refueling probes can make use of almost any Air Force, Navy, or NATO tanker aircraft available. This becomes especially easy if the aircraft is one of the big KC-10A Extenders, which can refuel aircraft with either boom or probe refueling systems. There also are Air Force HC-130 Hercules tanker aircraft that are assigned to search and rescue, as well as special operations units, and these can refuel either the Harriers or the CH-53E Super Stallions.

—Airborne Early Warning (AEW) Support—If there was any lesson that came out of the 1982 Falklands War, it was the overriding need for surface ships to have proper AEW coverage. Lack of a decent AEW platform for cuing of their limited force of Sea Harrier FRS.1 fighters probably cost the British several of the ships that they lost to Argentine air attack. Unfortunately, the Marine Corps has no such capability to augment their ACE aboard the *Wasp* (LHD-1), so they must make use of any AEW capability they can get. If they are near a CVBG, they can make use of AEW- information-data-linked E-2C Hawkeyes. In addition, data links on the ships of the ARG, as well as those on the new AV-8B Plus Harrier IIs, are compatible with a number of other such systems, including the force of E-3 Sentry AEW aircraft operated by the Air Force, NATO, and some of our other allies.

—Composite Wings—One of the important capabilities of the MEU (SOC) ACE is the ability to work jointly with air units of other services, or even other countries. As a result, don't be surprised if you don't someday see the use of Air Force composite wings like the 366th at Mountain Home AFB, Idaho, or the 23rd Wing at Pope AFB, North Carolina, with one of the MEU (SOC)s to support landing operations or provide cover for the ARG.

As you can see, it's easy for other units to team with or "plug into" the 26th MEU (SOC). Thanks to a robust series of command and control links aboard the ARG, the MEU (SOC) can provide connectivity for anything from an Air Force AEW aircraft to an airborne unit. Meanwhile, the staff of the Marine Corps Combat Development Command at Quantico, Virginia, is working hard to find new units and ways for the MEU (SOC) to play in the ever-growing world of joint operations.

- **Reconnaissance Support—**If there is anything that Colonel Battaglini wants more of on-cruise, it is high-quality, up-to-the-minute imagery of their AOR. If the imagery is satellite-based, it will come from the National Reconnaissance Office (NRO), with their fleet or orbital imaging satellites, and the Defense Airborne Reconnaissance Office (DARO), which manages UAVs and other airborne imaging programs. Both agencies are working hard to supply more imagery to an ever-expanding user base. In particular, plans are progressing to create a super-agency called the National Imaging Agency (NIMA), which will stand up on October 1st, 1996. NIMA will combine the services of NRO, DARO, the Central Imaging Office (CIO), the National Photographic Center (NPIC), and the Defense Mapping Agency (DMA), all under a single roof. In this way, the users of their products can more rapidly and efficiently obtain the wealth of assets they provide —everything from maps to real-time photography—when they need it.

The products themselves are growing in both variety and quality. As always, the NRO's satellite systems are producing a large volume of high-quality imagery, though efforts are under way to look at smaller, less expensive collection systems in the late 1990s.

Meanwhile DARO is making giant strides towards their vision of an integrated airborne reconnaissance architecture that does its job within today's limited budgets. The Pioneer UAV will continue in service for a few years. Because the Hunter UAV system that was designed to replace it has been terminated over high life-cycle costs, DARO has decided to move onto what is called the "Maneuver UAV," which is designed to provide Army and Marine unit commanders with a capability to obtain real-time video and other imagery.

In addition, the Predator program described in *Fighter Wing* is going great guns. The Air Force stood up its first Predator unit (the 11th Reconnaissance Squadron) at Nellis AFB, Nevada, just recently (these are roughly the same as the Gnat 750-series UAVs that did so well in their CIA-sponsored trials in Bosnia). The program is so successful that DARO is looking at buying and fielding more Predator systems to users.

In addition to the Predator and Pioneer programs, excellent progress is being made on longer endurance systems, like the Lockheed-Martin "Skunk Works"-produced Dark Star system. And even longer-range systems are being developed, as well as the data links, common control stations, and other equipment that will be needed to make the various UAV systems available to the widest possible base of users. DARO is also working on more traditional manned reconnaissance systems, including the introduction in 1997 of the new RF-18D Hornet equipped with the new Advanced Tactical Reconnaissance System (ATARS), and a new F-16-mounted recon pod system being introduced by the Virginia Air National Guard. The expectations are that by the year 2001, DARO's plans for completely remaking the airborne reconnaissance architecture of the U.S. military will be close to completion.

- **Intelligence Support**—In addition to the support provided by the various photographic and mapping agencies within the intelligence community, the 26th MEU (SOC) makes use of intelligence from several other agencies and organizations:

  —**National Security Agency (NSA)**— The NSA, which controls electronic and signals intelligence, is a significant supporting agency for an amphibious unit like the 26th MEU (SOC). Through the Ships Signals Exploitation Space (SSES), the ARG and MEU (SOC) commanders can tap into a variety of different signals and electronic intelligence sources, among these: RC-135 Rivet Joint,

ES-3 Shadow, and EP-3 Orion electronic intelligence aircraft and ferret satellites. Ground- and ship-based sensors (like the Classic Outboard ESM system) can also tap into all variety of different electronic signals, from SAM and air traffic radars to cellular phones and television signals.

–U.S. Space Command (USSPACECOM)–Based at Falcon AFB, Colorado, USSPACECOM provides space-based systems to support combat operations for all the military services. In addition to providing GPS navigation signals, communications support, weather reporting, and ballistic missile-warning, there are a whole range of new capabilities that will emerge in the next few years. These include integrated designation/communications/navigation/transponder systems, which will "net" individual Marines together on the digital battlefield of the 21st century.

–U.S. Department of Justice (DOJ)–Strange as it may sound, the U.S. DOJ and its assorted agencies are excellent sources of information for a variety of missions that might involve a MEU (SOC). The Federal Bureau of Investigation, the Bureau of Alcohol, Tobacco, and Firearms, the Drug Enforcement Agency, the U.S. Marshals Service, and other agencies offer useful information on everything from terrorist organizations to smuggling techniques. As a result, you frequently see DOJ and other government agencies (the Department of Energy, the Environmental Protection Agency, etc.) supporting unconventional operations by the Marines, as well as USSOCOM.

–Cable News Network (CNN)–Okay, let's all tell the truth here. Right now, CNN is the finest real-time intelligence-gathering service in the world. If you walk into the office of anyone who is really on the inside, you will inevitably find the television tuned to CNN. In the twenty years since Ted Turner launched his twenty-four-hour news network, CNN has brought most decision makers their first news of vital events as they are breaking. This kind of topical and timely coverage is the reason why intelligence staffs aboard ships are fighting so hard to obtain the stabilized satellite dishes needed to receive CNN. In this way, they can get the same real-time information as any other cable television subscriber!

CONOPS: The MEU (SOC) Way

Now let's look at how a MEU (SOC)/ARG operates. Returning to the rescue of our downed Air Force Captain, the dauntless Scott O'Grady, just how did that come together, and why did it work? To understand this is to understand how the MEU (SOC)/PHIBRON team works.

For Colonel Berndt and the personnel of the 24th, the rescue process started almost as soon as the young man was shot down in northwest Bosnia-Herzegovina. At the time, the 24th MEU (SOC) was embarked aboard PHIBRON 8–the

Kearsarge (LHD-3), *Pensacola* (LSD-38), and *Nashville* (LPD-13), under the command of Captain Jerry E. Schill. The group's job was to act as an offshore ready reserve force for the NATO forces that were enforcing the air and maritime embargo of Bosnia-Herzegovina. The 24th's duties also included combat search and rescue (CSAR), in the event that such a mission was required.

In the Marines, a CSAR mission is called a Tactical Recovery of Aircraft and Personnel, or TRAP. TRAP missions are something of an MEU (SOC) specialty, and are practiced regularly. The key to carrying out a TRAP, or any other MEU (SOC) special mission, is training and planning. *Really* fast planning. As for Scott O'Grady, on the night of June 2nd, 1995, the 24th MEU (SOC) was aware that Basher 52 was down, and that there was no confirmation that O'Grady was alive. All the staff of the 24th knew was that a rescue mission might be required. The commander of the 24th's GCE (the 3/8 BLT), Lieutenant Colonel Gunther (who was assigned the job of TRAP mission commander), quickly convened a crisis action team to begin the advance planning for a TRAP package of aircraft and personnel, should it be required. This done, the 24th waited and listened.

TRAP packages come in various shapes and sizes, with a variety of options available to the MEU (SOC) staff. For example, let's say that a helicopter from the ACE goes down as a result of a mechanical problem in neutral territory. And let's say that the aircraft is not too badly damaged, so it would be possible to repair the helicopter and fly it out. In that case, a small security team from the GCE, together with some maintenance personnel from the ACE, would fly out and establish a security perimeter around the downed aircraft. The TRAP force would then repair the aircraft and fly it home to fight another day.

The downing of Basher 52 was an entirely different problem. In that case, the shootdown was in an isolated area, over 30 nm/55 km from the coast, and in generally rough and mountainous terrain, well within reach of hostile Bosnian Serb forces. Given these parameters, and the expected threat level (there might still be an active SA-6 battery in the area), Gunther and Berndt decided to lay on what they call a "D" package. This was the largest of the five TRAP packages available to the 24th, and involved sending a pair of big CH-53E Super Stallions loaded with the mortar platoon from the 3/8 BLT's headquarters company. When I later asked why the mortar platoon was chosen for this duty, Chris Gunther answered, "They were available, and they did it during the workup." In other words, given the variety of tasks that the 24th was being tasked with at that time (possible evacuation of UNPROFOR personnel from Bosnia-Herzegovina, etc.), these personnel were not tasked for other duties, and they had trained specifically for the job. The CH-53Es from HMM-263 (the 24th's ACE unit) were chosen over the elderly CH-46Es Bullfrogs because of their greater range, speed, and lifting capability. In addition to the cargo helicopters, there would be an escort of AH-1W Cobras, and AV-8B Harrier IIs. All told, the rescue force, if it were ever needed, would have fifty-seven Marines and four Navy Corpsmen. By the morning of June 3rd, the personnel for the TRAP package had been alerted and the aircraft readied. The planned TRAP package looked something like this:

- **CH-53E Super Stallion Flight (2)**—The lead CH-53E was flown by Major William Tarbutton (the air mission commander) and Captain Paul Oldenburg. This helicopter would carry half of the mortar platoon commanded by First Lieutenant Martin Wetterauer, Lieutenant Colonel Chris Gunther (the BLT CO and mission commander), and two Navy corpsmen. The second Super Stallion was to be flown by Captain Paul Fortunato and Captain James Wright. This aircraft would carry the rest of the mortar platoon, two Navy corpsmen, Colonel Martin Berndt–the 24th MEU (SOC) CO, and Command Sergeant Major Angel Castro, Jr.

- **AH-1W Cobra Flight (2)**–The lead Cobra was to be flown by Major Nicholas Hall and Captain James Jenkins II. The second helicopter by Major Scott Mykleby (the escort flight leader) and Captain Ian Walsh.

- **AV-8B Harrier II Flight (4)**–This flight was led by Major Michael Ogden of VMA-231, and was made up of four aircraft, so that at least two would be available over the recovery site.

In addition to these forces, there would be mission spares (called "bump" aircraft) readied, as well as reinforcement units (known as Sparrowhawk and Bald Eagle), should the TRAP package encounter problems.

For the next six days, the situation remained quiet, as the pilots of the 31st FW flew over northwest Bosnia-Herzegovina, hoping to hear something from Basher 52. During this time, the TRAP package stayed on Alert 60 (an hour's notice), eating and sleeping on alert, hoping for the chance to go in and snatch the young Air Force officer out of harm's way. Down in the 24th's LFOC on the *Kearsarge* (LHD-3), the plan was refined, based upon the minimal information that was available at the time. On the night of June 7th/8th, O'Grady was located. Out on the *Kearsarge*, the staff of the 24th monitored the transmissions and began to get ready, even before the execution order arrived at 0300. Once this was given, Captain Schill tightened up the formation of PHIBRON 8's ships, and headed towards the Dalmatian Coast to be ready to launch the TRAP force. Immediately, the *Kearsarge*'s CO, Captain Chris Cole, ordered the ship to Flight Quarters, and things began to happen. Realizing that the TRAP team might be headed into an area where SAMs had recently been active, the MEU (SOC) requested, and Admiral Leighton Smith (the commander of NATO Forces, South) ordered, a support package of Air Force, Navy, and Marine aircraft to be launched. This included F-15s, an E-3 AEW aircraft, and Marine F/A-18Ds with AGM-88 HARM missiles, just in case the Serbian SAMs decided to engage. As it turned out, organizing this force took longer than expected, causing the TRAP force to remain airborne over the ARG until just before daybreak

Down in the 24th's LFOC, the decision to use the "D"- size TRAP package was reaffirmed, and the final steps necessary to ready the force were under way. Aircraft were fueled and armed. Weapons were test-fired and checked. And then they had what's called a "confirmation briefing." This took place down in the *Kearsarge*'s war room on the 02 Level just prior to the loading of the aircraft; and it's the final piece of what the Marines call "the rapid response planning process"–a planning sequence that allows a MEU (SOC) to commence the execution any of its preplanned special operations missions within just six hours of the reception of an execution order.

The confirmation briefing is the final coordination meeting for the officer and enlisted personnel of the ARG and MEU (SOC). And it is an amazing thing to watch. Taking only about fifteen to twenty minutes, it covers more than twenty different subjects—from weather and intelligence to radio call signs and aircraft weapons loadouts. The briefing can proceed so quickly because every movement and action of the TRAP mission has been practiced many times in training back in the U.S. and while afloat. In the confirmation briefing, the speakers execute what is called "briefing by exception": A speaker walks (rapidly!) to the front of the wardroom, slaps down a briefing transparency just long enough for the assembled crowd to see it, and speaks only about those operations that are not normal or running to plan. In short, you brief a particular topic in between thirty and sixty seconds. This procedure is not designed to be hasty or frivolous. Rather, it's done only when time can kill a mission. This was one of those times.

Following the briefing, the Marines moved up the ramp from the hanger deck to the flight deck. Each "stick" of Marines was led to their aircraft by Combat Cargo personnel from the ship and trooped aboard. When the loading was completed, the order was given for the helicopters to start engines. By 0505 hours, all of the helicopters were airborne, waiting for the word that the NATO support package was airborne and in position. At 0545, the "go" order was given. And at 0549, the TRAP force went "feet dry." At 0640 the lead Cobra, code named "Bolt," made contact with O'Grady, and ordered him to pop a smoke flare. Sighting the smoke, the Cobra crew dropped a flare of their own, and began to coach the lead CH-53E into a small clearing adjacent to the young pilot's position on the side of a hilly, rock-strewn pasture. Heavy fog blanketed the area, and caution was required to get the big choppers down. As soon as they hit the ground, Lieutenant Wetterauer and his mortarmen exited the helicopter to set up a defense perimeter and commence the search for O'Grady. The standard TRAP mission plan assumes that the person to be rescued is injured, so a security perimeter was established, just in case extra time was needed to carry O'Grady out. As it turned out, this was not required.

As the lead helicopter was unloading, and the second CH-53E was coming in to land, there was a small problem. A small fence was in the LZ, and the second chopper came down on top of it. This only caused a moment's delay, however. Moving forward a bit to clear the fence, Captains Fortunato and Wright set the CH-53E down and dropped the rear loading ramp. Before they could unload the chopper, though, Captain O'Grady came sprinting out of the underbrush, brandishing his radio and pistol as he headed into the CH-53E. After he was relieved of these items (for safety's sake!), the helicopter lifted off. Captain Fortunato then notified the air mission commander that O'Grady was safely aboard. This done, the four helicopters, with their Harrier escorts flying above, headed back to the coast at full speed and minimum altitude. Even during the burst of AAA and SAM fire, their planned procedures worked well. When the Cobras sighted the ground fire and SAMs, they immediately ordered an SAM break from the CH-53Es (an evasive turn while firing flares and chaff decoys) and continued on to the coast. With the rescue completed, TRAP rules advise that you avoid a fight in enemy territory, so the rescue force continued on with only a few return shots fired by a door gunner. By 0730, the TRAP force was back over the *Kearsarge* and safely home.

A reunion of 24th MEU (SOC) personnel with Air Force Captain Scott O'Grady (eighth from right in rear) in April of 1996. Included are Brigadier General Marty Bernet (far left), Lieutenant Colonel Chris Gunther (fifth from left), and Sergeant Major Angel Castro (fourth from left). The female officer (third from left) is Lieutenant General Carol Mutter, the first woman to achieve such rank.   *JOHN D. GRESHAM*

Captain O'Grady was then escorted from the flight deck down to the medical department, where it quickly became apparent he was in pretty good shape: He had a minor case of dehydration, his feet were a little beat up, and he had minor friction burns on his neck and face. Meanwhile, the Marines from the TRAP force turned in their unused ammunition, cleaned their weapons, went through debriefing, and headed down to breakfast. At the same time, the after-action reports were started...along with the preparations for the inevitable surge of press personnel. All before Colonel Berndt had his morning coffee!

Getting Ready: 26th MEU (SOC) Training and Operations

Once upon a time when you were a teenager, you probably dreamed of driving an automobile. In those days, making the quantum leap from walking or bicycling to bounding in a car from city to city or from state to state surely seemed comparable to getting the captain's chair on the starship *Enterprise*. Of course, driving a car didn't turn out to give you the freedom you hoped for. In fact, before anyone would let you loose in a machine that dangerous, you took driving classes and driving lessons in high school. Later, you went down to the Department of Motor Vehicles, took a written and visual exam, and finally took a personal examination that tested your driving skills in actual traffic conditions. All of this for the simple right to drive alone in traffic. Or is it so simple? Badly driven cars kill more Americans every year than we lost in all the years of the Vietnam War. To put this more practically: When we're out there on the freeways, we want the other guys in their big, fast machines to drive as well as we do. Reasonable people take the privilege of driving quite seriously.

Now, if an operation as easy as driving has to be so heavily monitored and regulated, you can imagine how the Marine Corps oversees the training and certification of a Marine Expeditionary Unit–Special Operations Capable (MEU–SOC). Just consider how much has to be done before one of these units can be sent cruising about the world, armed and dangerous. Up until now, we've been looking at the structure, personnel, equipment, and capabilities of a MEU (SOC). Seen through those windows, these are wonderful units. But what makes a collection of regular Marine formations *really* useful is training. More of the same kinds of training that created the esprit and ethos that made these people Marines in the first place.

To give you an idea of how this is done, let's follow the members of 26th MEU (SOC) as they prepare for their 1995/96 deployment to the Mediterranean Sea. I'll take you on some of the workup exercises, and try to give you a feel for the range of missions that MEU (SOC)s train for and how they are examined and certified ready to go. This cruise was to be no normal MEU (SOC) deployment (as if there is such a thing!). As the 26th was getting ready, the war in Bosnia-Herzegovina was coming to a head, and the 24th MEU (SOC) had just plucked Scott O'Grady out of harm's way. It was not hard to see that Colonel Battaglini and his Marines, as well as Captain Buchanan and his sailors, might be headed into the middle of a shooting war.

MEU (SOC) Missions

Today, MEU (SOC)s on both coasts and Okinawa are trained to a single set of standards and missions, which are constantly reviewed and examined to assess their validity in a changing and dangerous world. To support this effort, in 1989 the

Marine Corps implemented a set of standard MEU (SOC) training handbooks. These provide a common training syllabus for all MEU (SOC)s. The key to understanding the operations of a MEU (SOC) is to look at its various missions. A quick note about definitions: In the Marine vernacular, "assault" means to forcibly take an objective and hold it until relieved or reinforced. "Raid," on the other hand, means to enter an area, destroy or capture specified targets and equipment, and then return to wherever you started from.

Amphibious Assaults

This mission is the traditional amphibious/vertical-envelopment assault so fundamental to the Marine Corps ethos. In the case of a MEU (SOC), this mission could be executed on behalf of follow-on forces, such as Army airborne and/or fly-in units, or perhaps Marine MPSRONs/Army AWR-3 units. This is a forced-entry, "kick-in-the-door" kind of operation, and would likely be done very quickly (a matter of days) after the outbreak of a crisis. Meanwhile, the National Command Authorities would likely rush additional amphibious ships with extra Marines to beef up the MEU (SOC).

Amphibious Raids

In amphibious raids the MEU (SOC) would use its landing capability to move rapidly across a hostile beach to temporarily take an objective, and then render it useless to an enemy. Examples might include strikes at power plants, industrial areas, or military bases. Another possibility is the destruction of weapons facilitates capable of manufacturing or storing chemical, nuclear, or biological munitions. Airfields are also viable targets for an MEU (SOC) raid, as are ports.

Limited Objective Attack/Deception Raid

A specific kind of raid, the limited objective attack is defined as a short-duration raid or assault designed to divert the attention of an enemy away from a larger or more vital operation. This operation is designed to cause a lot of noise and "flash," after which the unit scoots away before the bad guys figure out which attack is the real one. Such a raid might be composed of a single big attack along a narrow front, or a series of smaller simultaneous operations spread over a wider area.

Maritime Reinforcement/Assault/Inspection

One of the many reasons for having a Navy and Marine Corps is to provide protection for the world's sea lanes. This is more than just a fancy-sounding way of saying, "Get off of my road." Maintaining freedom of navigation in the world's sea lanes is a shared responsibility of all naval powers. Therefore, one of the missions of the MEU (SOC)/ARG team is to help protect shipping against piracy and/or capture by terrorists, such as happened to the Italian cruise ship *Achille Lauro* in the 1980s. These maritime support missions come in several varieties: First, the MEU

(SOC) can provide a security or reinforcement detachment to protect a vessel during passage through troubled or threatening waters. Second, the MEU (SOC) can execute an assault type of mission, to take back a ship which has been captured by pirates or terrorists. Third (quite common in the last few years) is maritime inspection in support of an international embargo. In the Red Sea, Persian Gulf, and Adriatic, Marines have been the key to enforcing a number of maritime embargoes, like the one that helped strangle Iraq back in 1990/91.

Show of Force

Sometimes, when you want to make a point, you act like a "big dog" (large with a nasty growl), show your teeth, and let everyone in the neighborhood know you are *bad*. This mission is all about acting like a *big* dog. General Chuck Krulak will tell you that a traditional "Show of Force" is the single most valuable service that an ARG/MEU (SOC) force can perform. Nothing other than a modern version of what was once called "gunboat diplomacy," it is a unique and effective form of deterrence against small dictators and warlords who have more ambition than common sense. When an ARG or carrier battle group decides to park itself just outside the territorial waters of a nation, it sends a message that is both loud and clear. Stay put and keep to your own borders! It worked for Teddy Roosevelt, and it still works today.

Tactical Recovery of Aircraft and Personnel (TRAP)

Though TRAPs come in a variety of configurations—from simple repairs and recovery of damaged aircraft in a benign environment to full-blown raiding forces equipped to forcibly enter hostile territory to recover injured personnel—most TRAPs occur under peaceful conditions in friendly territory. All the same, MEU (SOC) Marines treat every TRAP like the O'Grady rescue, just in case.

Clandestine Recovery Operations

This mission is a variation of the TRAP mission, with the specialized requirement that it is done covertly with deniability. For instance, such an operation might be run in a short-of-war situation, where the National Command Authorities wish to avoid armed conflict. Under such conditions, stealth and patience will probably be more important than speed and firepower. In any case, if the mission is to be successful, the nation whose sovereignty is violated would have to remain blind and dumb.

Non-Combatant Evacuation Operations (NEOs)

In the last decade, evacuations have been probably the most common operations that MEU (SOC)s were called upon to execute. A crisis occurs in some far-off land like Liberia or Somalia—a civil war or some other event that breaks apart the fabric of local society. Nothing bad has happened to the Americans in the country yet, but clearly something can or will. With their organic transport helicopter force, hospital facilities aboard ship, and Marines for security, the MEU (SOC) becomes the perfect instrument to extract "non-combatants" from the danger. "Non-combatants" is the military term for civilians, be they tourists or embassy personnel (there are exceptions).

In addition to "civilian" NEOs, we are sometimes asked to evacuate military forces from a hopeless situation in a civil war or other kind of armed insurrection. A good example was the evacuation of the UN peacekeeping force from Mogadishu, Somalia, in 1994. The National Command Authorities usually request this kind of operation after an official diplomatic request. The NEO force then inserts itself into the danger area, makes contact with the unit to be evacuated, and secures a safe perimeter. Once the unit to be evacuated has been reached, transportation is arranged, and the operation is concluded as quickly as possible.

In-Extremis Hostage Rescue

Hostage rescue is probably the toughest operation that a MEU (SOC) can be asked to conduct. If this happens, it will be roughly like the planned rescue of the hostages in the U.S. Embassy in Tehran back in 1980. Though this kind of operation is usually the specialty of units like the Army's Delta Force, if time is critical, an MEU (SOC) may be the only unit that is forward-deployed. Marines therefore train for this mission, utilizing their big, long-legged CH-53E Super Stallions to provide transport. Should tanker support be available (such as a forward-deployed detachment of Marine KC-130Fs), then the mission can be run virtually non-stop thanks to in-flight refueling.

Security Operations

Security operations are like crowd control. You're trying to keep an area safe and operating normally. From the MEU (SOC) point of view, this kind of mission would normally involve the "beefing up" of an existing Marine or other military detachment. Usually this would be a Marine embassy detachment or airfield security force. The ill-fated deployment to Beirut in the 1980s was just such an operation. The difference today is that such a job would probably not be so open-ended, with much more of a "big dog" kind of presence out in the open, where everyone can see it.

Humanitarian Relief Operations

Humanitarian operations are becoming a significant priority in the post-Cold War world. Given what a force like a MEU (SOC)/ARG team is capable of delivering in the way or food, water, and medical supplies and services, such a unit is the perfect mobile relief force. Of late, we have seen a host of such operations worldwide. And soon, disaster-relief operations in our own country may become one of the major missions of the MEU (SOC)/ARG teams. As we have seen, when they are in port, the LHDs of the Atlantic Fleet represent the sixth largest hospital facility in the Commonwealth of Virginia. Should a major disaster such as a hurricane or earthquake strike a coastal community in the next few years, don't be surprised if you see a MEU (SOC)/ARG team leading the relief effort.

Civil Support/Training Operations

Back in the Cold War, we used to call this "winning hearts and minds." This rather broad category of mission covers a range of activities designed to build better relations between the U.S. and other governments and peoples. For example, joint training and exercises with local military forces help promote understanding and goodwill with our allies. Or, while making a port call, the ARG commander might open the medical department of an LHD or LHA, with the support of the MEU (SOC) medical department, to the local population to provide inoculations or dental services. Other options include assisting in the building of bridges, roads, and other basic infrastructure and services.

Military Operations in Urban Terrain (MOUT)

Infantry hates fighting in built-up urban areas. House-to-house fighting is quite dangerous. It can tear the guts out of an infantry force if it is not extremely well trained and very methodical. Marines, with their considerable experience rooting enemies out of caves and urban areas, have a healthy respect for such operations. The Corps' predeployment training program is designed to teach MEU (SOC) Marines how to take an edge into such situations. Called Training in an Urban Environment (TRUE), it covers everything from demolishing walls between buildings to proper movement through built-up urban areas.

Initial Terminal Guidance

Once upon a time this used to be called a pathfinding or pioneer mission. In today's context, the Initial Terminal Guidance mission is designed to provide navigational support to another, larger mission. Usually the task involves inserting onto a beach or into a helicopter landing zone (HLZ) a small team carrying specialized direction finding and navigation gear that assists incoming landing craft or helicopters in making a safe and accurate approach and landing. Even in an era when GPS allows for pinpoint accuracy with split-second timing, there is nothing like someone coaching you in personally.

Signals Intelligence (SIGINT)/Electronic Warfare (EW) Collection

It goes without saying that having a tap on your enemy's phone is desirable in a war. For this reason, the ships of PHIBRON 4 and the 26th MEU (SOC) have a robust capacity to conduct intelligence gathering of communications and other electronic signals. But sometimes more information is needed. Both the SSES on the ships and SIGINT/EW teams deployed by the MEU (SOC) can generate a vast amount of useful intelligence for decision makers from the tactical level to the National Command Authorities. While much of their equipment and techniques are highly classified, the 26th MEU (SOC)/PHIBRON 4 team can listen to virtually the entire electromagnetic spectrum.

Clandestine Reconnaissance and Surveillance/ Counterintelligence

Because of its heavy complement of Marine Force Reconnaissance personnel, SEALs, and ship-based sensors (both active and passive), a MEU (SOC) is an extremely capable force for collecting intelligence. Covert missions of this type might include insertion of teams into hostile territory, perhaps in short-of-war conditions. When the mission is complete, these teams can be picked up without the hostile forces becoming aware that they have been watched. As currently configured, the MEU (SOC) is set up to collect information in any number of other ways. Among these: It can observe terrorist groups (through national intelligence sources); it can monitor enemy road convoys with the Pioneer UAV detachment of the ARG; or it can go in and take human sources.

Seizure and/or Destruction of Offshore Platform Facilities

Over the last half century, offshore facilities that exploit the resources of a continental shelf have become quite common around the world. Oil production platforms of various configurations have also been commonly used by nations like Iran and Iraq as sensor and weapons platforms. Luckily, offshore operations are a MEU (SOC) specialty. To provide a capability to hold these platforms at risk, MEU (SOC)s are trained to assault and, if necessary, render such a facility unusable. By the way, doing this is quite simple. Damage the well head assembly in such a way that it requires complete replacement (which takes time and money), but does not cause a spill of raw crude oil into the environment. It is a touchy job, but one that the Marines have already executed successfully in combat, where it really counts.

Specialized Demolitions

In the Marine Corps there is an old saying: "There is no problem that can't be solved by an appropriately sized, placed, and fused charge of high explosive." It is true. Marines have a gift for blowing things up, which makes this kind of mission one of their favorites. As with offshore platform operations, the key is to destroy a particular target without damaging the rest of the neighborhood...or the neighbors.

Fire Support Coordination

A MEU (SOC) must be able to accurately call down fire on targets of interest. Modern fire support involves scouting, designating, and damage assessment for offshore ships and artillery, as well as attack helicopters and aircraft. In an age when many of the traditional Marine fire-support assets have been retired or cut back, proper use and direction of what is left will be crucial to the success of the overall Marine mission.

MEU (SOC) Training and Qualifications

It is easy to see that an ARG/MEU (SOC) team can accomplish a large number of missions. Still, this number is—and has to be—limited. The MEU (SOC) concept is successful because MEU (SOC) units stick to doing what they do well! Reaching the level of proficiency required for these missions is hard on a MEU (SOC)'s personnel, and expensive for the taxpayers. Nevertheless, few people who understand a MEU (SOC)'s capabilities would ever question the costs. Especially, someone like Scott O'Grady.

Preparing a MEU (SOC) for a cruise takes time. Each cruise lasts six months, and it takes three MEU (SOC)s/ARGs to keep one forward-deployed full-time. This is why three are on each coast. To support this requirement, the MEU (SOC)/ARG teams work on a fifteen-month cycle that looks like this:

- **Refit/Basic Refresher Training (Months 1 through 3)**—If there is a period of rest for the personnel of the various MEU (SOC)/ARG components, this is it. During this time the ships can squeeze in a dockyard period to repair and upgrade equipment and systems. This is also the time when new personnel rotate in to replace outgoing ones. Meanwhile, everyone else has an opportunity to take some leave and/or spend some time with the family. Life in an expeditionary unit is tough on the people in it; any time they can get away is treasured. When they're not on leave, or with their families (such time is all too short!), they get "back-to-basics," so they can have their primary skills down pat for the coming workup period.

- **MEU (SOC) Workup/Qualification Period (Months 4 through 9)**—During this period, the various MEU (SOC) component units come together and learn to work as a team. The ships of the ARG are then added to the training exercises, so that by the end of the process, the entire force functions as a team. Meanwhile, the MEU (SOC) goes through a complete workup and qualification process that is carefully supervised by specialists from the Marine Corps Special Operations Training Group (SOTG).

- **Deployment (Months 10 through 15)**—The whole process pays off in the deployment phase—when the MEU (SOC)/ARG teams are out on cruise. Because there are seven MEU (SOC)s in service (the 11th, 13th, and 15th on the West Coast, the 22nd, 24th, and 26th on the East Coast, and the 31st at Okinawa), two or three will be at sea at any given time. For the National Command Authorities, this translates to having one MEU (SOC) always in the Mediterranean, one always in the Western Pacific, and one occasionally in the Persian Gulf.

The key to making this all happen is the MEU (SOC) Workup/Qualification Period. For the MEU (SOC), this is the functional equivalent of a National Training Center (NTC) or Red/Green Flag rotation for an Army or Air Force unit—with the added dimension that it lasts six months! That is a long time to prepare for a cruise that will last half a year, and it takes a toll on the personnel and equipment. Still, the old saying goes, "The more you sweat in training, the less you bleed in war." It is true. The training and examination during this period are incredibly intense. Training and evaluation periods are round-the-clock, and it is rare for the personnel of the MEU (SOC) and ARG to get more than four to six hours of sleep a night. In fact, most Marines I spoke with would claim that the workup/qualification period is actually tougher than actual combat operations!

The actual standards and syllabus for this process are spelled out in a document called Marine Corps Order 3502, which was issued in 1995. It lays out the step-by-step procedure for taking a BLT, an HMM, an MSSG, and other Marine units, and turning them into a fully qualified MEU (SOC). At the end of the process, there is a final examination called a Special Operations Capability Exercise or SOCEX. To gain the (SOC) designation for their unit, the MEU must pass every single point in the book to the satisfaction of some very tough judges—both their regular evaluators and folks from the Marine Special Operations Training Group (SOTG), the keepers of the MEU (SOC) syllabus. According to Marines and sailors who have done it before, the qualification process is six months of pure hell, with two *really* bad weeks at the end!

Initial Phase (10 Weeks)

The initial training phase is designed to pull the various Marine and Navy component units together. The process is much like a pro football mini-camp at the start of training where rookies and veterans can get to know each other. Major events in this phase include:

- **ARG/MEU (SOC) Workshop**—A "101-level" course for the various component and ships staffs.
- **Special Skills Courses**—These classes are the specialized training courses that provide the essential technical skills for the various SOC-type missions.
- **Initial At-Sea Training**—This phase represents the first at-sea merger of the various Navy and Marine components. Much of the time is dedicated to "bread-and-butter" skills like quickly and safely loading helicopters and landing craft, as well as basic amphibious and helicopter assault techniques. There are also several training exercises during a workup, depending upon ship and training range availability.
- **Fire Support Coordination Exercise**—Since one of the most important and difficult skills required for the full range of MEU (SOC) missions is calling in supporting fire from ships, artillery, and aircraft, there is a special live-fire exercise run to hone these skills.

All of these activities are designed to provide the ARG and MEU (SOC) personnel with a foundation of skills and experience for the Intermediate Phase of training. Much like learning how to walk before you run, the Initial Phase gives you the confidence to do simple things, so that more difficult tasks are possible.

Intermediate Phase (8 Weeks)

The Intermediate Phase turns the basic MEU into a truly dangerous weapon, the MEU (SOC). The emphasis is on taking the teamwork and synergy developed in the Initial Phase and combining these with the missions spelled out in Marine Corps Order 3120.9. It is a tough period, over three months in length. The Marines and sailors of the various components and ships spend most of their time in the field or at sea. By the time the phase is over, the personnel of the ARG/MEU (SOC) team will be a fully functional—and keenly sharp—combat unit. The following events make up the Intermediate Phase:

- **Maritime Special Purpose Force (MSPF) Interoperability Training—** This training is for fifty or so members of the MEU (SOC)'s Maritime Special Purpose Force (MPSF—a specially formed team of Marine Force Reconnaissance personnel trained in the more extreme forms of special operations). These tasks include small boat and underwater insertions, demolitions, mountain warfare, and close-quarters fighting.

- **TRUE Training/Exercise—**TRUE training provides the opportunity for tactical training in unfamiliar urban environments. To enhance its realism and effectiveness, actual cities around the U.S. (such as San Francisco, New Orleans, etc.) are used to provide the Marines with a real-world environment to practice this difficult set of mission skills.

- **Marine Expeditionary Unit Exercise (MEUEX)—**The MEUEX is the first real opportunity for the ARG and MEU (SOC) commanders to evaluate how well their units are performing. With the assistance of the SOTG, they run the sailors and Marines through a week of continuous operations, one mission after another. It is the functional equivalent of running an O'Grady-type TRAP mission and an amphibious raid on alternate days.

- **Gas/Oil Platform and Maritime Interdiction Operational Training—**In former years only the West Coast MEU (SOC)s were trained for these maritime missions. Now all MEU (SOC) units are qualified to deal with them.

- **Long Range Night Raid Training—**Long a Marine specialty, raids against enemy targets are a valuable tool for national decision makers. This particular part of Intermediate Phase training emphasizes night raids with long transits to and from the target.

- **Intermediate At-Sea Training Phase—**There are one or more at-sea training periods, which are used to practice various missions. The exact mix of missions and training is at the discretion of the ARG/MEU (SOC) staff, and is limited mainly by ship and training range availability.

By the time the Intermediate Phase is completed, the ARG/MEU (SOC) team is almost ready for their final examination, the SOCEX. Before that begins, they are given a short break to catch their breath, do required maintenance, and fix any problems they can.

Final Phase (8 Weeks)

The Final Phase is long and brutal. During the two weeks of the SOCEX, the sailors and Marines of the MEU (SOC) must prove to the observers of the SOTG that they are qualified for certification as special operations capable.

- **Pre-Embarkation Maintenance Stand-Down**—A short, *realistic* stand-down for the sailors and Marines. The idea is to get their equipment loaded and ready as if they were leaving on an actual deployment.

- **ARG Advanced Amphibious Training**—Just prior to the SOCEX, key members of the ARG staff and ships' crews are given final training to assist with advanced amphibious warfare techniques. The training targets communications, navigation, fire-support operations, and many of the other procedures that make 'gator warfare so risky and dangerous.

- **FLEETEX and Special Operations Capable Exercise (SOCEX)**—The FLEETEX/SOCEX is the final examination and certification for the MEU (SOC) and ARG. Run over a period of days, the exercises consist of a series of no-notice missions, all of which require use of rapid planning and briefing techniques. Each of these missions must be executed within six hours of the reception of the alert order, with only minimum safety margins for weather and other conditions as an excuse for delay. Sometimes the missions are allowed to go to full execution. Other times, the MEU (SOC) is ordered to hold the mission at the start-up point for a period of hours or days, while other missions are run. After successfully completing the SOCEX, the MEU (SOC) and ARG are fully certified as capable of heading out on deployment, and are only a matter of weeks from being sent to a potential combat zone.

- **Pre-Overseas Movement**—Soon after the completion of the SOCEX, all the unit's vehicles, equipment, and personnel are staged, either to their home base (such as Camp Lejeune, North Carolina, or Camp Pendleton, California) of the MEU (SOC), or the port of embarkation (such as Norfolk, Virginia, or San Diego, California) onto the ships of the ARG.

- **Crisis Interaction Requirements Exercise**—One of the last bits of preparation for the leadership of the force is a tabletop war game. Keyed to fast-breaking situations, the exercise is designed to hone the crisis-response and management skills of the various leaders within the MEU (SOC), the ARG, and JSOC.

- **Area Commanders Brief**—The final act before deployment is a series of area briefings for the ARG and MEU (SOC) commanders. Usually these are run by the various agencies (Department of State, Joint Staff, Headquarters Marine Corps, CIA, DIA, NSA, NRO, etc.) in the Washington, D.C., area. Conducted just days before the ARG/MEU (SOC) deploys, the briefings are designed to give the unit's leadership an up-to-the-minute view of the area that they are headed in to.

A pair of HMM-264 helicopters prepare to take off from the after helicopter spots of USS *Wasp* (LHD-1).

JOHN D. GRESHAM

Getting Ready: The Summer of '95

During the summer of 1995, I made a series of visits to the 26th as well as PHI-BRON 4, and had the opportunity to watch the team get ready. It was an exciting set of experiences.

Onslow Bay, Off Camp Lejeune, North Carolina, June 16th, 1995

My first visit to the 26th MEU (SOC) and PHIBRON 4 came during the Intermediate Phase of their workup process. After a short early morning flight down from Andrews AFB, Maryland, to MCAS New River, North Carolina, I boarded a big CH-53E Super Stallion for the ride out to the USS *Wasp* (LHD-1). Donning a "Mickey Mouse" helmet with ear protection and a life preserver, I sat down on the web seating, and hung on. The weather, while warm and humid, was decidedly raw, with a stiff breeze coming in off the bay. All this was from the tail end of another summer storm, one of many the East Coast had recently endured. On the way out, the helicopter passed over the *Whidbey Island* and *Shreveport*, which were standing just a few thousand yards offshore. The ride took only about twenty minutes, and as the helicopter circled into the landing pattern I got my first look at *Wasp*. It's *big!* *Wasp* is to ordinary ships as Australia is to ordinary islands. Moments later, with a swing onto the landing spot, the helicopter thumped down, and I quickly exited. At the direction of one of the deck handlers, I moved to the starboard side of the flight deck, and entered a hatch on the port side of the huge island structure.

After removing my helmet and life preserver, I was greeted by Gunnery Sergeant Tim Schearer, the MEU (SOC) PAO, and Major Dennis Arinello, the S-4 (logistics) officer. Moving to a VIP arrival area, I was hit by a wave of cold air from the ship's incredible air-conditioning/Collective Protection System (CPS). After a round of introductions and quick admonitions about what not to touch, I was guided down into the 02 Level to one of the small wardrooms near the officers mess area. After an excellent dinner of shrimp stir-fried (*Wasp*'s mess specialists are quite good), I was taken to the Flag Briefing and Planning Room, which is adjacent to the

Landing Force Operations Center (LFOC), for a briefing by Colonel Battaglini and Captain Buchanan. There I was also introduced to Captain Raymond Duffy, the commanding officer of *Wasp*.

Ray Duffy is a jolly-looking surface warfare officer, who has spent most of his career in destroyers and amphibious ships. He is especially proud of his current ship, and rightly so, since the *Wasp*-class vessels are the largest surface combatants in the U.S. fleet. Backing him up was *Wasp*'s Executive Officer (XO) Captain Stan Greenawalt, a naval flight officer who previously commanded a squadron of S-3 Viking ASW aircraft down in Florida. Stan was the gent who watched over the ship for Captain Duffy, and had all the "heavy" jobs where the ship's personnel were concerned. A man of medium build, he kept his office and stateroom on the starboard side of the 02 Level always open, with coffee and wit flowing freely. Together they provided the wide range of skills necessary to run a ship as complex and versatile as the *Wasp*.

The briefing covered information on the MEU (SOC) and ARG and explained the ins and outs of the MEUEX I was about to observe. The exercise had already been going on for several days; I was to observe one of the simulated missions, a modified NEO of a small combat unit that had gotten itself on the wrong side of a peacekeeping "green line." They were encircled and *very* anxious. The MEU (SOC) was tasked to get them out. By 2000 hours (8:00 P.M.), the briefing had broken up, and I had a chance to look around the hangar. When I emerged onto the hangar deck, it was bathed in the sickly yellow sodium-vapor lighting used to preserve night vision. Tonight, most of the 26th Aviation Combat Element or ACE was up on the "roof," so that the majority of the hangar bay could be dedicated to laying out the equipment and weapons for the units involved with the mission in the morning. Along with the NEO team, other units of the MEU (SOC) were prepping their gear around the deck of the bay. One of these, a TRAP team, is kept on standby anytime that the ACE has aircraft in the air. This was only a small team, but provisions had been made to have larger Sparrowhawk (platoon-sized) and Bald Eagle (company-sized) units available, should they be required.

As I walked around, I was introduced to Lieutenant Colonel John Allen, the CO of the 26th's Ground Combat Element (GCE), BLT 2/6. John Allen stands in vivid contrast to Colonel Battaglini. While the 26th's CO is tall and lean, with a hard, intense gaze, Allen is shorter and more muscular, with a sunny, humorous nature that belies the concentration going on inside his head. He is *always* alert. If you watch his eyes, they are always moving, always taking note of details. With a friendly smile, he quietly suggested that I show up for the confirmation briefing that was to be held in the officers' mess at 2200 hours (10:00 P.M.), if I wanted to know more about what was to come in the morning.

I headed back up to the 02 Level, and I found a spot to sit on the port side of the mess area, while probably a hundred officers and NCOs came in and sat down. Most of them carried thermal mugs emblazoned with "USS *WASP* (LHD-1)" or "BLT 2/6, 26th MEU (SOC)." Moving over to the nearby drink area, they refilled the mugs—coffee for those on the late or mid-watches, fruit-flavored "bug juice" for those who might still have delusions of sleep on this late evening. Many carried notebooks, and some had briefing slides which were clearly intended for use with the

A pre-operations confirmation briefing in the officers wardroom of the USS *Wasp* (LHD-1). These meetings are highlighted by extreme brevity by the briefers to keep things short.

JOHN D. GRESHAM

overhead projector placed at the front of the mess. Along with that there was a white board, an easel with drawing pad, and a large projection screen.

Promptly at 2200, Colonel Battaglini, Captain Buchanan, Captain Duffy, and the various COs of the MEU (SOC) components marched in, and the briefing began. This was my first experience with the rapid-response briefing format; it was enlightening. Colonel Battaglini quickly laid out tomorrow morning's mission, then turned the floor over to a succession of fast-talking briefers. In less than an hour, the following topics were covered:

- **Weather**—Reported the air, sea, and ground meteorological conditions expected in the morning. At the moment a storm was raging topside. In spite of that, the prediction was for clear skies in the morning.
- **Operations**—An overview of the planned NEO, with a discussion of the forces to be employed and their planned movement to the objective area—the "Combat Town" facility at Camp Lejeune, used for urban warfare training.
- **Evacuation and Recovery Plan**—A short briefing to explain how the NEO forces would be recovered in the event of a failure. Contrary to Hollywood stereotypes, Marines are not "do-or-die," suicidal maniacs. On the contrary, they are for the most part highly professional, calm, and thoughtful. They always have a "Plan B," or even a "Plan C"!
- **Amphibious Operations (PHIBOPS)**—Run by the PHIBRON 4 operations staff, this briefing went over the main points of the landing craft missions that would support the planned NEO in the morning. This included landing of an armored task force and an evacuation force of 5-ton trucks and HMMWVs to remove the trapped unit from Combat Town. The evacuation force would be landed by LCACs from the *Wasp*, while the armored force of AAVs had already been brought ashore from the *Whidbey Island*.
- **Rules of Engagement (ROE)**—The ROE briefing explained the rules under which deadly force could be used. The MEU (SOC)'s normal policy on such matters is to educate all the Marines in the force on

appropriate application of the ROE to ensure that use of force is commensurate with the particular situation and the overall safety of the force. Because the NEO was part of a peacekeeping operation, the ROE for this mission restricted the MEU (SOC) to firing only if they were fired upon.

- **Mission CO's Briefing**—This briefing was conducted by Lieutenant Colonel Allen, who would command the forward elements of the rescue force. Colonel Battaglini would run the overall operation from his command console in the LFOC on *Wasp*.
- **Ground Security Force CO**—The ground security force, drawn from troops on *Wasp*, would be composed of a reinforced rifle company inserted by helicopter into LZs adjacent to Combat Town, and would move the evacuees via 5-ton trucks and LCACs to the *Shreveport*.
- **Task Force (TF) Mosby CO**—TF Mosby was an armored task force off the *Wasp* that had already landed by LCACs. It would provide reconnaissance and screening for the security force.
- **LHD Evacuation Plan**—Since storm and sea conditions might not allow evacuation to the *Shreveport* by landing craft, a backup or "bump" plan was developed to bring the evacuees out to the *Wasp* by helicopter. Details such as weapons safety and stowage, as well as containment of possible infectious diseases and other problems, were covered.
- **Fire Support Plan**—Since no gunfire support from the offshore ships was planned, contingency fire plans were put into place and made ready. The bulk of supporting fires for the NEO would come from the ACE's force of AH-1W Cobra attack helicopters, armed with anti-tank missiles, rockets, and 20mm cannon. The 81mm mortar platoon also would be on call if required.
- **GCE Communications Plan**—One of the most interesting parts of the Confirmation Briefing was the GCE communications plan, which was presented by designating the various radio and satellite communications channels. For example, no less than three satellite communications terminals (sharing one frequency) were to be dedicated to the morning's efforts.
- **Tactical Reconnaissance Plan**—Even before the start of the Confirmation Briefing, the MEU (SOC) had inserted reconnaissance elements into the Camp Lejeune area, which were feeding intelligence back to the Joint Intelligence Center on *Wasp*. The intelligence reports were fairly good; the force to be evacuated was staying put in Combat Town, and their opponents were behaving themselves.
- **Escort Flight Plan**—The commander of HMM-264's flight of four AH-1W Cobras laid out his plan for escorting the security force's transports into an LZs near Combat Town, and then providing security for the ground forces during the evacuation. Bump plans and the procedures for rearming and refueling the Cobras (aboard *Shreveport* if required) were laid out.

- **Air Boss/Department Plan**—The head of *Wasp*'s Air Department, Commander Frank Verhofstadt (also known as the "Air Boss"), laid out the air department plan for the following day. This included the side numbers of the primary and bump aircraft, as well as the spotting plan for the flight deck during various phases during the day.

- **Logistics Plan**—The MEU (SOC) S-4, Major Arinello, quickly briefed logistical support for the armored force already on the beach, as well as describing the loads of rations, ammunition, water, and other supplies the individual Marines, or "PAX" as they are called, would be carrying.

- **Air Mission Commander (AMC) Plan**—The AMC laid out the air plan, showing which units would be loaded aboard which aircraft, and then how the flights to and from the beach would go in in the morning. In particular, safety and divert plans were covered.

- **MEU (SOC) S-6 (Communications) Plan**—The detailed communications plan for the ARG and MEU (SOC) was laid out and checked for compatibility with the plan of the GCE.

- **TRAP Commander Plan**—Though no TRAP mission was yet required or anticipated, a platoon-sized TRAP team with two CH-46E Sea Knights would stand by just in case. Also, the contingency plans for the Sparrowhawk and Bald Eagle units were quickly covered.

- **MEU (SOC) Medical Plan**—The MEU (SOC) medical officer laid out his plans for handling the members of the evacuee unit, as well as any Marine/Navy casualties that might occur as a result of the NEO mission.

- **ARG/LHD Surgeon Plan**—The head of *Wasp*'s medical department ran down the status of his facilities, including available bed space, as well as the condition of the various operating theaters. As expected, all were ready and primed, with only minor bed cases currently residing aboard.

The entire briefing was finished in less than forty-five minutes, with the briefers only speaking if there had been a change from standing procedures. The speakers each spent an average of less than ninety seconds over their viewgraph slides. Finally, Colonel Battaglini and Captain Buchanan stood up to re-emphasize that this was an exercise and that safety was paramount. H-Hour for the security force to hit their LZs was set for 0900 the following morning, and then the briefing broke up.

At 2300 hours, Lights Out was sounded over the 1MC system, and *Wasp* took on a nocturnal air. I decided to join some of the MEU (SOC) and ARG staff for "mid-rats." You can tell a lot about a ship from the kind of mid-rats that they serve, and *Wasp* is pretty good. Some nights it's leftovers from dinner; other nights it's cold cuts and chips. But on nights when something special is in the wind, Captain Greenawalt usually orders up something special like "sliders" (*really* good cheeseburgers) and fries.

Following a short discussion over the mid-rats, I adjourned to a bunkroom for a few hours sleep. The motion of a ship at sea is quite soothing, and since flight operations were fairly light this evening, there was a minimum of noise from the flight deck just a few feet/meters above my head. Despite the killing heat outside, the temperature in the CPS citadel was almost too chilly. In the background were the noises that you hear aboard a warship at sea—the occasional low announcements from the

1MC system, the hum of generators and the air-conditioning/CPS system, and personnel walking through the passageways.

Aboard USS *Wasp* (LHD-1) in Onslow Bay, 0600 Hours, June 14th, 1995

Reveille came over the 1MC at 0600 (6:00 A.M.) that morning. Within seconds, there was a bustle of activity in the passageways. Having traveled light for this visit, I only had to freshen up and head to breakfast in the officers' mess to start the day. By 0800 hours, *Wasp* had gone to Flight Quarters, and had the stern gate down to launch the LCACs that had been loaded in the wee hours of the morning. Since we were only about 15 nm/27.4 km off Onslow Beach, the LCACs did not have to launch until about 0830. As they backed out of the well deck, flight operations on the deck above were temporarily suspended, so that the jet wash from the LCACs would not interfere with the helicopters taking off and landing just a few yards/meters above. Once clear of the ship, the three LCACs formed up and headed for Onslow Beach and the armored unit that was already there to escort them into Combat Town. I was slated to go in shortly with the helicopters.

The day was heating up rapidly, but the deck crews were kind enough to let me stroll about the flight deck for a few minutes before the helicopters started engines. Then, it was time to board the choppers for the ride to the LZ near Combat Town. Donning my helmet and life preserver, I strapped in and we lifted off. It was a beautiful summer day, and as we passed over *Whidbey Island* and *Shreveport*, I could see the LCACs making their runs into the beach. Then we were over the beach and the sand pines of the North Carolina coast. The chopper I was in, which was running about fifteen minutes ahead of the security force helos, landed in a quiet clearing. There I was greeted by PAOs from Camp Lejeune and driven in a van into Combat Town.

Combat Town, Camp Lejeune, North Carolina, 0900 Hours, June 14th, 1995

As I arrived in Combat Town, the PAOs warned me to stay in a nearby grove of pines, and to quietly observe what was happening. Precisely at 0900, I heard the distinctive sound of twin-rotor CH-46Es landing in a LZ about a thousand yards/meters away. Within a few minutes, the first scouts for the security force were moving forward to find the evacuee unit. The security force unit was built around "G" or Golf Company of BLT 2/6, which is commanded by Captain Andrew "Andy" Kennedy. As the security force formed a perimeter around Combat Town, Captain Kennedy made contact with the members of the evacuee unit, then made arrangements to get them on the trucks and back to the beach and safety. Around the perimeter, an opposing force (OPFOR), played by Marines from the 2nd Marine Division, harassed the security force. They even fired an occasional blank round to keep things interesting. Meanwhile, the security force commander had finished his coordination with the commander of the evacuee unit (also played by 2nd Marine Division personnel) on the procedures for the move. By this time, Lieutenant Colonel Allen had arrived with his headquarters detachment and set up a satellite

The command group of BLT 2/6 led by Lieutenant Colonel Allen (kneeling, second from left) confers with simulated evacuees in the "Combat Town" complex at Camp Lejeune, N.C.
JOHN D. GRESHAM

A pair of HMM-264 CH-46E Sea Knights land in a meadow at Camp Lejeune, N.C., to pick up simulated casualties during a 1995 exercise.
JOHN D. GRESHAM

communications relay back to Colonel Battaglini in the LFOC on *Wasp*. Overhead, a pair of Cobra attack helicopters prowled and watched. Soon the trucks and HMMWVs arrived, loaded up, and began to move out to the beach where the LCACs would take them back to the *Shreveport*. So far, everything had gone according to plan, and seemed to be proceeding well.

Then, a message came through that something was wrong in a nearby meadow. Quickly jumping into a van, the PAOs and I headed there to have a look at what was happening. The folks from the SOTG like things lively in training exercises. So they always throw in a few surprises—to imitate Clausewitz's "friction." What they had done was to orchestrate some "friction" in the meadow for the Marines of BLT 2/6. When I arrived, I saw that SOTG had arranged for one of the 5-ton trucks of the evacuation force to "suffer" an accident. The passengers were spread around the ground nearby, with prosthesis and makeup to make them appear severely injured. With the SOTG judges watching, things began to happen.

Within a few minutes, the first Marines from the security unit arrived. They instantly called into Lieutenant Colonel Allen that an emergency situation had developed, that personnel were "down" and suffering severe trauma, and that personnel from the MEU (SOC) medical team were needed at the site immediately. Since medical evacuation to the *Wasp* would be required, with at least three CH-46s needed to carry the load, John Allen quickly relayed the request to Colonel

Battaglini in the LFOC, and the helicopters were airborne within minutes. In the meantime, the Marines of Golf Battery (the unit of M198 155mm howitzers, already ashore from a previous mission) formed a security perimeter and applied first aid to the accident victims. A few minutes later Navy corpsmen arrived in a HMMWV, and things began to look up for the "injured."

Less than half hour after the first call from Lieutenant Colonel Allen, the three Sea Knights arrived in the meadow, escorted by a pair of AH-1Ws. As the Cobras moved to an overwatch position, the three transport choppers landed and made ready to take aboard their cargo of injured evacuees. That was when when a SOTG observer strode out to one of the CH-46Es and declared it "down" with a mechanical failure. Lieutenant Colonel Allen made another call back to *Wasp* LFOC, this time to ask for the TRAP team on Alert status, as well as a spare CH-46E to finish the evacuation of the injured.

While the new flight of Sea Knights flew in from the sea, the injured were triaged according to the severity of their injuries. The worst cases were loaded onto the two "good" CH-46s, and the choppers lifted off to take them to *Wasp*'s trauma center. Meanwhile, the Cobras continued to patrol over the meadow and keep the occasional OPFOR patrol away from the meadow HLZ. When the two new helicopters arrive the TRAP team and their mechanics leapt out, surrounded the "downed" bird, and got to work. Within half an hour, the "problem" would be "fixed" to the satisfaction of the SOTG observers, and the wounded bird would be allowed to return to the *Wasp*.

While this was going on, the remaining "injured" personnel were loaded onto the replacement medical evacuation chopper and flown out of the LZ. With the last of the SOTG-inflicted "friction" dealt with, Lieutenant Colonel Allen began to pull his forces together and withdraw them back to the safety of the sea and the ARG. After the trucks and HMMWVs were loaded onto the landing craft, together with the armored vehicles of the security force, all that was left was to get Golf Company aboard their helicopters and back to the *Wasp*. With the ever-present Cobras overhead, Captain Kennedy and his men returned to their LZ, boarded the helicopters, and headed home. One of the last units out was Lieutenant Colonel Allen and his command team, careful to make sure that nobody was left behind. It had been a good day.

The SOCEX final examination came the following month.

Final Exam: The SOCEX

The second week of July 1995 was terribly hot and muggy. It was the kind of heat designed to break men—and even Marines. Hot or not, for the members of the 26th MEU (SOC) and PHIBRON 4, it was final examination time. I didn't have a chance to observe this exercise, but word filtered back that they had performed "superbly" and were now ready to head out to relieve Marty Berndt and his 24th MEU (SOC) from their vigil in the Adriatic. But before that, they had one more hurdle to clear—an exercise tacked onto the end of their SOCEX called Joint Task Force Exercise 1995 (JTFEX-95).

Extra Credit: JTFEX-95

JTFEX-95 is a series of joint service exercises designed to try out operational warfighting concepts in contingency and expeditionary situations. The JTFEX-series was initiated in the fall of 1994, and the 26th MEU (SOC) and PHIBRON 4 were to be some of the key players in this edition. Unlike the NEO I observed in June, the 26th would not be working alone; it would be part of a larger joint service force, simulating an operation that could easily take place at the start to the kind of military deployment that we ran in the Persian Gulf in 1990.

JTFEX-95: The Scenario

The JTFEX-series exercises are run through the U.S. Atlantic Command (USACOM), which is headquartered in Norfolk, Virginia: Component units from each of the services are assembled in a joint task force (JTF), which is commanded by the 2nd/Atlantic Fleet battle staff aboard their command ship *Mount Whitney* (LCC-20). Component commanders are drawn from around USACOM to provide community leadership, and then a mission is assigned. For our JTFEX, the Navy would supply the *America* CVBG and PHIBRON 4, the Marines contributed the 26th MEU (SOC), the Army donated the 1st Battalion of the 325th Airborne Infantry Regiment (1/325th) from the 82nd Airborne Division at Fort Bragg, North Carolina, and the Air Force kicked in units from a variety of different bases, including F-15s from the 1st Fighter Wing at Langley AFB, Virginia; F-16s, A-10s, and C-130s from the 23rd Wing at Pope AFB, North Carolina, and even a couple of cells of B-1B bombers from Ellsworth AFB, South Dakota. This force would play out a hypothetical war game with an opposing (Red) force. They would have a specified period of time to achieve their objectives.

The scenario to be played out involved an invasion of an imaginary small country ("Kartuna") by a larger, more powerful neighbor ("Koronan"). In many details, it resembled the invasion of Kuwait....but with several additional challenges for the U.S. (Blue) forces—now known as Joint Task Force Eleven (JTF-11). For one thing, other than local land-based air support, there were no nearby bases for the Blue force to use. All the ground forces involved would either come from the sea or be flown in during the airdrop of the 1/325th. Next, the Red ("Koronan") forces were going to be anything but the automatons that the Iraqis had been during Desert Storm.

The Red forces were drawn from Marine, Air Force, and Navy units along the coast of the Southeastern United States, and they intended to fight like hell to keep the Blue forces out at sea. The OPFOR included a Marine regimental headquarters, a BLT (a sister unit of Lieutenant Colonel Allen's) heavily reinforced with additional armor, several squadrons of Marine F-18s out of MCAS Beaufort, South Carolina (simulating Mirage F-1 fighter bombers equipped with AM-39 Exocet anti-ship missiles), several squadrons of helicopters (acting the part of Super Pumas loaded with Exocets), and an assortment of small frigates, submarines, and patrol craft from the naval base at Norfolk, Virginia. JTF-11's object was to liberate the Kartunan homeland and destroy the ability of the Koronans to threaten their neighbors.

The area for this matchup was a region bounded by the Camp Lejeune reservation and some other parts of coastal North Carolina. This was both good and bad for the Blue forces. On the one hand, it meant that everyone on both sides knew the ins and outs of the planned battlespace well. On the other hand, it was an extremely small place to fight a war; there weren't many maneuver possibilities for the 26th MEU (SOC) and the 1/325th. Also, the Koronan forces knew they were coming, and would consequently be alert. The exercise would start on July 18th, 1995, and run some four days.

PHIBRON 4, off the Virginia Capes, Tuesday, July 18th, 1995

The day started for me on the steaming ramp at NAS Norfolk, Virginia, boarding an HC-6 UH-46D for the ride out to PHIBRON 4 and the *Wasp*. As I rode out over the Virginia Capes, I talked with some personnel from the 26th MEU (SOC) command group who gave me some background on the coming exercise and the challenges the unit was facing. The 26th had only just finished up their SOCEX a couple of days earlier, and their biggest challenge was that they were jumping into JTFEX-95 before they could take a breather. Because of the round-the-clock planning schedule, the command group was showing fatigue from almost two weeks of continuous operation. In addition, there'd been almost no opportunity following the SOCEX to pull maintenance on equipment, vehicles, and aircraft. Maintenance crews were working frantically to make their machines ready. The operations were to start that evening.

As we entered the landing pattern of the *Wasp*, she was already steaming south for the waters of Onslow Bay, with *Whidbey Island* and *Shreveport* in a tactical (triangle) formation. The force was doing over 20 kt/36.6 kph with a bone in their teeth. A few miles ahead, JTF-11 had already started the air campaign against the Koronan forces, with strikes by CVW-1 off the *America* and various Air Force units against air and naval targets, including some "SCUD" sites in the Koronan "homeland." The air units would have to work smart and fast, for the invasion of the Kartunan homeland was scheduled for the morning of the 21st.

After the helicopter thumped down on the deck, I was met by the friendly faces of Major Arinello and Gunnery Sergeant Shearer and escorted to my stateroom on the 02 Level. As I stowed my gear, they explained that I would have full run of the ship, and would be able to go almost anywhere, and do almost anything I might desire. I intended to make the most of the opportunity. After a break for lunch, the first major event was the confirmation briefing for the 26th's initial mission of the JTFEX—insertion of their reconnaissance and surveillance (R&S) elements into Camp Lejeune. The 26th needed to develop an intelligence picture of what the Koronan ground forces were up to.

Compared with my earlier experience, this briefing was leisurely; it ran over a period of about two hours. Here is a short version: Using three CH-53E Super Stallions from HMM-264, the MEU (SOC) was going to covertly insert fifty-two PAX in ten different teams around the Camp Lejeune reservation that evening. A couple of problems were foreseen: For one, the weather was looking marginal. Tropical Storm Chantal had been beating the hell out of the Atlantic, and was still a threat to our north. Chantal was forcing a cold front down on top of our planned launch position that evening, and weather conditions might get dicey as a result.

Lieutenant Colonel John Allen (third from left) and the BLT 2/6 plan operations during JTFEX-95 in August of 1995.

JOHN D. GRESHAM

There was also the matter of the Red (Koronan) forces. The Koronan ground component was composed of a BLT from the 6th Marines, heavily reinforced with armor and artillery. Though the OPFOR had no organic helicopters, their armor overmatch was about two-to-one compared to what Lieutenant Colonel Allen and BLT 2/6 could bring to bear. In addition, the Koronan ground force was commanded by a Marine lieutenant colonel who was reputedly smart and aggressive. To counter all of that, Colonel Battaglini and Lieutenant Colonel Allen had given their personnel carte blanche to their own forces to conduct deception operations and generally screw with the minds of their opponents.

As for the R&S mission itself, the job of the various teams was to position themselves at strategic points around Camp Lejeune and pass their observations back to the JIC aboard *Wasp*. Nine of the teams would be "eyes"-observation-capable, while the tenth would include a radio-intelligence capability for intercepting enemy short-range tactical communications. It was hoped that these—together with intelligence assets from JTF-11, the *America* CVBG, and national sources—would shine some light through the "fog of war" that always obscures force-on-force engagements. Some of these other intelligence assets included the ships' SSES spaces, the PHIBRON's Pioneer UAVs from the *Shreveport*, TARPS imagery from the VF-102 F-14 Tomcats, ES-3 Shadow ELINT/SIGINT aircraft, as well as several new systems that were being tested in this exercise.

For the R&S units, their ROEs were simple: If possible, *don't* engage enemy forces. They were clandestine reconnaissance teams; their job was to avoid detection by the Red security forces. They could use force only in self-defense. This meant they would be allowed to lay simulated Claymore mines, but they could not use incendiary weapons. According to the insertion plan, the ten teams would board the three CH-53s at 2200 hours, and lift off at 2215. The flight would take over seventy minutes (we were still several hundred miles away from Onslow Bay). The choppers would fly in formation at low level, and would use every deceptive trick available to keep the locations of the teams secret from the Red forces. Should an evacuation be required, a TRAP team would be in continuous standby, and ready to pick up any team from any LZ that they could access.

Marines of the 26th MEU (SOC)s reconnaissance teams embark on an HMM-264 CH-53E Super Stallion on the afternoon of July 18th, 1995. This mission was in support of JTFEX-95, which was running at the time.

JOHN D. GRESHAM

After the briefing broke up, I wandered up topside to get some air. While the flight deck is usually restricted, there is a wide catwalk along the starboard side of the island where the rules are relaxed; it is a favorite among the crew. This is wonderful place to sit and watch the sea. So I found a folding chair and sat awhile. Alongside *Wasp* was a fleet oiler, which was shooting messenger lines across the space between the ships to set up for a refueling. At the same time, UH-46Ds were shuttling back and forth from the oiler to the ships of the ARG, lifting and delivering pallets of food and aircraft parts and whatever else the ships needed. All this seemed so bizarre it was almost unnatural—like watching hippos dancing. Bizarre or not, the ability to refuel and resupply at sea sets a great power apart from those nations with only coastal defense forces. These operations went on for over an hour, and only the coming darkness and dinner caused me to break away and head back inside.

USS *Wasp*, 200 nm/366 km Northwest of Camp Lejeune, North Carolina, 2100 Hours, July 18th, 1995

At 2100 hours, I joined Lieutenant Colonel Allen on the hangar deck to talk with the members of the various R&S teams that were preparing to head upstairs and board their helicopters. For this mission HMM-264 to made ready all four of their CH-53E Super Stallions, so there would be a bump aircraft in case one went "down." As I walked around the hangar bay, the teams were checking their weapons and other equipment, particularly their communications gear. This included a number of satellite-radio and HF sets, which were designed to provide secure communications back to the *Wasp*. Every team had at least one GPS receiver. Some had Trimble PLGR units, and the rest had the newer handheld Rockwell SLRGs.

At 2145 hours, the 1MC announced Flight Quarters, and things began to pick up. I walked up the ramp from the hangar deck to the island and waited there with the fifty-two R&S team members, quietly sweating in the subdued lighting. As I waited for the engines to start, Colonel Battaglini moved quietly up the ramp, talking softly to his Marines, encouraging them to keep tough and focused on what was clearly going to be a long and hot four day mission in the bush. The order to start engines came at

2200, the teams loaded up, and then the Super Stallions held, awaiting final clearance from the Air Boss. As I watched from the island, I could see the blue flares of static electricity flying off the rotor blades of the CH-53s, looking like something out of a science-fiction movie. Then, at 2215, the three choppers lifted off, immediately dousing their normal red and green navigation lights (they have infrared and subdued green ones for clandestine operations), formed a stepped formation, and headed southwest for Camp Lejeune. As quiet returned to *Wasp*'s flight deck, I headed down to the officers' mess for mid-rats and what I expected to be a fairly early evening.

That all changed just eight minutes later at 2223 hours. Flying low over the water to avoid radar detection by the air traffic control radars at MCAS New River, the three Super Stallions ran smack into a gift from Tropical Storm Chantal. The cold front had moved in over the warm water of the Gulf Stream, and a thick bank of fog had come up with no warning. Suddenly flying blind on night-vision goggles is a very dangerous situation, and peacetime rules require a quick, prescribed response. The three crews went into a pre-planned separation maneuver, formed up again north of the fog bank, and immediately aborted the insertion mission. All of this was done without radio transmissions, to avoid revealing to the Red forces radio-interception units that anything untoward had happened. Less than a half hour after liftoff, the whole force was back aboard the *Wasp*, pleased that they had safely handled the emergency, but angry that the MEU (SOC)'s entire intelligence-collection plan had just gone into the scrap heap.

Meanwhile, the normally calm demeanor of Battaglini and Allen was showing some cracks. I quickly followed them down to the LFOC, where they sat their staffs down and began to make plans to rebuild as much of the R&S plan as they could. Some damage, they knew, could not be undone: In addition to the situational awareness they would give up because of the absence of the R&S team, they would lose a full day of supporting fires from air strikes and offshore destroyers. At 0200, while everyone was still tensely trying to make the best of a tangled situation, I excused myself back to my cabin to get some sleep. "Friction" had again struck a MEU (SOC) mission. Things were going to get very interesting in the three days left in the JTFEX-95 exercise.

USS *Wasp*, Wednesday, July 19th, 1995

By reveille at 0600, the folks in the LFOC had come up with a plan to restart the stillborn R&S effort of the night before. Overnight, they had put together an unconventional insertion plan based on the fact that Camp Lejeune was their home base and they knew how it worked. All told, there are over thirty thousand Marines based there, which means that men moving across the base in full kit is as common as the sun coming up. It also turned out that the 26th had left a "stay-behind" counterintelligence team ashore after the completion of the SOCEX, and this was to be used to support the new insertion plan. So, after a few calls on a cellular phone, arrangements were made to re-run the CH-53E insertion mission of the night before that afternoon. In addition, the ashore team was to conduct a covert observation of the Red headquarters and go through their office trash, looking for documents related to the coming operation. These would be FAXed out to the *Wasp* via a secure link.

Finally, the Pioneer UAV detachment on the *Shreveport* was ordered to run as many sorties as possible to obtain naval gunfire targets for the USS *Scott* (DDG-995).

By 1800, these measures had been put into effect, and dinner was attended by an extremely tired pool of officers in the wardroom. By now, the BLT and MEU (SOC) staffs had been up for almost thirty-six hours straight, and they still had one more big event to go before this evening was finished—the dress rehearsal for the operational confirmation briefing that would be held the following morning. This briefing would provide a detailed look at the Friday morning assault on the Kartunan homeland. Held at 2000 hours, the briefing went over every detail of the planned "invasion." And it was a complete bust...mostly because the tired young officers hadn't been able to put the necessary time and coordination into their briefing slides. When they were done, Jim Battaglini, a man of few words, stood up and made his displeasure clear. "Get it right for tomorrow," he commanded. At the morning briefing, the ground and amphibious forces component commanders from JTF-11 were scheduled to fly over from the *Mount Whitney* to review the invasion plan, and he wanted it right. After suggesting that the young officers work out their problems over mid-rats, he left to head back to the LFOC. Then the young officers headed back to their staterooms to retrieve their notes and laptop computers.

Returning to the wardroom at 2315, they discovered that mess specialists had pulled out all the stops...in the form of hot ham-and-cheese-melt sandwiches and a small mountain of French fries. Soon you could feel the energy and morale level of the group change as they munched their way through the coordination problems that had plagued their briefings. As the group broke around 0100 to get some sack time, I wandered down to the LFOC to see how things were going. During the evening intelligence briefing I'd noticed some disturbing trends in the air campaign, and I wanted to talk to John Allen about them. I wasn't the only one to pick up on this situation. In fact, by the time I found Allen, Colonel Battaglini had already started to deal with it. He had called Allen and the ACE commander, Lieutenant Colonel "Peso" Kerrick, for a short talk, and at their invitation I joined them.

Directed by the JTF-11 staff aboard *Mount Whitney*, the air campaign against the Red forces had so far been a mixed affair. While the Koronan naval forces had been decimated, their air force had suffered less than 30% attrition in over two days of operations. Worse yet, the simulated force of Exocet-armed Mirages and Super Pumas was making a nuisance of itself, and had just scored a hypothetical hit against the nuclear-powered cruiser *South Carolina* (CGN-37). Though the missile warhead was assessed to have been a "dud," the battle group commander was extremely upset. Predictably, he was demanding better protection for his ships. What happened was the JTF-11 staff had allowed their air units to be drawn into a personal duel with the Koronan air force, and were failing to keep up with their operational objectives. For instance, several planned air strikes against ground targets had yet to be executed. And this meant that on Friday morning Lieutenant Colonel Allen's BLT might walk into a fight against a force that overmatched his in armor and artillery, and was dug in on the very objectives he was required to take. While Battaglini, Allen, and Kerrick put together a plan to deal with this situation, I headed back to my own stateroom, wondering how all these "friction" elements were going to effect what was happening in less than twenty-four hours on the North Carolina coast.

USS *Wasp*, 50 nm/91.4 km West of Onslow Beach, 0600 Hours, July 20th, 1995

At 0630 the next morning, I was sitting across from Lieutenant Colonel Allen, and he was showing a thin smile; he had gotten some sleep, and things were looking decidedly better than the previous night. For starters, his R&S teams were reporting in and finally delivering the kind of targeting data he needed to knock back some of the Red forces. In addition, the JTF-11 ground component CO, General Keane, seemed to have finally "persuaded" the JTF-11 staff to remember some of *his* mission objectives, and there had been air strikes against the planned targets ashore. There were also some excellent results from the surveillance of Red force's garbage. John Allen wasn't the only one looking on top of the world. All around the mess area, you could feel new energy. The 26th had just eighteen hours until the invasion...and there was a feeling that they might pull it off.

By 0900, the officers' mess had been reconfigured to support the mass confirmation briefing for the invasion. It was set to begin when the ground and other component commanders, who were flying over from *Mount Whitney*, made their appearance. The ground CO for JTF-11 was General John M. Keane, the commander of the Army's famed 101st Air Assault Division. "Sadly," he didn't exactly make it—another example of "friction." After the simulated Exocet attack on *South Carolina* the previous evening, the fleet AAW coordinator was convinced that he needed to provide a tighter defensive screen for the JTF-11 naval forces. So he upped the alert level and ROE of the picket ships armed with surface-to-air missiles (SAMs) to "Warning Yellow—Weapons Hold," meaning that attack by enemy air units was expected. A Blue ship detecting a confirmed unfriendly aircraft should shoot it down immediately—the equivalent of shooting first and asking questions later. When the SH-3 Sea King helicopter carrying the component commanders and their staff flew from *Mount Whitney* to *Wasp*, its electronic Identification Friend or Foe (IFF) transponder was mistakenly turned to the Off position. One of the escorting picket ships therefore shot it down with a simulated SAM. If the ship's AAW coordinator had been at a lower alert level, he would probably have taken the time to check the JTF-11 Air Tasking Order to see if the helicopter was a "friendly." But in his desire to avoid a strike by hostile forces, he screwed up seriously. The result: When the various commanders and their staffs arrived on *Wasp*, they were greeted with the news that they were "dead." They were notably unhappy by the time that they arrived in the officers' wardroom as simulated corpses.

The briefing began, and things went considerably better than they did the night before. The plan for the invasion of the Kartunan homeland was clearly laid out: At 0000 hours (midnight) that night, elements of the 26th MEU (SOC) would land inshore from Onslow Beach and along the New River inlet. The key to success was the capture of a causeway bridge and several strategic road junctions. This was to be accomplished in some innovative ways. *Whidbey Island* and *Shreveport* would be landing their armored task force in the inlet. The task force would use the AAVs as riverine gunboats to dominate this natural barrier through the middle of Camp Lejeune. Following them up the inlet would be a rifle company in the rubber raider craft. The company would take the northern part of the inlet. Another company

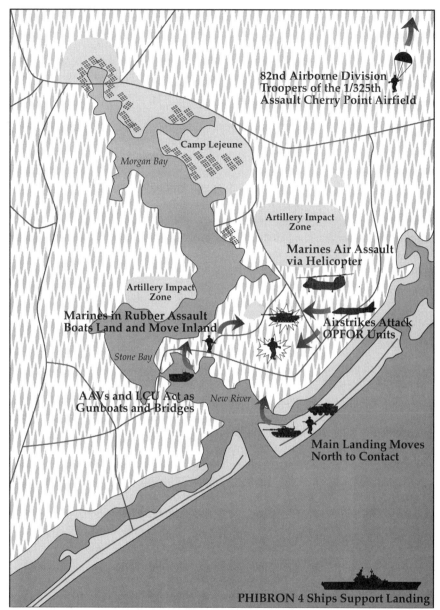

The invasion of Camp Lejeune during the August 1995 JTFEX-95 exercise.

JACK RYAN ENTERPRISES, LTD., BY LAURA ALPHER

would then be landed inland by helicopter to block the approaches to the landing beach near the point of the inlet. When this was done, the rest of the heavy equipment would be brought ashore on the LCACs.

As this operation proceeded, the 1/325th from the 82nd Airborne would parachute onto an airfield a few miles/kilometers inland, to provide a base for follow-on fly-in units. This would be followed by a helicopter insertion into a LZ to support the amphibious landings. There would also be a series of deception operations—such as the temporary unloading of the portable toilets into a dummy LZ—to encourage the Red forces to believe that the landing would be occurring at the easternside of the training area. With luck, the Red forces would be drawn there. By the way, don't laugh at the portable toilets scam. Even though we were in a "war," EPA and DoD regulations about waste disposal still applied.

After the briefing broke up, I headed over to the LFOC to see how the war was going. When I arrived, it was clear that the 26th was now playing "their" game with the Koronan forces. The Red force command and control capabilities were down to less than 50% effectiveness, their navy was out, and the JTF-11 staff was successfully using their airpower: The OPFOR air forces were also down under 50%. Just to make sure that enough sorties went against ground targets, Colonel Battaglini had arranged for extra air strikes by VMA-231 AV-8B Harrier IIs from MCAS Cherry Point. The Harriers had been left home for this exercise and the SOCEX to prepare for the coming deployment, but were called in now to provide the 26th with some "Marine" airpower that they could depend upon! By noon, the VMA-231 Harriers had flown their first strikes. I spent the afternoon visiting the LFOC and spending time on the portside island catwalk enjoying the peace before the storm. After dinner, I watched *Wasp*'s combat cargo crews load the LCACs and helicopters with Dennis Arinello for the first wave that would leave at 0000 hours.

Well Deck of the USS *Wasp*, 0000 Hours, July 21st, 1995

The LCACs would go in on the first wave of the invasion. These craft would carry the LAR/CAAT team to the beach as a covering force for the units that would follow. If all went well, the full combat power of the 26th would be ashore before sunrise (about 0600 hours) and then off-load the rest of the equipment ASAP after that. There was more than just a desire for combat efficiency involved by now; this landing meant that however JTFEX-95 came out, they would be home shortly. As the earsplitting whine from the LCACs faded into the distance, you could hear the 1MC announce Flight Quarters. Then the first wave of four AH-1W Cobra attack helicopters was launched, along with a UH-1N Huey with John Allen and his staff

Marines of the 26th MEU (SOC) come ashore in rubber raider craft during a 1995 exercise at Camp Lejeune, N.C.
JOHN D. GRESHAM

aboard. Things were getting busy, and would stay that way through the night. But for now, I went back to my stateroom to catch a few winks before reveille.

USS *Wasp,* 0600 Hours, July 21st, 1995

I was already awake when the 1MC blared its wake-up greeting at 0600, a sign that I was getting into the routine aboard ship. After a quick breakfast, I headed down to the LFOC one last time to see how things were going. It turned out that I was too late. By midday, the main Red force units had been engaged and defeated by BLT 2/6. The battle would continue for another twenty-four hours to completely secure the objectives, but it was a total victory for the 26th MEU (SOC). Everything was over before the sun set the next day. Despite problems with the close air support at H-Hour, and a breakdown in communications with the R&S teams, all the landings had gone as planned. Early on contact with the Koronan forces had been surprisingly light. We learned later that many of the Red force artillery pieces, tanks, and other armored vehicles had been knocked out by the last-minute air strikes and offshore destroyer gunfire. The airdrop by the 82nd Airborne's 1/325th had also gone well.

By 0800, it was time to get ready to board a CH-46 for a ride into the landing beach area. Before I left, I took a small side trip to the logistics center, where an extremely fatigued Dennis Arinello was trying to stay awake for the six hours he needed to get the ships unloaded. Wishing him good luck and a good night's sleep, I headed upstairs, and boarded the Sea Knight for the ride in. As I passed over *Whidbey Island* and *Shreveport,* I could see the LCACs and one of the LCUs moving in and out from the beach to unload the ships of the ARG. As the Bullfrog landed, I was picked up again by the Camp Lejeune PAOs. With the taking of the airfield by the airborne troopers and the seizure of the port facility and beach by the Marines, a hypothetical liberation of Kartuna would now be possible.

By noon the next day, the JTFEX-95 observers would issue a "change of mission" order, and the exercise would be concluded. While it had not been pretty, the 26th had performed superbly, adapting well to the many problems thrown at them. Best of all, the 26th MEU (SOC) and PHIBRON 4 could now concentrate on getting ready for their deployment to the Mediterranean, some five weeks away. Before I left, I made a promise to Colonel Battaglini to visit his unit while they were deployed.

Monday, August 28th, 1995, Norfolk Naval Station and Amphibious Base, Little Creek, Virginia

Less than a month after the end of JTFEX-95, I was back at Camp Lejeune, N.C. to witness the culmination of six months work for Jim Battaglini, C.C. Buchanan, John Allen, and all the other members of the 26th MEU (SOC)/PHIBRON 4 team: deployment. It was planned to occur just prior to Labor Day. Deployment is a process that sees the men and women of the ARG and the MEU (SOC) severing their ties from the land and their families, and boarding their "second home," the ships of the ARG. The day dawned rainy and decidedly ugly as the warships of the USS *America* (CV-66) carrier battle group (CVBG) and PHIBRON

4 got underway. A heavy thunderstorm was moving up from the south, and just getting under way was becoming quite a challenge. For the ships of the ARG, it involved heading out over the underwater automobile tunnels of the Chesapeake Bay, taking a hard right at the Virginia Capes, and heading south for Onslow Bay. During the transit, Captain Buchanan had the ships' crews stow everything possible away, because in the morning, they would be taking aboard the entire aircraft, vehicle, equipment, and personnel load of the 26th MEU (SOC).

Tuesday, August 29th, 1995, 0500 Hours, Marine Corps Air Station, New River, North Carolina

Deployment day for the 26th MEU (SOC) started early, even before the sun rose. In the hangar bay of HMM-264, Lieutenant Colonel Kerrick and his Marines had risen early to be the first element of the 26th to be loaded aboard the ships. Seeing that this was to be the largest ACE ever deployed by an MEU (SOC), some thought had gone into the effort, and now HMM-264 would get a chance to see if their plan would work. Around the HMM-264 hangar that morning, Marines and their families began the ritual of separation, usually over Egg McMuffins and coffee. Wives, girlfriends, parents, and children tried (sometime unsuccessfully) to hold back the tears that come with the start of a six-month cruise. It is a gut-wrenching thing to watch, and brings home the price that we ask of the sailors and Marines that serve our interests around the world.

Unlike the day before, August 29th had dawned clear and cool, a perfect summer day in North Carolina. As the first pink glow of sunrise appeared in the eastern sky over Onslow Bay, things swung into high gear. At 0545, the order was given for engine start of the first group of aircraft that would be launched. This would be a flight of three CH-46E Sea Knights that would start the cruise aboard the *Shreveport* (LPD-12), which was proceeding separately across the Atlantic. The helicopters began their taxi roll at 0613, and were airborne just five minutes later. At almost the same moment, the six AV-8B Harrier IIs of VMA-231 started launching from MCAS Cherry Point, N.C., some miles to the north. The idea was that the six Harriers would be taken aboard the *Wasp* (LHD-1) first, and lashed down to their parking spots aft of the island. Then, the other helicopters of the ACE would be brought aboard and carefully tucked into every space that could be found.

For the next hour, helicopters continued to leave MCAS New River in threes and fours, gradually emptying the ramp in front of the hangar. By 0715, quiet had returned to the HMM-264 ramp, and the crowd of ground and maintenance crews made their good-byes to their loved ones, loaded their gear onto trucks, loaded themselves into buses, and headed up to Morehead City for the boat ride out to the *Wasp* and *Shreveport*.

Tuesday, August 29th, 1995, 0800 Hours, BLT 2/6 Headquarters and Barracks Area, Camp Lejeune, North Carolina

Back at Camp Lejeune, Lieutenant Colonel Allen and his headquarters team were doing their own version of what had just occurred at MCAS New River. Down at the barracks for BLT 2/6, the various companies were sorting themselves out, and loading up. Amid crying women and children, last hugs and kisses, the Marines loaded up onto their buses, and began the trip up to Morehead City, where they would ride out to the *Wasp* and *Shreveport*. When the last of the buses was loaded, Lieutenant Colonel Allen walked over to his office one last time, and loaded up his briefcase. Wishing good luck to the remaining office staff closing up BLT 2/6 headquarters that day, he happily grabbed his bags and headed down the stairs, commander of his own battalion for one last cruise before heading up to Washington, D.C., to become General Krulak's aide in the Spring of 1996. All around Camp Lejeune, there was the bustle that comes with deployment day for a unit. Over in the headquarters of the 26th MSSG, Lieutenant Colonel Cooper had already pre-loaded much of his equipment, personnel, and supplies on the ships up in Norfolk, so this day was a little less manic for him than his GCE and ACE counterparts.

Tuesday, August 29th, 1995, 0900 Hours, 26th MEU (SOC) Headquarters, Camp Lejeune, North Carolina

Among the last of the components of the 26th to deploy was the headquarters. Behind the headquarters building, four large charter buses were being loaded along with some of the special communications equipment that the 26th MEU (SOC) would take with it. Sergeant Major Creech was busy kicking butts, and generally making life easier on officers, who had their own families to deal with. At 0955, the move to the ships kicked into high gear when one of the HMM-264 UH-1N Iroquois helicopters landed in front of the headquarters building to pick up Colonel Battaglini. Wishing us a hearty farewell, as well as an invitation to visit the 26th "on cruise," Jim Battaglini climbed aboard for the start of his first MEU (SOC) command deployment. Having already said good-bye to his teenage son, he was able to get on with the business at hand with a minimum of distractions, and you could see the confidence and pride that he had in himself and his Marines. Around the back of the headquarters, the last of the farewells were going on, and the buses were starting up. Our good friend from the 26th's S-4 (logistics) shop, Major Dennis Arinello, was saying good-bye to his wife Kathy and his kids, doing his best to set a good

Marines of the 26th MEU (SOC)'s headquarters unit ride out to the USS *Wasp* (LHD-1) on an ACU-2 LCU. They were preparing to leave for the Mediterranean on August 29th, 1995.
JOHN D. GRESHAM

example. Then, with a final set of waves, the bus convoy pulled out of the base, and headed north to Morehead City.

Tuesday, August 29th, 1995, 1100 Hours, Morehead City Harbor, North Carolina

Around noon, the bus convoy pulled into Morehead City. Pulling off to a large concrete beaching ramp, the buses unloaded, and the headquarters personnel joined other members of the unit for their ride out to the *Wasp*. This job was being done by a quartet of LCUs from ACU-2. In the distance, we could see *Shreveport* completing her loading farther up the harbor. Supervising the effort on the ramp was Captain C.C. Buchanan in his ever-present blue coveralls. Right now, he was as happy a man as could be imagined, because the loading of *his* ARG was going perfectly, and everything was on schedule. It was, by any standards, a perfect summer day in the sun. After a short wait, we were ordered aboard one of the LCUs, and headed out on a short journey to the *Wasp*. With us were members of the detachment which would control the landing craft and beachmaster parties for the ARG. Pulling along steadily, we soon pulled alongside *Wasp*. As we did, helicopters from the ACE were still coming aboard and being stowed, giving the flight deck the look of a power line full of birds.

As the LCU beached in the well deck, a chief warned us that we would need to be back aboard in thirty minutes if we did not want an all-expense-paid trip to the Adriatic! Properly forewarned, I helped Dennis Arinello with his baggage, and started the long climb up the loading ramps and ladders to his cabin on the O2 level. We slowly trekked around the ship, as over 1,400 other Marines were doing, and could see the transition going on between the land and the "second home" of the ship. Emotions were easing and calm determination seemed to be settling over the Marines and sailors all over the ship. Despite the favorable conditions this day, they were under no illusions as to what the sea could do to them if things got rough.

Then it was time to leave. Bidding Dennis and the others good luck and farewell, we headed back to the well deck. Getting back to the LCU just in time, we headed back to shore. As we did, *Shreveport* passed us on the way out of the harbor, headed rapidly out into the Atlantic. She looked like a gypsy wagon, loaded to the

An ACU-2 LCU enters the flooded well deck on August 29th, 1995. The landing craft was transporting personnel and their gear to the ship just prior to deploying to the Mediterranean.
JOHN D. GRESHAM

gunnels with men, vehicles, equipment, and the three of the CH-46s that we had watched take off from New River just eight hours earlier. By the time that we reached the loading ramp, the job was almost done. Before sundown that afternoon, *Whidbey Island* had joined up with her, and they headed east, over the horizon, to start their 1995/96 deployment. It was hard not to shed a tear, and wish that we were going alone with them. It had been a long, sweltering summer, and we had come to know these people so well.

Thursday, September 21st, 1995, 1100 Hours, Camp Lejeune, North Carolina

The final act of the 26th's outbound deployment cycle came some three weeks later with the return of now-Brigadier General Marty Berndt's 24th MEU (SOC). They were coming home flush from their rescue of Captain O'Grady some three months earlier. The process, almost the reverse of how a deployment begins, is something you have to see to believe. Each unit is staged into their barracks, where an open-air picnic is laid on. Everywhere, bedsheet banners decorated the building and fences around Camp Lejeune, proclaiming the joy and relief of family members waiting for their Marines to come home.

We chose to join in the reunion of the Marines of the 3/8 BLT, which had made up the GCE of the 24th, led by their commander, Lieutenant Colonel Chris Gunther. Their return was a triumph. What said it all was when Gunther, a veteran of over twenty years in the Corps, saw and hugged his wife and kids for the first time in six months. At moments like this, you feel almost guilty about intruding, but the sight is so compelling that you just have to watch. For the next couple of hours, there was a feeling that was like a decompression. With the pressure of a six-month cruise behind them, the Marines began to become human beings again. When things calmed down, we had a few minutes to visit with Lieutenant Colonel Gunther and

Lieutenant Colonel Chris Gunther, the commander of BLT 3/8 at the moment of his reunion with his family, following the memorable 1995 Mediterranean cruise of the 24th MEU (SOC).
JOHN D. GRESHAM

discuss the deployment. He confirmed that the handover to the 26th had gone well, although not exactly to plan. Normally, the two units would meet at the naval base in Rota, Spain, and spend a couple of days conducting equipment exchanges and data transfers. This time, though, the handoff had been done while under way, and the 24th's port visit had been dedicated to getting ready to come home.

Wednesday, February 14th, 1996, Naval Station Rota, Spain

I kept my promise...but only at the last possible moment.

The previous day, the ARG had "chopped" out of the Mediterranean Sea and 6th Fleet command and had started the long voyage home. But before they could do that, they had to stop and clean up after a hard six months on cruise. The stop was at the Spanish Naval Base at Rota (near Cadiz), on the Atlantic coast just north of Gibraltar. The U.S. Navy uses Rota as a rest and inspection stop for units coming home from Europe. Here all the equipment can be washed down, everyone can rest for a few days before the Atlantic crossing, and U.S. Department of Agriculture inspectors can check for pests or unwanted plants.

Wasp was moored on the north side of the bay, with *Shreveport* and *Whidbey Island* on the south side. All three ships had their vehicles out on the concrete piers; sailors and Marines were washing them down with freshwater. On the nearby beaches, the LCACs and LCUs were beached, also getting cleaned up after a busy cruise. In between was a sizable chunk of the Spanish Navy, including their small aircraft carrier, *Principe de Asturias*. Off the coast, the *America* battle group was exercising with a British force based around HMS *Invincible*. Aircraft came and went from the Naval Air Station. There was a buzz in the air from all the activity. As I marched up the brow, I was greeted by a host of smiles. Keeping promises, even little ones, means a lot to military personnel.

Thursday, February 15th, 1996, Naval Station Rota, Spain

The next day after dinner, I was invited to join Colonel Battaglini, Lieutenant Colonel Allen, and other members of the staff for a detailed briefing on the deployment. It should be noted that I have left out some details that relate to operations security issues, but I think you will understand the basic story. The 1995/96 cruise started with a series of joint international exercises around the Mediterranean. These included:

- **COOPERATIVE PARTNER**–Shreveport and her embarked units conducted this exercise with the armed forces of Bulgaria between September 14th and 18th, 1995.
- **ATLAS HINGE**–At the same time as Cooperative Partner (September 17th thru 21st, 1995), *Wasp* and *Whidbey Island* ran a series of force-on-force engagements with elements of the Tunisian military. This operation proved the validity of Colonel Battaglini's decision to include the platoon of M1A1 heavy tanks in the TO&E of the 26th. Of particular note was a counterattack conducted by the M1A1s at a critical point

Marines of the 26th MEU (SOC) trek through the Negev Desert in Israel during a 1995 exercise.
OFFICIAL U.S. MARINE CORPS PHOTO

in one engagement; it really surprised the Tunisian forces. A Tunisian comment was: "We didn't know you had those things!"

- **RESCUE EAGLE II**–This was the second in a series of mountain/TRAP exercises that have been conducted in Albania. Run between October 2nd and 14th, 1995, Rescue Eagle II saw Marine units off *Wasp* obtaining valuable high-altitude and small-unit infantry training for the MEU (SOC).

- **ODYSSEUS**–Simultaneous with Rescue Eagle II (October 3rd thru 13th, 1995), Odysseus was run with the armed forces of Greece. Marines aboard *Shreveport* and *Whidbey Island* provided the forces for Odysseus.

- **ISRAEL** –Early in the fall, the entire ARG/MEU (SOC) came together for a live-fire training exercise with the Israeli Defense Forces in the Negev Desert. Almost two weeks long (October 22nd to November 7th, 1995), this was one of the larger exercises that the force participated in. Following this, the force was given a short port liberty...which had to be cut short because of the tragic assassination of Israeli Prime Minister Rabin. In fact, several members of the armored task force embarked on *Whidbey Island* were just a few blocks away drinking beer in an open-air cafe when Mr. Rabin was killed.

- **BRIGHT STAR 95**–One of the longest-running exercises in the world today, Bright Star provides forces assigned to U.S. Central Command an opportunity to exercise in their AOR. It was based out of Cairo West Airfield in Egypt, and the whole of PHIBRON 4 and the 26th MEU (SOC) were involved, along with numerous other U.S. and allied units. Bright Star 95 ran between November 10th and 17th, 1995, and was highly successful.

- **ALEXANDER THE GREAT**–Following BRIGHT STAR (November 22nd thru 28th, 1995), *Shreveport* and *Whidbey Island* ran another exercise with the armed forces of Greece.

Chairman of the Joint Chiefs of Staff, General John Shalikashvili, visits with a Marine of the 26th MEU (SOC) off of the Albanian Coast. The Chairman was visiting the area following Operation Rescue Eagle II in October of 1995. *JOHN D. GRESHAM*

Despite the hectic exercise schedule, there was a real-world crisis to deal with in Bosnia-Herzegovina, and the 26th MEU (SOC) and PHIBRON 4 were actively involved in it. During the run-up to the Dayton Peace Agreement and the introduction of the NATO Implementation Force (IFOR), HMM-264's six AV-8B Harrier IIs flew some ninety-nine missions in support of Operation Deny Flight (sixty-three sorties) and Decisive Endeavor (thirty-six sorties) in the Balkans. The MEU (SOC)/ARG was then alerted that their services would be required during the coming IFOR operations in Bosnia-Herzegovina. They were to be ready for any contingency.

The Balkans get cold in the winter, and preparing the force for action involved a major effort. Special cold-weather clothing and rations were delivered to the 26th by mid-November. Because of the multi-national makeup of the IFOR (United States, France, Great Britain, etc.), special consideration had to be given to communications. Numerous hookups were planned around the NATO communications systems. Best of all, the personnel of the MEU (SOC) and ARG now had an Internet link that allowed them to send and receive daily E-mail from home. Along with the obvious matériel upgrades for operating in the Balkans, the staff of the MEU (SOC) ordered additional mine detectors, chains for the vehicles, and a small augmentation force of personnel with skills that might be required for the IFOR mission.

When the preparation was done, the 26th MEU (SOC) and PHIBRON 4 became the IFOR Theater Reserve Force. This meant that during the two months the IFOR ground forces were setting up in their positions on the ground, the ARG would be steaming in wide "doughnut" patterns around the Adriatic. For the rest of the cruise, Colonel Battaglini had to keep his personnel ready and alert. A rigorous drill and exercise program helped, but boredom slowly began to take over. The enlisted personnel started calling the force "the Maytag MEU" (after the terminally bored Maytag repair man in commercials back home). But they worked hard to stay sharp. All the classroom time spent studying ROE, mine detection and clearance, cold-weather operations, and counter-sniper tactics helped. By early February 1996, it was finally time to come home. They handed off to the 22nd MEU (SOC) at sea, and now they were at Rota in the final stages of washdown and reloading the ships. The next day, they would leave at noon for home.

Friday, February 16th, 1996, Naval Station Rota, Spain

By 1000 on Friday morning, Captains Duffy and Buchanan were knocking at my stateroom door. If I wasn't on the dock soon, they told me, I would be riding home the long way! Grabbing my bags, I headed down to the vehicle deck and the brow. Captain Buchanan was not kidding either: At 1200 sharp, all three ships of the ARG weighed anchor, pulled up lines, and promptly headed past the breakwater and out to sea. In less than a two weeks, the MEU and the ARG would have their home-comings at Camp Lejeune, New River, Little Creek, and Norfolk. Once home, they would start the ritual of preparing for their next cruise, planned to start in November of 1996. Colonel Battaglini would give up command of the 26th in the spring of 1996 to become an aide to the Secretary of the Navy, John Dalton. John Allen was headed up to the Commandant's Office at the Pentagon as the Commandant's aide. And after several years, Dennis Arinello was leaving the 26th for a shore assignment.

As for the ships of the ARG, *Wasp* headed into dry dock for her first major overhaul since being commissioned. For the 1996/97 cruise of the 26th MEU (SOC), PHIBRON 8, comprising the USS *Nassau* (LHA-4), USS *Ponce* (LPD-15), and USS *Pensacola* (LSD-36), would handle the job of transportation. Captain Buchanan planned to retire in 1997, Captain Duffy went to Washington to chair a promotion board and attend the National Defense University, and Stan Greenawalt relieved Ray Duffy as CO of *Wasp* in April of 1996.

In May 1996, it all began again.

In earlier chapters, I have shown you what a Marine Expeditionary Unit–Special Operations Capable, a MEU (SOC), can do in combination with its Amphibious Ready Group (ARG). Now we'll sketch out a couple of alternative futures to examine how a MEU (SOC)/ARG team might operate in the early part of the 21st century. The MEU (SOC)s will tackle two "major regional contingencies." Follow along as we explore some near-term possibilities.

Operation Chilly Dog: Iran, 2006

Back in the 1960s, the Shah of Iran, Mohammad Reza Pahlavi, knew that someday the oil would run out. (He was wiser than most rulers in the region.) "Petroleum," he once said, "is a noble material, too valuable to burn." So he envisioned a national electrical grid powered by a series of clean, modern nuclear plants. The French were doing the same thing, and he admired everything French. He also knew that possession of nuclear technology brought prestige that would enhance Iran's position as a regional power. It had worked for Israel. He also admired the Israelis. The sleepy Persian Gulf port of Bushehr made an ideal site for the first plant. The Bushehr peninsula was a solid, isolated block of rock, standing out along the generally flat, barren, central Persian Gulf coast. Nature had intended it to be an island, but ages ago silt had filled in the narrow channel, and a road built on an elevated embankment led to the town. Power lines from the nuclear plant would run alongside the road and up through the mountains, supplying the great inland city of Shiraz with cheap, abundant electricity.

In 1979 the Islamic Revolution came; the Ayatollahs threw the Shah out of the country, and the foreign engineers and construction crews departed soon afterward. The Ayatollahs may have been fanatical, but they weren't crazy. They remembered what had happened to Saddam's ambitious Osirak nuclear power plant, smashed into rubble by a few Israeli bombs. The Shah's nuclear dreams were abandoned, and intelligence officers in the West nicknamed the site "Dead Dog." With the passing years, war came and went. And oil continued to flow. But the Shah was right; it would not flow forever. A new generation of Iranian technocrats rose into positions of power, and they rediscovered the Shah's vision. Russia offered nuclear reactors on advantageous barter-trade terms. Nuclear technology brought prestige, enhancing Iran's position as a regional power. It had worked for Israel.

Iranian Army Officer Training School, March 1991

The young officers in the cadet program were a privileged elite; they were permitted to watch the Gulf War and its aftermath on CNN. Those who understood a little English translated for the rest who spoke Farsi, but the images spoke for themselves. It was a gut-wrenching experience to watch the destruction in four days of the hated Iraqi army which had defied the mobilized might of the Islamic Republic for eight grinding, bloody years of attrition warfare. Every young man in the room had lost friends or relatives in battle against Saddam's Revolutionary Guards' armored divisions...and now Saddam's armor evaporated like snowflakes in the hot desert sun.

It was a bitter joy they found in the humiliation of their enemy. For the victory which should have been *their* victory was being won by an even more hated enemy, the Great Satan, America. The junior officers were the best and the brightest of their generation. But it didn't take much to see the writing on the wall. If the Americans can do this to Saddam, *what* could they do to us? They listened attentively to lectures by officials from the Ministry of Islamic Guidance. The Great Satan's victory, it was said, had been bought with the oil revenues of the corrupt Gulf sheikdoms. The godless Russians had given the Americans the secrets of Saddam's defenses. Iraq had only collapsed because the martyrdom of a million faithful Iranians had fatally weakened his regime. After evening prayers, the junior officers gathered in the dorm, arguing late into the night. The mandatory time for lights out came and went, but no one could sleep. They resolved that whatever it took, they would understand the causes of Iraq's defeat, and they would ensure that their nation never suffered the same fate.

Sub-Lieutenant Gholam Hassanzadeh did not have to wait long. Before the Islamic Revolution, he had studied physics at Teheran University for a year. He spoke good English and fluent Arabic. His first assignment was debriefing a planeload of Iraqi nuclear technicians who had escaped to Iran after their prototype isotope-separation plant was reduced to rubble by U.S. BGM-109 Tomahawk missiles. They would be working for the Islamic Republic of Iran now. The technicians, all educated men and good Muslims, had little love for Saddam. They had escaped only minutes ahead of the Mukhabarat secret police that Saddam had dispatched to execute them, to keep them from telling the Americans what they knew.

Gholam took an instant liking to these men, uprooted from homes and families by the winds of war. Their accommodations were harsh, little better than prison barracks, but Gholam had grown up in a culture where hospitality toward the stranger was not only a religious obligation, but a fine art. He did what little he could to make their exile more comfortable. They reciprocated with a torrent of information. His reports were read with growing interest by top Government officials. One caught their special interest. In it he outlined a plan for an Iranian nuclear deterrent force. Gholam swiftly made captain, and then major. In a few years, he was given the leadership of the team that managed secret nuclear labs that were building a true Islamic bomb.

International Hotel, Bushehr, Iran, August 8th, 2006

The humidity was near 100 percent, and the temperature was about the same as body heat. He half expected it would cool off after sunset, but then remembered he was in the Persian Gulf and that it was August. An air-conditioner sat mockingly in the hotel room's window, but salt fog had corroded it into junk years ago. He hated this place almost as much as the local people hated him—he was a symbol of the West, the Infidel, the Enemy. Hans Ulrich, Senior Technical Inspector for the International Atomic Energy Agency (IAEA), dreamed about the alpine glaciers of his native Switzerland as he sat unhappily in the stifling room that passed for luxury accommodations in Bushehr.

Tomorrow he would complete the solemn, high-tech ritual of placing inspection seals on meticulously weighed and measured fuel rods of the Bushehr #1 Reactor Unit, a Russian VVER-440. He hated working around Russian reactors. He knew, of course, that this one was a pressurized water type, a safer, more modern design than that horrible graphite pile of crap at Chernobyl. Still, it was a sloppy piece of work by his standards, and that offended every neuron in the finely machined clockwork of his Swiss brain. In a few hours of work, he would accumulate almost an entire year's permissible radiation exposure. Then he would face the struggle of getting back to IAEA headquarters in Vienna with a quarter ton of inspection equipment from a country where every official, cab driver, and schoolchild regarded him as an enemy spy. As Ulrich continued to sweat, he went back to writing out his report in longhand. He would have used his laptop data slate, but it had gone into thermal shutdown an hour before, and was useless to him now. He hated trusting his thoughts to a sweat-stained notepad, but it would have to do for now.

As he sweated, he took a swallow of warm orange juice and sat back. He was thinking about the sealed cases of uranium cores that he had seen in the plant's secure storage area. He had been here just six months earlier to certify the refueling of the #2 reactor, and the spent fuel rods from that operation were still in their containers. When he asked why they had not been shipped out for reprocessing, he was told that the tons of rod assemblies from the first refueling had been held over to save on costs. In that way only one shipload would be required. While technically not a violation of the rules, it was not good management. As much as 75 kg/ 165 lb of weapons-grade plutonium might be mixed in with the witches' brew of radioactive isotopes in those cases. Until they were safely at a certified reprocessing facility, raw material for at least a dozen nuclear weapons lay in Iranian hands.

USS *Abraham Lincoln* (CVN-72) in the Gulf of Oman, August 14th, 2006

At least once a week, an elderly F-14 equipped with a TARPS reconnaissance pod made a low-level run from the carrier around the Persian Gulf's north coast, keeping carefully outside Iranian air space. If anything nasty was happening onshore, the NRO's imaging and radar satellites would pick it up almost immediately. Nevertheless, it was good training for the naval aviators and the ES-3A Shadow crews farther out in the Gulf, who expectantly monitored the electromagnetic spectrum, hoping the Iranian radars would light off some new frequency or pulse modulation. Thanks to some trick of the Gulf's freakish aerial refraction, this week's imagery was particularly good.

As he studied the fine detail in high magnification on the workstation, Lieutenant JG Jeff Harris, a photographic intelligence analyst assigned to the carrier's air wing, saw something odd about a new pair of oil platforms under construction off Bushehr. His fingers danced across the keyboard as he opened a new window on the screen and called up precise 3-D renderings of typical Persian Gulf drilling and production platforms and then rotated the images for side-by-side comparison. Something was definitely different. The steel lattice at the center of each platform was much too light to support the massive structure of a drill rig. Staring for a few moments, he reached around to a small classified safe, dialed the combination, extracted a CD-ROM, and loaded it into the workstation's drive. As the program displayed various pieces of equipment, it became clear there was nothing in the wildcatter's world that could be mounted there. But the grid pattern would fit the dimensions of a vertical-launch canister cluster for a Russian SA-N-9 surface-to-air missile (SAM) system. Those round fittings at each corner of the new platforms made no sense as mountings for any kind of drilling equipment. But they were exactly the right size and shape as mounts for CADS-1 gun/SAM mounts. And what looked like racks for drill pipe might be mounts for Chinese CS-802 surface-to-surface missiles (SSM). Rubbing his eyes, he rose to pour another cup of coffee, then picked up a phone to call his department head. While he waited for the intelligence chief to arrive, he suddenly realized that these platforms had been built to protect something. He pulled more CD-ROMs from the security safe, and began to think.

Defense Intelligence Agency Headquarters, Bolling AFB, near Washington, D.C., August 22nd, 2006

In all the vast bureaucratic labyrinth of the American intelligence community, you probably would not have found a bigger collection of prima donnas than the Counterproliferation Coordinating Committee. The Committee, of course, did not officially exist. Its funding was buried in an obscure Interior Department line item that covered long-forgotten uranium mining subsidies to a holding company in Utah. Attendees for the every-Tuesday-morning meetings were drawn from the CIA, various imagery agencies, all four military services, the Department of Energy labs, a sprinkling of academic physicists and engineers, an FBI deputy director, and whatever senior analyst the State Department could spare that week. All of them had the right "tickets" (Special-Access security clearances). The older guys tended to be "Kremlinologists," long-time Russian specialists who had acquired profound cynicism and paranoia during long, bitter years of being outwitted by the KGB and its successors. The younger guys tended to be East Asia specialists, who built entire careers on the interpretation of enigmatic scraps of data from the bizarre information vacuum of North Korea. The absence of Middle East specialists might have seemed startling, unless you understood the politics of the intelligence community. In the aftermath of 1991 Gulf War, it was discovered that Iraq had operated several vast, parallel nuclear weapons programs right under the high-tech nose of countless billions of dollars worth of U.S. spy satellites. This was one of the great intelligence failures of the century; it sent a clear message to a new generation of intelligence officers. Stay away from anything connected with nuclear proliferation in the Middle East; it's not career-enhancing. Anyhow, it's the Israelis' responsibility. They can collect HUMINT (Human Intelligence) over there; we can't. Besides, they don't like anyone else messing around on their turf.

This Tuesday morning, the agenda began with a presentation by Dr. Rob Kennelly, a young nuclear engineer from the Department of Energy's Oak Ridge National Laboratory. He was describing new developments in laser-based gaseous isotope separation to extract plutonium from spent reactor fuel. Conventional extraction of weapons-grade fissionable material required construction of vast industrial complexes that were impossible to hide. But this laser plasma technique could be scaled down to machine-shop dimensions; a complete facility might be concealed on the grounds of a nuclear power plant. The only "signature" would be diversion of megawatts of power to drive banks of high-energy lasers. At the completion of the presentation, the committee chairman politely thanked the engineer, dismissing him with a nod toward the door. He wasn't cleared for the afternoon's agenda. But he wasn't ignored. During the lunch break, the lieutenant colonel who represented the Marine Corps took the young man aside and bombarded him with questions.

Russian Embassy, Teheran, Iran, August 25th, 2006

Yuri Andreevich Rogov was carried on the official Embassy roster as Senior Science and Technology Attaché, but of course he reported directly to the station chief of the SVR, the successor of the KGB. Like its Soviet predecessor, the SVR selected its officers carefully and trained them rigorously. Fluent in Turkish, Farsi, and Arabic, Rogov operated with smooth confidence and impeccable courtesy in an often suspicious, hostile, and unpredictable country. He had stood and gazed in wonder at the ruins of Persepolis, and marveled that these people had built a cultured and efficient world empire back when his own distant Slavic ancestors lived in reed huts and slogged through the Pripyat marshes as hunters and gatherers.

Another characteristic the Russian Republic shared with its Communist predecessor was that it demanded long hours and hard work from its Foreign Service officers. One of Rogov's many duties was maintaining contact with the hundred or so technicians down at Bushehr. Officially, they were independent contract hires, working directly for Iran's Energy Ministry. Some of them had Ukrainian or Kazakhstan passports, but the project was based on an agreement between Russia and Iran, and the Iranians expected the Russian Embassy to keep the men happy, healthy, and out of trouble. No women were permitted, an irrational and unfortunate local custom, Rogov thought.

Several times a year, the various team leaders flew up to Teheran for an extended debriefing with Rogov. This was also an opportunity to update and check the contingency plans for an emergency evacuation of the crew in the event of a military or political crisis. Rogov's contacts in Bahrain and Dubai maintained a fleet of small speedboats, normally employed in the lucrative trans-Gulf smuggling trade, but instantly on call if it became necessary to clear out. Lev Davidovich Telfian was an ethnic Armenian, but his people had lived in Novorossiysk for generations, and he was a Russian citizen. His job at the plant was Training Manager, which gave him unusual freedom to move about, and a wide range of close personal contacts with Iranians on the site. He was one of Rogov's best informants. Unfortunately, this trip Telfian had had nothing new to report regarding the SVR's highest collection priority, the Chinese and North Koreans who occupied a separate, heavily guarded compound on the military base adjacent to the plant. As Rogov sipped his tea and re-read the report, he knew that there was something here. Or perhaps, the absence of something? He decided to send an Eyes Only message to SVR headquarters, and let them try to sort it out.

Iranian Ministry of Machinery Automobile Plant #3, near Bandar al Abbas, September 5th, 2006

Wendy Kwan sat uncomfortably in the director's lobby of Iran's newest automobile plant, carefully sipping a cup of tea. One of CNN's top foreign correspondents, she was here to interview the Iranian Minister of Machinery, as well as the factory manager. The American trade embargo, dating back to the 1990s Clinton Administration, had been tough on Iranians, but their response had surprised Western observers. Rather than knuckling under, they had launched a modest industrialization program, which had grown dramatically in the last few years. Plant #3 was the prototype, based on the latest Japanese flexible manufacturing techniques. It was configured to produce everything from new automobiles to tractors and other heavy equipment. It could also produce combat vehicles. Though her current beat was financial news, Wendy had started as a Far East correspondent fifteen years before, and she knew more than she wanted to know about military hardware. The interview went almost too well. Both men and their assistants (bodyguards?) were pleasant. This put her ill at ease. She was not on some Wall Street trading floor. Here, she could disappear at the whim of one of these men. Today, however, no problems developed. Following the interview, they escorted her and the camera crew on a factory tour.

It was impressive. She was surprised by the sophistication of the technology. The Iranians had made deals with companies in Russia and the former Soviet Republics for start-up capital in trade for a guaranteed flow of equipment at favorable prices. All around, brand-new robots and computer workstations were being installed. Every piece of equipment, she was told, was tied into a central computer, which held a complete design database for every product built on the line. As they passed the Engineering Department, she noted that the doors all had cipher locks, and that a plainclothes guard seemed to be checking workers' ID badges before they went in. Rather odd for an automobile plant, she thought. Then, as she was finishing up her close-out shot for the story, she saw on the edge of her vision someone who made her blood run cold. Trying not to look again, she calmly told the crew to pack up and head back to the airport. Only after they had returned to Bahrain that evening did she even allow herself to think about the man she had seen.

Wendy had first seen him ten years earlier. In those days, Professor Kim Ha Soon had been a top physicist in North Korea's now-dead nuclear weapons program. And he had led the delegation that negotiated away that program for energy and food supplies at the end of the Cold War. Wendy had covered those talks with CNN's field team. She never forgot the war scare that swept the Korean Peninsula back in 1994 and 1995. She had seen him then, and remembered rumors around the press pool. Not only was Kim the brains behind their uranium enrichment program, he had also devised the deception and cover plan that had hid the Korean effort for years. Now he was at an Iranian automobile plant, coming out of a security zone, talking with the plant director. She decided to make a quick detour through Washington, D.C., on the way home, and to take the master tape with her. Her old Georgetown University roommate was now an Army major working for the Defense Nuclear Agency at Fort Belvoir. Wendy thought she might be able to feed this to someone who could make use of it.

Iranian Ministry of Machinery Headquarters, Teheran, Iran, September 15th, 2006

The Iranian Minister of Machinery sat in a high-backed chair and looked over a thick file folder of material about the "Special Machinery" project at Bushehr. So far security had held, and with only three months to completion, there appeared to be nothing to be concerned about. The CNN interview had shown only what he had wanted them to show, and his own performance had been both soothing and convincing. The image was exactly what he wanted—that his ministry was merely overseeing a plucky country's industrial program, trying to overcome the shackles of an unjust embargo. His own lack of military service (he had trained as a mechanical engineer in France) meant that he probably was not known beyond a thin file at CIA headquarters. He had never been politically active, and was considered rather boring in most trade circles. He was, he thought with a thin smile, the perfect cover for a nuclear weapons program.

The smallest details of security had been considered. For example, graduate students at several Iranian universities published scientific papers on nuclear physics under the names of key scientists in the program, so that their absence would not be noticed by Western scientists. Best of all, it was a small program, with just the two facilities at Bushehr and Bandar al Abbas on the coast 320 miles/512 kilometers to the southwest. Thanks to the new laser-plasma isotope separation process and a secure central computer database, less than 250 personnel were involved.

A folder on the desk held the time line for the final three months of the first production run—a dozen boosted fission weapons with a nominal fifty-kiloton yield, based on an implosion design using plutonium. Half would arm a squadron of intermediate-range ballistic missiles (IRBMs), and the other six would become warheads of Russian-supplied AS-19 cruise missiles, for air-launch by Iran's SU-24 Fencer fighter bombers. These weapons would allow Iran to deter any aggression from the Americans or their Arab lackeys in the Gulf while even more powerful weapons and delivery systems were developed by his ministry.

It had taken a long time. Almost fifteen years earlier, he had read the papers written by his good friend, now-Colonel Gholam Hassanzadeh. Armed with these, he had gone to an old mentor in the Defense Ministry with the proposal for a careful and discreet program to build nuclear warheads and delivery systems. It would take time and patience, but the plan would yield results. The Defense Ministry had entrusted him with industrial responsibility for the project, while Colonel Hassanzadeh handled security. That made them two of the most important men in Iran.

Now the project was about to bear fruit. He looked at the time lines with satisfaction, and mentally reviewed the schedules. Final assembly of the weapons was timed for the American holidays at the end of the year, when their attention would be focused on that bizarre form of football they worshipped more than their God. Over the Thanksgiving weekend, the components for the warheads would be moved from the fabrication shop at Automobile Plant #3 to the nuclear plant at Bushehr, where the plutonium was being extracted from the last batch of fuel rods from the twin reactors.

Starting on Christmas Eve, twelve warheads would be assembled in a special facility at Bushehr, over a period of seven days. Finally, the warheads would be brought back to the auto plant for mating to the IRBMs and AS-19s, with delivery to operational units the following day. Once the weapons were deployed, there would be a declaration that Iran was a nuclear power and would no longer submit to unfair treaties or agreements imposed by Western powers. From that moment, they would be the regional superpower. The Iranian people would again be able to seek their destiny, without interference by outsiders.

Russian Embassy, Teheran, Iran, September 26th, 2006

To Yuri Andreevich Rogov, sitting in his embassy office, the CD-ROM in his hand felt like a disk of deadly plutonium. It might as well have been, for it held the very documents and diagrams that the Machinery Minister had been reviewing the night before. The disk had been smuggled out of Bushehr in an audio CD case, labeled as Armenian folk music. Someone had copied an actual audio CD, adding written data to the outside tracks. The disk had been passed to Telfian covertly by one of the Pakistani technicians in the secure area, while they had been in the cafeteria together. Telfian had had no idea what it was at first—not until he inserted the disk into his multimedia laptop computer to listen to it, and accidentally found the data files. Telfian had then used a special code phrase whereby the embassy could request his recall on a phony family emergency. He gave the disk to Rogov, then returned the next day to Bushehr to maintain the cover for the brave Pakistani. Now Rogov had the problem of getting the disk back to SVR headquarters in Moscow. There was really only one way. He made reservations on an Aeroflot flight home in two days, so as not to appear too eager.

Defense Intelligence Agency Headquarters, Bolling AFB, near Washington, D.C., September 30th, 2006

The chairman of the Counterproliferation Coordinating Committee brought the meeting to order, and quickly summarized the data the Russians had forwarded that morning. Combined with other bits and pieces that had come in, they now had a full picture of how Iran planned to join the nuclear "club." The documents detailed an exquisite deception and security plan. The Iranians had purchased daily 1-meter-resolution commercial satellite imagery covering every base in the Western world that supported special operations forces. The list read like a mailing roster for a snake-eaters convention. Fort Bragg, North Carolina; Hurlbert Field, Florida; the

SEALs training base at Coronado, California. Even the garrison and training facilities for the British SAS and German GSG-9. They had arranged for Iranian nationals to emigrate to each nation and set up businesses, usually things like dry cleaners and pizza-takeout shops, just outside the bases themselves. The Iranian agents reported home though a complex E-mail path over the Internet using encrypted messages. It was an almost perfect system, and it would be noticed immediately if one of the agents were arrested. The result was that special operations units which could neutralize the Iranian weapons program were covered with an Iranian surveillance blanket, making surprise impossible.

What made the situation worse for the intelligence types was that they had done their job. Thanks to their efforts to bring together the intelligence community and build relationships with past enemies, they had achieved an intelligence coup. Yet because of the Iranians' patience and care, it seemed as if nothing could be done. But unless they did something radical soon, the balance of power in the Middle East was about to take a dangerous tilt. The Marine lieutenant colonel broke the gloom with a comment about the Iranian surveillance list. Nowhere on it was even one U.S. Marine Corps base.

Fleet Marine Force Atlantic Headquarters, Naval Station Norfolk, Virginia, October 5th, 2006

Dr. Kennelly and Lieutenant Harris were both wondering what they were doing in the secure conference room of the Fleet Marine Force Atlantic (FMFLANT) headquarters this fine day. The heat and humidity of summer had finally broken, and you could almost feel fall in the air. In the room with them were a number of Navy, Air Force, and Marine officers, none over the rank of colonel or captain. Precisely at 0800, the brigadier who served as deputy commander FMFLANT rose and went to the podium. He pushed a button to display a briefing slide onto the large-screen projector to the side of him.

"Ladies and gentleman, we have here an opportunity to excel...."

Everyone in the room tensed up, knowing exactly what that kind of invitation meant. As he outlined the situation at Bushehr and Bandar al Abbas, you could feel the anxiety in the room rise. Dr. Kennelly wondered if this was how it felt back in 1949 when the first Soviet A-bomb test was announced. If the data was correct, a new nuclear power would be born in just three months. The fact that he had contributed to the discovery just made him sicker. What the general had to say next stunned them even more:

"Your job is to stop this program, and bring home irrefutable evidence of what the Iranians have been up to."

He flipped through his charts, and the officers made furious notes on the hard-copy charts that had been supplied to them. A Marine colonel spoke next.

"Sir, am I to understand that the 22nd MEU (SOC) is the only element of the force that will actually be in the Gulf itself?"

The reply came quickly. "Yes, Colonel, you'll relieve the 31st as scheduled, with the extra training and support that we have described here previously. Other than that, we want nothing of this operation to ever touch ground in the region. We're trying to provide complete deniability for the Saudis and our other friends.

The President, the Congressional leadership, and the Joint Chiefs are all behind this one, and they want it to go off smoothly. Any questions?"

"How about the name of the operation sir?"

The General replied with a smile, "Back in the 90s, the old-time intelligence analysts called this plant 'the Dead Dog.' When we get done with it, it's going to be a Chilly Dog!"

There was a long pause, after which the colonel replied, "Semper Fi, *sir!*"

Aboard USS *Bataan* (LHD-5), off the North Carolina Coast, November 1st, 2006

"All right, ladies and gentlemen, this is our final confirmation brief before we do this run-through for the last time. Are we clear on all the important points?" Colonel Mike Newman was going over the last of his briefing slides.

The young captain commanding Charlie Company replied, "Yes, sir. The last time showed that we're good on time and tasks, but we need to work on order and flow?"

"That's right, Jimmy. It's not so much that you're doing anything wrong; it's just that I want to see you guys flowing like black ink through the compound mock-ups. There's nothing we can do about being noticed eventually. I just want to delay the inevitable as long as possible, so the diversion force can really get the attention of that battalion on the north side of the access road." He stopped, and then his face wrinkled into a thinly veiled grin. "I want them giving their full attention to defending their own barracks," he continued, "and not bothering with a few guys in black jumpsuits." He finished with:"Let's do it right this last time, and put it into the can, folks!"

The last run-through was nearly perfect, good enough to satisfy Colonel Newman and the SOTG observers. With this part of the preparation completed, and the procedures for the disposal of the defensive oil platforms dealt with, they would be ready to deploy in early December.

Warhead Assembly Room, Bushehr, Iran, December 4th, 2006

The Machinery Minister looked around with satisfaction at the twelve warhead assembly bays that were being finished. The movement of parts from the automobile factory had gone without incident, and the last phase of the plutonium extraction process had begun on time. In three weeks, a dozen nuclear weapons would take shape in this room, and there was nothing that the infidels or anyone else could do about it. That morning, he had received an intelligence briefing from his assistant at the ministry. The young man had a gift for this work, and amazingly, did absolutely nothing that was illegal in any country. The 1-meter-satellite imagery was acquired from a half-dozen different providers from France to the People's Republic of China. Data on movements by military units was also available over the Internet; it was as good as what most intelligence analysts saw in their morning briefings.

There was absolutely no indication of anything unusual at the bases where enemy special forces were plying their trade. In fact, there was a steady decline of military activity by the U.S. and her allies around the world. Even the U.S. Air Force, with its boast of "global reach," had been cutting back. The only matter of note that would be happening in the next month was a handover between two token Marine units in the Gulf. Nothing to worry about: only a single battalion aboard three ships with a couple of escorts. The carrier battle group based around the USS *Constellation* (CV-64) would be operating out in the Arabian Sea, and would not enter the Persian Gulf on this cruise. It was going to work.

Onslow Beach, Camp Lejeune, North Carolina, December 7th, 2006

It was deployment day, and Captain Bill Hansen had the double problem of saying good-bye to his own wife and baby daughter, and getting his company of fifteen amphibious tractors aboard the USS *Trenton* (LPD-14). He would have the honor of taking the first unit of AAAVs out on cruise. He knew the real reason for this honor. Unlike others who had taken new systems to sea for the first time, he knew he would be taking this one into battle in just a few weeks. Luckily, the new vehicle had proven pretty reliable in field trials, and he had four contractor technicians to keep them in good shape.

His concentration was broken suddenly by the buzz of twin turboprops, and he looked up just in time to see Lieutenant Colonel Colleen Taskins banking her MV-22B Osprey to the north, followed by three other Ospreys from HMM-263. She had a fifteen-minute flight ahead, and then a landing aboard *Bataan*. He smiled, because Taskins faced the same problem he did. Though the Osprey had been in service for a few years, this would probably be its first combat trial. Lieutenant Colonel Taskins had been chosen as the first woman to command a Marine combat helicopter unit; now she would likely be the first female to command a Marine unit in actual combat. Not that this was a problem: Inside the pixie-faced lady who could turn the head of every male Marine in the MEU (SOC) was the heart of a warrior. He also knew that if something went wrong at Bushehr, she would be the first one in the air to come pick them up. Shaking the thought off, he climbed into his AAAV, and ordered the driver to head into the surf.

Reactor Control Room, Bushehr, Iran, December 15th, 2006

Lev Davidovich Telfian was nervous. A few days earlier he had been visited by Rogov, from the embassy in Tehran. The visit, sponsored by the Iranians, was one of many to industrial plants employing contract Russian personnel. He and Rogov had gone walking along the waterfront, beyond the ears of Iranian security. Rogov had quietly advised him to be ready for "something," possibly even "anything." Then he'd headed back to Teheran. Since that time Telfian had taken to carrying his personal effects with him. His computer, passport, and hard currency were carefully stashed in his briefcase, along with a clean pair of socks, underwear, and a toothbrush. He explained this to the security guards as an accommodation to the plant managers who were asking him to work extra shifts, which in fact they were.

Now he was doing his turn in the rotation as the midnight-to eight supervisor. He would continue on this schedule until New Year's Eve. After that, the Iranians had offered all the foreign workers a paid three-month vacation. He was ready for it. Even though he had grown up in the former Soviet Union, where you learned to suppress all outward signs of fear, suspicion, or thought, the stress of staying calm every day was immense. Nevertheless, he noted with curiosity that after running at top capacity for six months, the twin reactors were now at only 66%. He just knew it had to do with the CD-ROM the Pakistani had given him. Something bad was happening, and it was about to get much worse. He wondered if he would survive.

Oval Office, the White House, Washington, D.C., December 17th, 2006

The National Christmas Tree lighting had gone well earlier that evening, and the press was already speculating about the forthcoming State of the Union message. Expected to be another one-term President, the Chief Executive had stunned the world with a last-minute victory over his rival, a senator from Washington State with a penchant for bribery and adultery that offended the electorate's infatuation with morality. Now, as the President looked out over the balcony towards the park, he wondered if winning was going to be worth it. Like so many other men who had sat in this office, he had entrusted foreign policy to others while he dealt with the challenge of the budgetary shambles left by an angry Congress.

The result was that he was betting the future on a military adventure that his advisors said was foolhardy. He had to make good his own failure to pay attention to

Map of the raid on the Iranian nuclear weapons assembly facility at Bushehr.

JACK RYAN ENTERPRISES, LTD., BY LAURA ALPHER

the troubles of the world; this desperate venture was the best, last chance. Making a pre-New Year's resolution, he promised that if Chilly Dog came off, he would gut the National Security Council and State Department and start fresh. He sipped the bourbon in his glass slowly, exhaled, and looked out over the city once again, wondering if God listened to political deathbed wishes and confessions.

Hangar Deck, USS *Bataan* (LHD-5), about 40 nm./73 km. Southwest of Bushehr, 0000 Hours, December 28th, 2006

It was game time for Operation Chilly Dog, and Colonel Mike Newman had never felt more alive. All through his career, he had wanted to lead Marines on an important combat mission, and now this skinny kid from Wisconsin was about to do just that. It was almost enough to keep him from losing his mid-rats right here in *Bataan*'s hangar bay. The worst part of the whole thing was that he would not get more than a few yards/meters from this very point. Given the complexity of Chilly Dog, he would fight this battle from a console in the Landing Force Operations Center (LFOC) on the ship's 02 Level. For the first time in his career, he would not lead from the front, and he felt guilty. It made no sense, of course, because unlike John Howard at Pegasus Bridge in 1944 and Dan Shomron at Entebbe in 1976, the only way that this mission could be coordinated in space and time was with the electronic tools of cyberspace.

Around the hangar bay, over five-hundred Marines were checking weapons and equipment. The sickly yellow-orange sodium lights cast a surrealistic glow over the scene. As he walked from group to group, encouraging them to muted shouts of "Oooh-rah!" and "Semper Fi, sir!" he watched Marines apply desert camouflage paint to their war faces. Over near the port elevator bay stood the most important group of all, the media/observer team. There had been much debate about bringing them along, but in the end, the need to justify the action to the world community had won out. Dr. Kennelly from Oak Ridge was talking with Hans Ulrich from the IAEA. He smiled at the thought that these two bookish men had given them the first "tickle" on Bushehr. Both were looking decidedly uncomfortable in desert "cammies." Wendy Kwan and her CNN crew were mingling with the DoD Combat Camera team that would document the event for the world. She had been offered the opportunity as a reward for fingering Professor Kim Ha Soon at the automobile plant. Now she and her crew would get the rest of the story. Colonel Newman grimly smiled, and hoped that she would live to collect the Emmy that would inevitably be hers—if she survived.

Suddenly, Captain Fred Rainbow, commander of *Bataan*, ordered Battle Stations with an old-style bugle call, and then followed it up with the Marine Band's classic recording of *The Marines Hymn*. The troops on the hangar bay immediately struck a brace, and sang along at the top of their lungs. It was almost too much for Colonel Newman, who wondered how many of the young men and women in this bay he would have to write letters for tomorrow.

He went up to the LFOC on the 02 Level and sat down at his battle console, motionless until a thermal mug of coffee laced with cocoa thumped down in front of him. He looked up to see Lieutenant j.g. Jeff Harris, who had been transferred to his intelligence staff following his discovery of the two defensive platforms near Bushehr. "Showtime, sir," was the comment from the young officer, who showed a pensive smile. His errand to his colonel done, he sat at the console beside Newman's, where he would monitor the sensor feeds from the UAVs that had just launched from the flight deck of *Trenton* (LPD-14). The call to flight quarters brought Newman back to reality, and he said a silent prayer as he watched his Marines troop aboard the helicopters over the deck television monitor. Ten minutes later, they lifted off into the inky night.

Defense Platform #2, 10 nm/18.3 km West of Bushehr, Iran, 0200 Hours, December 28th, 2006

The duty officer of the platform stood over a radar operator monitoring the formation of ships to the south. There had been some launching and landing of helicopters and Harrier jump jets, but this was entirely normal for the enemy, who loved to fly at night, like bats. The sensors of the heavily armed platform detected nothing unusual, and he picked up the telephone to report in to the security center at Bushehr. The fiber-optic data link to the center at the Bushehr Airport ensured that communications to the mainland were not subject to the vulnerability of radio transmission which could be jammed or intercepted by an enemy. As he finished his hourly check-in call, he moved over to the teapot to pour himself a cup of Persian brew. It would never touch his lips.

At precisely 0201 hours, an AV-8B Plus Harrier II from VMA-231 fired a salvo of four AGM-88 High Speed Anti-Radiation Missiles (HARMs). A few hundred feet above the platform, their warheads detonated almost simultaneously, spraying thousands of armor-piercing tungsten cubes which shredded exposed antennas and weapons canisters. Within a minute of the missile strike, a raider craft carrying a four-man SEAL team cut the armored fiber-optic cable back to the mainland and flashed a signal to an MV-22B Osprey. By 0204, over twenty Marines from the 22nd MEU (SOC)'s Maritime Special Purpose Force (MSPF) had fast-roped down to the deck and cleared out the seven survivors of the missile attack. The dazed prisoners were taken aboard the Osprey and sent back to the *Bataan*. At the same time, an identical force was taking out the other platform some miles to the south. The outer layer of Bushehr's defenses had just been eliminated, and the Iranians did not even know it. In a few more minutes, they would not care.

Road Causeway between the Town of Bushehr and the Power Plant, Iran, 0205 Hours, December 28th, 2006

The Marine Force Reconnaissance platoon had been in place for two days reporting back over a secure satellite link to Colonel Newman in the LFOC. Now they had just cut the phone lines to the power plant and prepared the causeway for demolition, should anyone try and come down the road. They were armed with Javelin anti-tank missiles to maul anyone who tried. This platoon was one of two covering access routes from the town of Bushehr, and the sergeants leading them prayed that the extraction plan worked as planned. The alternative was a very long walk to Pakistan.

Bushehr Airport, Bushehr, Iran, 0205 Hours, December 28th, 2006

The loss of signals from the data links was noticed immediately by Security Control at Bushehr airport. Like military personnel everywhere, the duty section called the maintenance section and poured another cup of tea to stay awake. Overhead, four bat-shaped B-2A Spirit stealth bombers from the 509th Wing at Whitman AFB, Missouri, silently took position for what had to be a perfect strike. They had staged out of Anderson AFB on Guam, refueling from KC-10A Extenders based at Diego Garcia. At 2007 hours, sixteen GBU-29 Joint Direct Attack Munitions (JDAMs) precision-guided bombs dropped from each bomber's weapons bays. Each bomb was guided by a GPS receiver to fall within five meters of a pre-surveyed aim point. The most important targets got a pair of bombs, the rest received a single JDAM. The first weapons struck the hardened concrete of the security center as planned, slicing through overheads with the penetrating power of 2,000-lb/909.1-kg warheads. Within thirty seconds, the command center, post office, telephone exchange, runways, hardened aircraft shelters full of MiG-29s, and other targets around Bushehr had been neutralized.

Two minutes behind the B-2s came eight B-1B Lancers from the 7th Wing at Dyess AFB, Texas, also launched from Anderson AFB and refueled from KC-10As at Diego Garcia. Their targets were two battalions of troops in barracks adjacent to Bushehr airport. Each unloaded twelve AGM-154 Joint Standoff Weapons (JSOWs) from their weapons bays, well outside Iranian airspace. Following a two-minute gliding flight, the ninety-six JSOWs, guided by onboard GPS receivers, unloaded their payloads of BLU-97/B Combined Effects Munitions (CEMs). They blanketed over a hundred acres of troop billeting and vehicle-parking areas with thousands of CEMs, and the effects were horrific. The two minutes since the bombs from the B-2 strike had given the troops time to throw on their boots, grab their weapons, and rush outside to be shredded into hamburger by exploding cluster munitions. After a few minutes, the Bushehr garrison could no longer defend itself, much less the power plant to the south.

Ra's-e Hhalileh Mud Flats, Southeast of Bushehr Power Plant, Iran, 0210 Hours, December 28th, 2006

Captain Hansen and his fifteen AAAVs were crawling across the mud flats south of the power plant. They had swum ashore minutes earlier, having crawled out of *Trenton's* well deck, some 25 nm/45.7 km offshore. Hansen had seen the flashes from the bombs in Bushehr, and was waiting for the radio signal that would send his pack of armored vehicles into a headlong cavalry charge. Nothing like this had been seen since Eagle Troop of the 2nd Armored Cavalry Regiment had charged an Iraqi brigade at the Battle of 73 Easting back in 1991. He just hoped that he was not leading his men into another Battle of the Little Bighorn.

The snipers of the 22nd MEU (SOC)'s BLT had broken into four teams, each armed with a Barrett .50-caliber sniper rifle. Each team sat in a spider hole, a mile from the guard posts of the power plant. As the spotters sighted the guards in the corner towers and passed them to the shooters, they awaited a signal at 0210 to go into action. The signal for each team came over a miniature satellite communications terminal, and all four fired their first shots within seconds of each other. Each weapon spat out a total of ten rounds, taking out the guards, radar and communications antennas, and power lines. Within a minute, all four teams flashed their "success" code back to Colonel Newman in *Bataan*'s LFOC. With a murmur of, "Dear Lord, don't let me screw up," Hansen ordered the AAAVs into action.

The fifteen AAAVs spread out in a wide line and charged forward at over 40 mph/65.5 kph across the mud. When they came within 1,500 meters of the garrison perimeter, they opened up with 25mm cannons, spewing high-explosive incendiary (HEI) shells into the compound. Buildings began to burn and soldiers ran about wildly. Ragged return fire fell around the fast-moving AAAVs. Captain Hansen's men fired an occasional Javelin missile against anyone who got too accurate. The vehicles churned up the mud east of the compound, generally raising hell and making noise. Captain Hansen hoped it would be enough diversion to cover the rest of the Marines coming in from the sea.

Delivery Pier and Ramp, Bushehr Nuclear Power Plant, Iran, 0215 Hours, December 28th, 2006

The rigid raider craft had disembarked from an LCAC about 10 nm/18.3 km off the coast, and had come the rest of the way on muffled engines. When Colonel Newman sent out his "go" signal, the raiders made their dash for the loading ramp at full speed. They came ashore at almost the instant Captain Hansen's attack began, and were covered almost perfectly. Out of the rigid raiders came Charlie Company, which split into three teams. One platoon disposed of the guards at the security posts, and then set up a security cordon inside the razor-wire fence, just in case the Iranian guard force remembered its *real* job. The rest of the company headed into the plant for the hard part, the assault on the assembly and reactor control rooms. As they went in, a pair of LCACs headed into the dock area, loaded with trucks, LAVs, and other equipment.

Ra's-e Hhalileh Mud Flats, South of Bushehr Power Plant, Iran, 0220 Hours, December 28th, 2006

Captain Hansen had ordered his AAAVs into defilade behind some low rises, to reload their 25mm ammunition and Javelin launchers and draw the Iranians out of their barracks. It worked like a charm. About four hundred Iranians moved out of their compound, escorted by light trucks and scout vehicles armed with machine guns and recoilless rifles. They had closed within one thousand yards/meters of his line of vehicles when he made a radio call. Seconds later, two pairs of AH-1W Cobra attack helicopters rose on either flank of the Iranians and opened fire with 20mm cannons and 2.75-in. rocket pods. At the same time, the AAAVs began to fire again. It was a slaughter. Under fire from three directions, the Iranians could not even retreat. In a matter of moments, white flags began to appear, and Hansen was forced to order a "cease-fire." He then ordered the Cobras to hold them there, and ordered his vehicle towards the power plant, the sea, and, he hoped, safety. He had to make sure the rest of the security battalion was kept busy, but he doubted there was much left to occupy.

Over the Power Plant, Bushehr, Iran, 0222 Hours, December 28th, 2006

Lieutenant Colonel Colleen Taskins thumbed the "tilt" control on the throttle column and flared her MV-22B Osprey to a hover over the roof of the weapons assembly building. As the aircraft shuddered to a halt, she jammed the intercom button, and called, "Let 'em go, Chief!" Just behind her, the crew chief lowered the rear ramp and Marines began to fast-rope out of the side exits and the rear ramp. In less than thirty seconds, all twenty were on the roof, working their way down into the building. Scanning left and right, she saw that the other five *Ospreys* of her flight had offloaded their Marines. Punching the radio transmit button, she ordered them to head back over the Gulf to orbit and wait. She would return for the pickup in under an hour.

Weapons Assembly Room, Bushehr Power Plant, Iran, 0223 Hours December 28th, 2006

The warning klaxon sounded, and the security reaction team rushed to the access doors. It did no good. The guards had hardly made it to their posts when the lights went out and the doors were blown open by small shaped charges. Combined with some flash-bang grenades, the effect was intended to render those inside temporarily deaf and blind, unable to respond. It worked pretty well, with only two guards requiring some non-lethal projectiles to take them down. The use of less-than-deadly force was not so much in the interests of humanity, as to minimize dust and contaminants in the almost surgically clean room. Within seconds, the Marines had the room secured, and Lieutenant Colonel Tom Shaw, the commander of the 3/8 BLT, the 24th's GCE, strode in to take charge of the scene.

What he found was a white-painted, high ceilinged room that looked like a cross between an automobile service center and an operating room. Twelve assembly bays were located around the perimeter, each with a partially assembled warhead, or "physics package," sitting on an assembly stand. Off to the side of each assembly bay was a rolling rack of parts and sub-assemblies. As he surveyed the prisoners, he noticed three older men standing off to one side of the cluster of dejected personnel. He ordered two of his Marines to take charge of them and ensure they were on the first evacuation flight back to the *Bataan*. He then went outside and called Colonel Newman in the LFOC to tell him to get the "penthouse" cells of the ship's brig ready for three special prisoners, the Iranian Minister of Machinery, Colonel Gholam Hassanzadeh, and Professor Kim Ha Soon of North Korea. He wondered how the United Nations would deal with these three, but decided to leave that to those with better-looking suits than his. Right now, though, he had a more pressing problem to deal with. "Safeing" a live nuclear power plant.

A VMA-231 AV-8B Plus Harrier II Over the Bushehr Nuclear Power Plant, Bushehr, Iran, 0230 Hours, December 28th, 2006

From the cockpit of Spade-1, Major Terry "Pirate" Kidd could see almost everything happening below through his night-vision goggles and the multi-function display of his FLIR targeting pod. He was flying at 12,000 ft/3,667.7 m. as lead ship of a two Harrier flight assigned to cover Chilly Dog against interference by Iranian forces. Each aircraft carried pairs of Sidewinder and AMRAAM air-to-air missiles, two CBU-87 cluster bombs, a pair of AGM-65G Maverick air-to-ground missiles, and a GAU-12 25mm gun pod. He had listened to Spade-3 and -4 taking out the defense platforms with their HARMs, and was now using his APG-65 radar to track the movements of the HMM-263 helicopters.

As he and his wingman orbited back to the west, he saw the LCACs unloading vehicles and other equipment at the dock, and he smiled as Lieutenant Colonel Shaw called in the success code for capturing the Iranian nuclear weapons and personnel. All that was left was to take care of the reactor itself, and then to get everyone back out into international waters. Thus far, Chilly Dog had gone perfectly, with only two Marines from Charlie Company suffering minor wounds from stray Iranian fire at the power plant.

Then it happened. One of the AH-1W attack helicopters came too close to the Iranian garrison compound, and a trio of shoulder-fired SA-16 missiles lanced out towards the Cobra. The helicopter evaded two of the missiles through a combination of maneuver and decoy flares, but the last missile hit home on the tail boom. Though it was heavily damaged, the pilot managed to get it to the ground, but not before he and the gunner both suffered sprained backs and ankles. They managed to crawl away from the wreckage (which, thankfully, didn't burn), calling on their rescue radios for a TRAP mission. Pirate immediately called for the standby TRAP team: a CH-53E and two Harriers. The TRAP team Marines were standing by on the hangar deck ready to go, and Colonel Newman in the LFOC indicated that they would be on station in twenty minutes. Until then, Pirate and Spade-2 would provide cover for the two downed Marines.

The first problem was to suppress the continuing ground fire from the Iranian compound. Kidd locked up his FLIR onto an air conditioning unit on the top of the nearest barracks and slaved his radar to provide a good delivery solution. He ordered his wingman in Spade-2 to hit the other end of the compound, and they dove on the complex, releasing their CBU-87s. Kidd tried to put out of his mind that he had probably just killed a hundred or more Iranian soldiers. Combat was like that. In the end, what kept him focused was the fact that he was doing it for two brother Marines who were down and hurt. His mind clearer, he turned the Harrier flight around in a wide starboard turn, and headed back to the crash site.

Reactor Control Room, Bushehr Nuclear Power Plant, Iran, 0250 Hours, December 28th, 2006

The last act of the Chilly Dog assault plan was "safeing" the reactor plant. This meant finding a way to rapidly shut down the plant, and then to make it incapable of producing plutonium. The solution had been found in an IAEA report on a Czech nuclear plant that was a near twin of the Bushehr facility.

When you perform an emergency shutdown of a nuclear pile, called a SCRAM, there is a *lot* of latent heat left in the reactor. Even with the cooling pumps working full speed, the IAEA specialists figured that it would take three to four days for the plant to go "cold" to the point where it could be completely shut down. Destroying the huge water-cooling towers was out of the question. Damaging the control rod assembly was also ruled out, since it would require opening the radioactive reactor pressure vessel. The experts therefore decided that the safest course of action would be to eliminate the ability to restart the reactor by taking out the control rod electronics and consoles, once the reactor had been SCRAMed and backup generators started to maintain the cooling pumps' vital flow of water. This would require access to the main control room of the plant, and that was easier said than done.

Just as the laws of physics dictate the design of a nuclear reactor core, regardless of the owner's ideology, the laws of small-arms fire and human psychology dictate the design of a reactor control complex. Security is a fundamental design criterion. To be certified as safe to operate, a reactor control complex must pass a rigorous security-threat evaluation, just as its overall design, systems redundancy, documentation, and operator training must be evaluated by appropriate experts. Over the years, a great deal of high-tech wizardry had been proposed for ensuring the safety of reactor controls against a well-armed and well-organized terrorist attack. Entry locks keyed to retinal patterns, fingerprints, or brain-wave spectrums of authorized personnel. Passageways that can be instantly filled with sticky foam, or debilitating gas.

At Bushehr, though, physical security relied on the tried and true system of steel doors with firing ports, and men with automatic weapons behind them. These defenses were deployed in depth, with a labyrinth of right-angle turns that created "man-trap" corridors with kill zones swept by fire from two directions.

But anything defended by men with guns can be taken by men with guns. The variables are hard to quantify, but they include training, small-unit cohesion, special weapons and tactics, and something indefinable that falls somewhere between uncommon valor and common craziness. The Marines of the 22nd MEU (SOC) had practiced this drill many times, often taking the role of the "aggressor" forces in exercises staged with the cooperation of the Department of Energy at a variety of active and decommissioned nuclear plants.

The main outer gate resembled a bank vault door; indeed it had been installed by the same firm that supplied most of the vaults for the better-known Swiss banks. In initial planning for the mission, Major Shaw of VMA-231 had proposed cracking the gate with the formidable armor-piercing warheads of precision-guided Maverick missiles; but the problem of targeting in the confusion of the ground battle, the proximity of friendly troops, and the risk of collateral damage to the plant had ruled this out.

In the end it came down to the practiced eye and hand of Lance Corporal Drew Richardson, an AT-4 missile gunner in the Heavy Weapons Platoon of Charlie Company. Repeated direct hits with shoulder-fired rockets left the massive steel door twisted and hanging from its hinges. Two Marines managed to loop a steel cable around the wreckage, and the powerful winch on an LAV combat engineer vehicle, landed by one of the LCACs, pulled it clear. Inside the door, the passageway made a right-angle turn, and the darkened corridor was under fire from both ends.

The security team had ordered the control crew, including the foreign contract workers, behind an armored door, and prepared to defend the room against the Marines that they now knew were inside the perimeter of the power plant. They had given up trying to call for outside help long ago, the phone lines having been cut by the Force Reconnaissance teams and the airwaves jammed by a LAV electronic combat vehicle brought ashore by the LCACs. There was little they could do except defend the room to the death, which was exactly what they intended to do.

The drill for the forced-entry team was not subtle. One man would toss smoke grenades around the corner, while a pair of AT-4 gunners, each wearing a respirator and lightweight FLIR goggles, rolled out onto the floor, firing into the next barrier. The team leader used a thermal viewer with a right-angle periscope to determine the results of each shot. This had to be repeated several times before the last guard posts were silenced and the final steel door to the reactor control room was breached with a demolition charge.

The assault force poured through the opening, to take the control room crew into temporary custody. The night-duty crew inside the control room consisted of about a dozen technicians. Some had been deafened by the blast, and a few were cut by splinters, but they had had the sense to stay away from the door when they heard the first muffled explosions. As the Marines cautiously entered the room, they quickly secured the technicians, binding their wrists with plastic handcuffs, and separating the contract workers from the Iranians. The native technicians were trooped outside to a holding area, while the three foreigners were kept in the room.

Among them was Lev Davidovich Telfian, who had wisely donned earplugs and goggles before the engagement began. He had decided as soon as the first warning had sounded to lay low, taking no action that might be construed as favoring either side. While he hoped that the Marines would evacuate him, he feared that they might just as well leave him behind. He was relieved when the young lieutenant commanding the assault team came forward and greeted him personally with a warm handshake.

AH-1W Crash Site South of Bushehr Nuclear Power Plant, Iran, 0255 Hours, December 28th, 2006

The two Cobra crewmen had taken cover behind a large rock, and were monitoring the status of the TRAP team's progress on their rescue radios. Colonel Newman was better than his word, and the big CH-53E with its security team arrived eighteen minutes after the emergency call. With two fresh Harriers flying top cover, the Super Stallion touched down, and a platoon of Marines fanned out to surround the site. While four Navy corpsmen saw to the injuries of the Cobra crewmen, four more Marines approached the wreck of the AH-1W, secured the classified and crypto components, and set demolition charges to ensure nothing useful to the Iranians was left behind. Five minutes later, mission accomplished, the CH-53E lifted off. The charges detonated, turning the wreck into a blazing fireball of jet fuel and ammunition.

Reactor Control Room, Bushehr Nuclear Power Plant, Iran, 0310 Hours, December 28th, 2006

The control room was becoming crowded with all the witnesses and the CNN camera crew. Colonel Newman had made it clear that this phase of Chilly Dog would be documented to the smallest detail. For Wendy Kwan, looking somewhat less than glamorous in a desert camouflage Kevlar helmet and battle-dress uniform, it was the most exhilarating and frightening experience of her life. She watched as the Marine technical team leader pressed the red SCRAM buttons for each reactor, setting off a chorus of alarms. Each move was supervised by Hans Ulrich, Professor Kennelly, and a Russian she did not know. After the alarms and warning indicators had been turned off, and the backup generators had automatically kicked in to keep the cooling circulation pumps running, the Marines went to work.

They rapidly dismantled the control rod assembly panel, leaving only cable ends clipped off, their connectors removed for good measure. The racks of control electronics were given the same treatment and wheeled out of the room. Finally, a pair of Marines with sticky-foam guns arrived. They filled the control conduits to the reactors with the quick-setting foam, making it impossible to restore the plant's control circuits without extensive demolition work. When this was done, the room was evacuated, and it was time to go home. Ten minutes later, at the suggestion of Lieutenant Colonel Shaw, the Iranian technicians returned to the control room, taking over the job of monitoring coolant flow to the rapidly faltering reactor.

LFOC, USS *Bataan* (LHD-5), 40 nm/73 km West of Bushehr, Iran, 0315 Hours, December 28th, 2006

Criminals say that breaking into a bank is hard, but getting away is harder. It was now time for the 22nd MEU (SOC) to get the hell out of Iran. While they had done immense damage, their luck could not hold forever. Already there were seven casualties, and additional delay on the Iranian coast would only cause more. First to go were the LCACs, with their load of equipment from the plant and reprocessing facility, as well as the heavy vehicles and trucks. The partially assembled warheads followed in a pair of CH-53Es, along with the prisoners from the assembly room in MV-22Bs. Charlie Company in their rigid raiders left next, escorted by the three surviving Cobra gunships. A single CH-53E, covered by Major Kidd's two Harriers, moved around the battlefield, retrieving sniper and Force Reconnaissance teams. Captain Hansen and his AAAVs withdrew through the mud flats, to begin their high-speed swim back to *Trenton* (LPD-14). Last out was Lieutenant Colonel Shaw aboard Lieutenant Colonel Taskins's Osprey. Five minutes later, the only sounds to be heard at the Bushehr nuclear power plant were the hum of the backup generators and circulation pumps and the sputtering explosions of ammunition cooking off in the barracks across the road.

USS *Bataan* (LHD-5), 0415 Hours, December 28th, 2006

The Air Boss spent a busy twenty minutes getting LCACs and aircraft aboard; the elevators had never been worked so hard in so short a time. First aboard were the Harriers, which were rapidly rearmed, refueled, and launched to provide combat air patrol (CAP) during the critical hours to come. The nuclear material was loaded into shielded containers and sealed for shipment. The prisoners were processed into three groups. The "special" tags were assigned to key leaders and technicians, who went straight to the ship's brig and a round-the-clock suicide watch. Minor personnel were restricted to a chain-link-fenced area in the hangar bay, until they could be returned to Iran through the Red Cross. Finally, there were evacuees like Lev Davidovich Telfian, who were given a ration of medical bourbon, a hot breakfast, and a stateroom to sleep off their adventures. He shared it with the Pakistani technician who had passed him the CD-ROM, and both slept well for the first time in months. For the Marines, Captain Rainbow had laid on a special meal of steak and eggs, followed by a quick cleaning and stowing of weapons before they hit their bunks. When *Trenton* rejoined the formation, the ARG and its escorts laid on 24 kt/44 kph, and headed for the Straits of Hormuz. There they would pick up a CAP of F-14 Tomcats and F/A-18 Hornets from *Constellation*, and would head out into the open ocean and Diego Garcia, where they would off-load their cargo and passengers.

USS *Constellation* (CV-64) Battle Group, Arabian Sea, 0430 Hours, December 28th, 2006

As things calmed down in the Persian Gulf, there was one final act to Chilly Dog. At 0430 hours, two Aegis cruisers and a pair of *Spruance*-class destroyers began to launch a strike by 124 BGM-109 Tomahawk missiles against the automobile factory at Bandar al Abbas and missile batteries in the Straits of Hormuz. After an overland flight from the Arabian Sea side of Iran, they leveled their targets, fully 88% of the missiles striking their designated aim-points. With this, Chilly Dog came to an end. The political fallout around the world, though, would last for months.

Joint Session of the U.S. Congress, Washington, D.C., January 18th, 2007

The final words of the President's 2005 State of the Union Address were simple, as all good speeches should be. "Ladies and gentlemen, I summarize the results of the Bushehr Raid in this way. We have decisively ended a clear violation of the Nuclear Non-Proliferation Treaty, as well as a threat to the stability of Southwest Asia. Even more important, the individuals who perpetrated this violation are to be tried shortly for crimes against humanity. Already, we have seen the fall of the Islamic Revolutionary government in Iran and the beginning of a thaw between ourselves and the people of that troubled land. We offer the hand of friendship and commerce to the people of Iran, and the hope that the terrible fire that we have contained will never again raise its head in the Persian Gulf.

"I also wish to thank the men and women who conducted this action. We live in a new millennium, and unless we choose different paths for ourselves and our world, the human race will not see another. Luckily, we have good people out there who stand on the walls, and guard them for us as we sleep. I never want to be without a military to watch over our interests, and protect them. God bless them, and God bless the United States of America!"

Officers' Mess, USS *Bataan* (LHD-5), Western Mediterranean, January 18th, 2007

Lev Davidovich Telfian watched the State of the Union address with a smile on his face, knowing that he had helped in bringing a happy end to this situation. He was still aboard the *Bataan*, where he would be safer until the memories of the Iranians were less fresh. Telfian had been thinking about what he wanted to do next, and he had several offers. One came from the IAEA to work on an inspection team in South America and Africa. Another one had been extended from the U.S., to work as a consultant for the Defense Nuclear Agency on counter-proliferation issues. Even the SVR was making him an offer, as an intelligence analyst at Moscow headquarters. That last one held few attractions for him. Perhaps the Americans. At least they were working to take the damned bombs apart. After being so close to the nuclear genie for so long, perhaps it was time to try and force it back into the bottle.

Operation Tropic Fury: The Liberation of Brunei, September 2008

Limbang Valley, Brunei, September 2nd, 2008

This morning, the Sultan of Brunei would dedicate a new clinic for the hill tribes of the upper Limbang Valley. The Royal helicopter, a luxurious Sikorsky S-76, snaked up the valley from the South China Sea toward the rain-forested mountains of northern Borneo. It was only a twenty-minute flight from the Palace, and the airspace over the entire district had been cleared of other traffic. Sometimes the Sultan, an enthusiastic and reasonably skilled pilot, liked to take the controls himself, but today he was content to sit back and browse through the electronic edition of the *Wall Street Journal* on his new Toshiba data pad. The lead story was, in fact, about his plan for partition and management of newly surveyed oil fields in the South China Sea.

They centered around the Spratlys, a few barren reefs that had ambitions to be islands at low tide. The new oil pools were probably the biggest at-sea petroleum find since the North Sea fields back in the 1970s. Unfortunately, the nations surrounding this oil discovery were nowhere as reasonable as Great Britain and Norway when they partitioned the North Sea fields. Half a dozen nations had claims over the new oil find, and few of these nations could be described as "reasonable." To the east lay the Philippines, where a share of the oil revenue might relieve an exploding population's chronic poverty. To the west, the Communist governments of China and Vietnam coveted oil to fuel their economies and earn hard currency from petroleum exports. To the north, Taiwan, still claiming to be the "true" Government of China, felt entitled to a piece of China's share. But the real trouble lay to the south, where Malaysia, Indonesia, Singapore, and Brunei all had claims on the new fields, and some of them were willing to fight for a larger share.

Tiny Brunei, with the wells that produced fabled North Borneo crude, the world's purest oil, was the richest nation per capita in the world. This provoked envy among neighbors, particularly Malaysia, with its growing population, simmering ethnic tensions, and lack of oil resources. The Malays had been full of threats this past summer. The reason was the upcoming United Nations conference at the end of October which would settle, once and for all, the development plans for the South China Sea oil fields. Malaysia had joined Indonesia in a coalition for the coming conference, and was trying to entice Singapore. The same offer had been extended to the Sultan, but he had politely declined the invitation.

He would place a proposal on the table to create a multi-national non-profit corporation which would invest the oil income into a regional development fund to build schools, roads, and other infrastructure so badly needed by the peoples of the region. The Sultan knew that the leaders of the other countries did not share his vision, and this was why he wanted to put his ideas on the table at the UN. The article in the *Wall Street Journal* spelled out the details of the plan—as well as the first reactions to it, which had come quickly. Malaysia and Indonesia had denounced it. Vietnam and China had remained ominously silent. But Singapore, Taiwan, and the Philippines had all endorsed the idea, and this gave him hope. He smiled and sat back, composing his thoughts for the clinic dedication.

The bright, newly painted yellow circled "H" of the clinic's medevac landing pad was just coming into view around a bend in the river. Today's duty pilot for the Royal Flight was a retired British Fleet Air Arm commander with thousands of hours logged in just about anything with rotors. He had also been trained in escape and evasion. It didn't do him much good. He caught the flash on the ground and instinctively pulled into a hard break to starboard. The data pad flew out of the Sultan's grip and bounced off the Plexiglas windscreen.

The pilot's move was too late. The seeker head of the first shoulder-fired missile had locked onto the hot metal of the turbine exhaust even as other missiles lanced upward from the opposite side of the valley. The high-explosive warhead detonated on impact with the port engine, shredding fuel lines, hydraulic tubing, and control cables. A single missile hit might have resulted in a survivable crash, but the second hit turned the tough and graceful chopper into a flying cloud of flaming wreckage. By the time the stunned VIPs and the medical team from the clinic arrived at the crash site, His Royal Highness, the Sultan of Brunei, by some estimates the world's richest human, was identifiable only from dental records.

The Palace, Bandar Seri Begawan, Brunei, September 2nd, 2008

Crown Prince Omar Bolkiah, twenty-six years old, was on the tennis court with an instructor when the elderly, respectful, and impeccably discreet palace chamberlain arrived with the news of his father's death. Omar was unsure which of his innumerable half-brothers had engineered the assassination, though he had a reasonable suspicion which foreign power had provided the hit team, and he knew that his own life would not be worth a ringgit if he were found anywhere within the fifty-acre Palace compound. Twenty minutes later, veiled and shrouded in women's garb and surrounded by a gaggle of his favorite sister's servants, he slipped out of a little-used riverfront exit and boarded a small boat. Within an hour, dressed in the plain white uniform of a junior naval ensign, he was embarked on the rusty but trusty missile patrol boat *Pejuang*, listening to the throb of the twin diesels as she slipped out of Muara harbor, bound for the treacherous shoals of Louisa Reef.

The young Prince ("No, now I have to start thinking of myself as the Sultan," he thought) had many worries, but pursuit was not one of them. There were men he could trust in the Navy. As the sun went down over the South China Sea in another magnificent tropical blaze of glory, every other patrol boat in the Royal Brunei Navy swung gently at her moorings, polished and scrubbed to the best Portsmouth standards, and thoroughly sabotaged. Within days, some very junior mechanic's mates would pay with their lives for their loyalty to their Prince.

•

British Embassy, Washington, D.C., September 5th, 2008

The package had arrived on a trans-Pacific red-eye flight into Dulles from Singapore under the diplomatic seal of Her Majesty's Foreign Service. It was picked up by a car from the British Embassy and escorted by two Secret Service Chevrolet Blazers. That was unusual, but at that hour, there was no one around to take any particular notice. The men gathered to examine the package came from a variety of military, diplomatic, and intelligence services. The Americans had better teeth. The British wore better-fitting suits. They had all been through this drill before. A fine linen tablecloth was flung over the exquisitely inlaid conference table; then the work began. Some of the tropical hardwood trees from which the table was made had been logged over a hundred years before, not far from the crash site. The package was opened without ceremony, and the charred and blackened shards of metal were passed from hand to hand for examination and judgment.

"Our Special Air Service lads picked these bits up the night after the crash. Rather a lot of confusion on the scene, as you can imagine. They had the devil of a job getting in and out without being spotted."

"No question," one of the Americans finally said. "This is a Chinese copy of the Stinger."

The missiles that had downed the Sultan's helicopter were, therefore, untraceable. You could pick one up in any Third World arms bazaar for a few thousand deutschmarks. The next question, asked by the President's National Security Advisor, was aimed at the British ambassador.

"Mister Ambassador, what is the position of the British Government on this matter?"

"My Prime Minister is, as you know, in a very difficult position. British Shell and Lloyds are the primary guarantors of more than a trillion British pounds of investment in both Malaysia and Brunei. Potential revenues from those two countries represent many times that amount. As might be imagined, British industry is putting huge pressure on our government to do absolutely nothing and accept this new arrangement as a fait acompli. The reality is that what we have here is nothing more than the rape of a small country by a larger and more powerful neighbor, just like Kuwait in 1990. Thus, while we will not be seen taking active measures, rest assured that we will support any initiative by your Government to restore the status quo." The ambassador then extended his hand to seal the latest of many such back-channel deals between the United Kingdom and her former American colony. Once again, the "special relationship" had been reaffirmed.

Off Louisa Reef, South China Sea, 0400 Hours, September 6th, 2008

Commander Chu Hsiang-kuo raised *Hai Lung's* periscope and slewed it around the horizon with a practiced flick of the wrist. There was the Bruneian patrol boat, a few hundred yards/meters to the south, just where he had been told to expect it. "Helm, come to course one hundred eighty degrees, slow to five knots and prepare to surface." Chu clicked the stopwatch button on his Rolex watch, a gift from an uncle who owned a major Taiwanese electronics firm. He planned to spend no more than three minutes on the surface, and had drilled his crew for days to shave every possible second off the tricky rendezvous and pickup. There were too many Mainland Chinese patrol planes about to allow him the luxury of loitering on the surface. Two minutes and forty nine seconds later the hatch clanged shut and His Royal Highness, the Sultan of Brunei, was a guest aboard the Republic of China's best submarine. The traditional formalities of piping a head of state on board were dispensed with; Omar simply gave Commander Chu a bear hug. The Prince was finally safe, though the patrol boat crew would have to hot-bunk in the Taiwanese sub's cramped accommodations for a while. As *Hai Lung* (*Sea Dragon*) dove, the patrol boat *Pejuang* wallowed deeper on the glassy sea and slowly capsized. Scuttling charges might have attracted unwanted attention. The Mainland Chinese had wired these waters for sound, and this last act of Brunei's Navy was played out quietly.

The Palace, Bandar Seri Begawan, Brunei, September 6th, 2008

Surrounded by thuggish bodyguards, twenty-one-year-old Prince Abdelrahman, brother of the missing Crown Prince, looked distinctly uncomfortable in the uniform of a field marshal. It would be his first press conference. Although his handlers had thoroughly drilled and briefed him in the four days since the assassination, the "live" satellite feed, in English with simultaneous translation into Malay, Mandarin Chinese, and several regional dialects, had a seven-second delay; and a senior Malaysian intelligence officer, covered as an audio technician, was standing by the "kill" switch, just in case Abdelrahman said anything particularly stupid.

He coughed and stuttered out, "In the name of Allah, the merciful, the compassionate, I, Prince Abdelrahman Bolkiah, Sultan of Brunei, have the sad duty to inform my people and the world of the events that have shattered the peace and tranquillity of our country during the past week. We have uncovered proof that our late father, the Sultan, was treacherously murdered in a plot by our half-brother, the former Crown Prince Omar, who has fled the country. We will pursue this criminal by every possible means and bring him to justice. Our government will regard it as a most serious breach of international law if any foreign power gives sanctuary to this criminal.

"Even as We exert every effort to avenge our father's murder, we must take thought for the future of our people. For over a hundred years, this Sultanate has been a vestige of colonialism and a geopolitical anomaly." He paused for a sip of water. The English phrases would be a mouthful for the poor translators. "We have consulted with representatives of our people and our faith." He nodded toward the hard-line Islamic fundamentalist imams who had taken control of the local ulema, the collective interpreters of Muslim religious law.

"We have therefore determined that Brunei will formally request admission to the Federation of Malaysia. We have received assurances from His Excellency, the Prime Minister of Malaysia, that the traditional prerogatives of the Sultanate and the customs, culture, and traditions of our people will be fully respected. Also, in the forthcoming international conference on territorial waters in the South China Sea, Brunei's historic claims will be represented with the full power of the Malaysian Federation. Our military forces will be merged into the Malaysian armed forces, and the Brunei dollar will be withdrawn from circulation and exchanged for the Malaysian ringgit at a very favorable rate. Foreign embassies will be given every assistance in relocating their facilities and staff to Kuala Lumpur, and we invite all the nations with which we have enjoyed friendly diplomatic relations to maintain an appropriate consulate here in Bandar Seri Begawan." He finished with the words "Peace be upon you." There were no questions. Everyone agreed that, for his first press conference, the kid had followed the script pretty well.

Headquarters, Fleet Marine Force, Pacific, Pearl Harbor, Hawaii, September 7th, 2008

Lieutenant General Sidney Bear, USMC, was not a subtle man. Built like his name, he carried an old Naval Academy nickname of "Teddy," reflecting his kind and gentle nature. But at times he had a temper. Now was one of those times. As commander of Marine Forces, Pacific (FMFPAC), he was responsible for all Marine Corps activities in the Pacific Theater, and he had problems, both big and small. The American decision not to recognize the new Sultan and ignore the order to relocate the U.S. Embassy in BSB had caught U.S. Pacific Command by surprise. The general's first concern was for his own, of course, the squad of Marine guards at the U.S. Embassy in BSB. He quickly set up a video teleconference over a secure satellite link. The military attaché at the embassy was an Air Force lieutenant colonel, but the general was relieved to see that the security detachment was led by an experienced gunnery sergeant. This was probably the first time the gunny had ever talked to a three-star general via the jerky image and fuzzy audio of an encrypted video phone, but his confidence and professionalism came across loud and clear.

"We've had a crowd of people at the gate this morning lined up to apply for visas, sir, but otherwise it's been business as usual."

"Gunny, I'm counting on you to be my eyes and ears until we can get you some reinforcements. Have your men keep a low profile. If they storm the embassy, let 'em have it. It's not worth dying for. We're gonna get you out of there real soon, but until then you're my eyes and ears on the spot. Anything unusual happens, you get on the horn to my Ops officer, ASAP. Understood?"

"Semper Fi, sir!"

No further explanation was required.

The White House, Washington, D.C., 1000 Hours, September 8th, 2008

The Secretary of Defense brought over a vanload of wall charts, slides, high-resolution satellite imagery, and documents to brief the President of the United States about the situation in Brunei. Then, the Secretary of State discussed the regional and global ramifications of the crisis. Finally, the National Security Advisor and the Chief of Staff explained it to him in simple language. These preliminaries over, the President made phone calls to London, Paris, and Moscow, and it was decided. The change of government in Brunei was an illegal coup d'etat. The policy of the United States was not to recognize any change in the international status of the Sultanate, and to seek to restore the reign of his rightful successor, the Crown Prince Omar Bolkiah.

Somebody recalled a phrase from the early 90s. "This will not stand."

A political solution through the United Nations Security Council would be pressed, but NSA analysis of the message traffic out of Beijing made it clear that a Chinese veto could be expected. That left only one alternative. The Secretary of Defense called the Chairman of the Joint Chiefs. The Chairman called CINPAC. CINCPAC called the FMFPAC. A planning cell was activated in a dingy basement office under security so tight that only a half-dozen officers were fully "read-in" on the time, the place, and the objective. Wheels began to turn.

Headquarters, The 7th Gurkha Rifles, Seria, Brunei, September 9th, 2008

For decades Brunei Shell Petroleum had entrusted the security of its oil fields to the small, brown, and very capable hands of the Gurkhas. A Nepalese hill tribe, the Gurkhas enjoyed a unique relationship with the British Crown, combining elements of honor, tradition, mutual admiration, and direct cash payment. Maintaining a regiment of nine hundred Gurkhas cost the Sultan fully five million British pounds a year, and it was worth every penny. Nobody messed with the Sultan's oil fields. No professional soldier in the world ever wanted to go up against Gurkhas.

It was a delicate situation. Recruited and trained for generations by the British Army, the Gurkhas had been hired by Brunei to defend its oil fields, and there was no doubt in anyone's mind that as long as one of them remained alive, they would do exactly that. Colonel Rai stood 5 ft 4 in./1.6 m tall and weighed about 105 lb/47.6 kg, soaking wet. He was fifty-two years old, and could still decapitate a water buffalo with one stroke of his razor-sharp *kukri,* the curved fighting knife that represented the mystical center of the Gurkha warrior tradition. He rarely wore his full-dress uniform; his days were mostly spent on patrol with his men, or with the handful of foreign special forces officers who were favored with the privilege of jungle training with the Gurkhas. But today, every crease was as sharp as a *kukri,* and every bit of brass gleamed like gold, because he was receiving a special guest, a personal envoy from his own Hindu monarch, the King of Nepal. Tea was poured, gifts were exchanged, and there was polite small talk while an orderly cleared the table.

"His Majesty desires the presence of your regiment in Katmandu for an important ceremony," the envoy said.

"We are not worthy of such an honor, and duty requires our presence here in Brunei. Surely His Majesty understands," Rai said.

"The 14th Gurkha Rifles will rotate in temporarily to perform your duties. The British Prime Minister has graciously offered the use of Royal Air Force transports to fly you and your men directly to Nepal at no cost."

The warrior and the diplomat made eye contact. Faint smiles flickered across their impassive faces. Little was said and much was understood.

"Please convey to His Majesty my deepest gratitude for this honor."

By the end of the week, the 7th Rifles were out of the country, and for some unaccountable reason, they wound up in Manila, billeted in the same hotel as the Crown Prince Omar Bolkiah. At the same time, the 14th Gurkhas were held up in transit. Problems with paperwork, it was said. Diplomatic channels hummed with profuse apologies, while Malaysian authorities scrambled to recruit temporary security guards. For now, though, the new Sultan had only a Malaysian shield.

Prime Minister's Residence, Kuala Lumpur, Malaysia, 1430 hours, September 10th, 2008

It was intolerable. The Prime Minster was not a patient man. He had devoted a long career to building up his fragile nation into a respected regional economic and military power. And now the insolent American task force, steaming provocatively though his territorial seas, was requesting that Malaysian patrol planes keep a distance of at least 50 nm/91.4 km to avoid "unfortunate incidents." In response, he had summoned the American ambassador and browbeaten the man for half an hour. The bland diplomatic replies about "freedom of navigation" and "precautionary measures" had only infuriated him more. Malays could be a hot-tempered people. *Amok* is a Malay word, and the Prime Minister was just about ready to run amok. As soon as the American had been dismissed, the Prime Minister grabbed the red phone that connected him directly to the Armed Forces Chief of Staff. He would give them an incident to remember.

Above the South China Sea, 1500 Hours, September 10th, 2008

There was a time when a flight of four vintage MiG-29 Fulcrum-Cs flying top cover for a gaggle of four shiny new F/A-18C Hornets might have seemed bizarre. In the New World Order, though, any mix of aircraft was possible. The Malaysian Air Force had stretched its limited budget by driving hard bargains, East and West, and the result was this formation. Squadron Leader Edward Tawau, call sign Red Dragon, nervously thumbed the stick's radar-mode select switch between air-search and surface-search modes. He didn't like this mission one bit. His orders were to fly directly over the ships of the American task force at low altitude, cracking sonic booms just above their mastheads. At the pre-dawn briefing, the Wing Commander had assured the pilots that the Americans would back off as soon as they understood that Malaysia was serious about enforcing its sovereignty.

The Wing Commander (call sign Blue Python) had been born into a princely family of one of the little sultanates that made up the Malay Federation and trained by the RAF. He was contemptuous of Americans, a strange people, wholly without courtesy and lacking in any sense of family honor or obligation. On the other hand, the Squadron Leader's parents had met in a factory that assembled circuit boards for an American computer company, and he had learned to fly his F/A-18C Hornet in Florida. He might not understand Americans, but he was not likely to underestimate them.

"Red Dragon, this is Blue Python," the radio crackled. "Surface ships bearing one hundred degrees at seventy-five miles." Tawau briefly flicked the switch to surface-search mode to confirm the contact, then went radar-silent. No point in giving the Americans any extra advance warning. They would be over the target in ten minutes. Five minutes later, his radar-warning receiver alarm sounded. One of the American escorts had just locked him up with a fire-control radar. This game of chicken was getting serious!

Combat Information Center (CIC), USS *Bon Homme Richard* (LHD-6), 1505 Hours, September 10th, 2008

Captain Mike Anderson had seen the incoming flock of fighters at almost 120 nm/220 km, and had already taken measures to deal with the threat. Two AV-8B Plus Harrier IIs loaded with Sidewinders and AMRAAMs on "Plus Five" alert had been launched when the force had hit 100 nm/183 km, and two more were being readied for launch. Then, over the communications net, he heard the PHIBRON commander, a rear admiral, order, "Warning Yellow, Weapons Hold," to the force and its escorts. This meant that an attack was expected, and that weapons could be fired in the event of a hostile act. What was headed their way looked like big trouble, and Anderson ordered General Quarters. It was going to be an exciting few minutes.

Red Python Flight, 1508 Hours, September 10th, 2008

At 65 nm/119 km, Squadron Leader Tawau heard an American-accented voice over the Guard channel of his radio warning them to veer off and maintain at least 50 nm/91.4 km distance from the force. Through his headphones, he heard the Wing Commander snort his contempt in response and order the aircraft to continue. It was getting ugly. Tawau decided to check the air situation, and was not surprised to see a pair of unidentified contacts closing in from the side. It got even worse a minute later. After crossing of the 50-nm/91.4-km line, his radar-warning receiver blared, showing a pair of air intercept radars to port. He wanted to order his flight to turn around, but as he moved his finger to press the microphone button, two of his F/A-18s exploded into blazing fireballs, victims of what had to be the fabled AIM-120 AMRAAM missiles of the American fighters. Then, through the top of his canopy, he saw two MiGs explode in the same way. Over the radio, he heard the Wing Commander call for him to close on the ships and shout, "Weapons free!" on the squadron net. Feeling growing anger at this stupidity, but unable to defy an order, he ordered the surviving Hornet in his flight to follow him, selected afterburner, turned on his jammer, and put the nose over into a dive heading for the amphibious ships. He never saw the Wing Commander's aircraft disintegrate into a ball of fire from a Sidewinder hit and the surviving MiG-29 running for home. He was following the last legal order he had been given, bad as it was.

Minutes later, as the indistinct shapes of the task force appeared as dots on the horizon, he saw the flash and smoke trail that indicated a SAM launch from one of the escorts. Both F/A-18Cs commenced evasive maneuvers heading for the deck. As he did, the SAM arched down towards him, detonating above and behind. Shredded by the fragments from the warhead, his Hornet began to break up, and he activated his ejection seat. Seeing his flight leader shot down roused the other young Malaysian pilot to fury, and he continued down to just above the wave tops. Heading along the last bearing to the enemy task force, he flew until one of the big amphibious ships suddenly loomed right in front of him. Arming his Gatling gun, he lined up on the ship and set up a shallow dive for a strafing run....

Aboard the USS *Germantown* (LSD-42), 1513 Hours, September 10th, 2008

When the F/A-18s had continued on a course that looked like a classic attack profile, the ARG commander had given orders that enabled the fire-control computers to engage as soon as hostile aircraft came within weapons range. But no one could have predicted that one F/A-18 pilot was crazy enough to close to strafing range at wave-top level, and no one could have predicted how badly a long burst of 20mm shells could shred the bridge of an amphibious transport. It did the young Hornet pilot little good in any case. One of the *Germantown*'s 20mm Mk 16 Phalanx CIWS mounts shattered the F/A-18, sending it careening into the sea, killing the pilot.

Pentagon Press Room, 0800 Hours, September 10th, 2008

"At approximately 2:00 AM Eastern Time this morning, eight Malaysian aircraft were detected approaching a U.S. Navy task force in international waters, transiting the South China Sea. After ignoring repeated requests on recognized international frequencies to maintain a safe distance, these aircraft were engaged by defensive systems. We believe that seven were shot down and one turned back. One aircraft closed on USS *Germantown* and opened fire with its cannon before being shot down. Twenty-six sailors and Marines on board *Germantown* were killed, and eighteen were seriously injured.

"Air-sea rescue helicopters from the task force are still searching for survivors from the Malaysian aircraft. The Secretary has asked me to emphasize that the United States does *not* regard this incident as an act of war. Let me repeat, we are not at war with Malaysia. We are attempting to defuse a volatile situation in a troubled part of the world. At an appropriate time we will seek, through diplomatic means, a formal apology for the loss of American lives and compensation from Malaysia for the damage to our ship. Meanwhile, the United States intends to closely watch the situation in occupied Brunei and will continue to uphold the principle of freedom of navigation in international waters, as we have done for over two hundred years."

A map of the South China Sea theater during the opening moves of the invasion of Brunei.

JACK RYAN ENTERPRISES, LTD., BY LAURA ALPHER

When the Department of Defense press officer sat down, the State Department press officer took the microphone, cleared her throat, and read the paper that had been handed to her a few minutes earlier.

"Until the situation is clarified, the State Department has advised Americans in Malaysia or occupied Brunei to leave the country by the first available means. Also, American passports will not be valid for travel to Malaysia or occupied Brunei. The President has issued an Executive Order freezing all assets of Malaysia and Brunei in American financial institutions. Our Ambassador to the United Nations has asked for an emergency meeting of the Security Council tomorrow morning. Thank you, ladies and gentlemen. No questions, please."

United Nations, New York City, September 11th, 2008

RESOLUTION 1446

The Security Council,

Grieved by the death of His Royal Highness the Sultan of Brunei, under unexplained circumstances,

Deeply concerned that the annexation of Brunei by the Federation of Malaysia is being implemented without regard to the freely expressed desires of the Bruneian people,

Alarmed by recent naval incidents in the South China Sea involving armed clashes between forces of the Federation of Malaysia and the United States of America,

Acting under Articles 39 and 40 of the Charter of the United Nations,

Demands the immediate and unconditional withdrawal of Malaysian military forces from the territory of Brunei,

Calls upon the Federation of Malaysia, the United States of America, the Sultanate of Brunei, member States of the Association of Southeast Asian Nations, and all other concerned States to begin intensive negotiations for the peaceful resolution of their differences,

Decides to meet again as necessary to consider further steps to ensure compliance with this resolution.

The resolution carried, with fourteen for, one against (Indonesia), and two abstentions (China, Japan). If China had vetoed the resolution, Malaysia's annexation of Brunei would have been a done deal, and the fragile "New World Order" could like it or lump it. Diplomatic pressure cut no mustard with the Chinese Communists. But in the days before the vote, chief executives of the major Western and Japanese banks and oil companies had called their Chinese contacts with a simple, back-channel message. If the takeover of Brunei went unchallenged, there would be no credits for any offshore oil development in the South China Sea, regardless of territorial claims by any power in the region. The Communist Chinese may have been true believers in Marxism-Leninism, but they weren't stupid.

Aboard USS *Bon Homme Richard* (LHD-6), PHIBRON 11, September 12th, 2008

Channel 6 of the Fleet Broadcast Satellite Net was carrying CNN, and the staff of the ARG had gathered in the wardroom at this absurd hour to watch the live feed from UN Headquarters halfway around the globe. The betting was about even. Half the officers figured Malaysia would back down, given the lesson that had been taught to Iraq some fifteen years earlier. Half expected immediate orders from CINCPAC to begin planning for the liberation of Brunei.

Colleen Taskins had not made Colonel in the Marine Corps by hoping for the best. She was the first female MEU (SOC) commander in the history of the Corps. She fully expected the worst, and she was about to get it on her first command cruise. The 31st MEU (SOC) and PHIBRON 11 had no orders yet to retake Brunei, but good commanders anticipate events, and she was trying to do that now. She gathered her staff for a late-night planning session. When the orders came down the chop chain from CINCPAC, her Marines and sailors would be ready.

American Embassy, Manila, the Philippines, September 14th, 2008

Crown Prince Omar Bolkiah and the colonel of the 7th Gurkhas were seated in the embassy conference room, being briefed on the plan for the liberation of Brunei from Malaysia. The young man thought it odd that others would talk so clinically in front of him about fighting for his country and his people. But Colonel Rai had counseled the young Prince that this was the way of soldiers. Although the Americans were talking about his country as if it were a chessboard, they had every intention of giving it back. This, the colonel said, was exactly what they had done back in 1991 for the Al Sabah family in Kuwait, and they would do it now for him.

On the large-screen projector appeared a series of viewgraphs, with smaller insert screens in the corners for each of the major participants in the briefing. One of them was dedicated to the impassive face of the Crown Prince, while the others showed the Chairman of the Joint Chiefs, the CinC Pacific Forces, and the commander of the 31st MEU (SOC), Colonel Taskins. The Prince wondered about entrusting his country to this pixie-faced woman, but she seemed to know her business, and the others on the screen were showing her respect.

The Americans called the forthcoming operation Tropic Fury. He wondered how this one would be remembered—as a triumphant liberation, like Desert Storm, or an abysmal failure like Eagle Claw, the raid to rescue American hostages in Iran. But Tropic Fury looked like it had a chance. Colonel Rai called it a "rock soup" approach, which meant that they would start with very little and try to feed more into the effort if the initial assault worked. He was amazed that the Americans seemed to have thought about this kind of problem so thoroughly, and then remembered how they had been humiliated in the 1970s. The Americans' ability to enforce their will was based upon long experience in such affairs, and he promised himself that he would learn more than a new tennis swing from his time here in Manila.

White House Briefing Room, Washington, D.C., September 15th, 2008

"Ladies and gentlemen, the President of the United States!"

The press room was jammed to capacity for a major policy announcement on what was becoming known as the South China Sea Crisis. Along with the usual Presidential media personnel were the Secretaries of State and Defense with an easel full of briefing charts. The TV lights were running hot when the President arrived, and he moved quickly into his presentation. After a short introduction reviewing events of the last few days, he got to the point.

"...therefore, the United States, in conjunction with the United Nations, is declaring a complete military and economic embargo of Malaysia. The Government of Malaysia has until midnight tonight, Eastern Daylight Time, to clear all air and maritime traffic to and from Brunei, or the U.S. will use force to enforce the embargo. In addition, Malaysia has just five days to withdrawal from the territory of Brunei and allow the return of the Sultan, or measures will be taken to evict them. This is the only warning that will be given, and there will be no negotiations. We have not created this situation. Malaysia has. Now let them solve it, or we will do it for them. That completes my statement. The Secretaries of State and Defense will now field any questions you might have. Good day, ladies and gentlemen."

He turned and headed offstage to cries of, "Mr. President???" from a hundred reporters.

USS *Bon Homme Richard* (LHD-6), Somewhere in the South China Sea, 1100 hours, September 16th, 2008

"Jeez, it's like a whole city of oil tanks. How are we supposed to fight in that?" the Lieutenant said. From the piers of Kuala Belait to the wellheads and pumping stations of Seria, 20 mi/32 km east, the coastal strip was a continuous landscape of immensely valuable and extremely flammable petroleum facilities, punctuated by the flames of a few flare stacks, in fields where it was too much trouble to collect the natural gas for liquefaction.

"With these, Lieutenant," Major Bill Hansen said, tossing a small but surprisingly heavy, round flat bag on the table.

"Excuse me, sir. My Marines are going into combat against guys who have live ammo, and we're supposed to shoot back with beanbags?"

"Non-lethal projectiles, Lieutenant. They're called Flexible Batons, and don't underestimate these things—they'll knock down a horse at twenty paces. And we're going to use these in the shotguns and grenade launchers until we're at least five hundred yards inland from the oil facilities."

The kid was too young to remember the fires of Kuwait. The Major had been there, and he never wanted to see anything like that hellish landscape of smoke and flame again. He patiently explained the rules of engagement for fighting in an oil field. The lieutenant would lead his company through the basics of combat-shotgun refresher training in the morning.

"Anyhow, we don't think you'll be going up against real soldiers in your LZ," the major said. "The oil company security guards are basically rent-a-cops, and we're trying to convince Shell to pull them out in any case."

"Well, sir, I've met some pretty bad-ass rent-a-cops."

"I'm sure you have, Lieutenant. I'm sure you have...." The kid had grown up on the mean streets in South LA. He had probably been shoplifting from liquor stores at an age when the major was learning to tie knots in the Cub Scouts. Now he was a Marine, one of *his* Marines. He was pleased to have such a man under his command.

South China Sea, 40 nm/73.2 km Northwest of Natuna Island, September 17th, 2008

The pride of the Malaysian Navy, *Sri Inderapura* had been launched at San Diego, California, in 1971 as the U.S. Navy's Tank Landing Ship *Spartanburg County* (LST-1192). Decommissioned in 1994 due to changing doctrine and force reductions, the five-thousand-ton vessel had been snapped up enthusiastically by Malaysia as an ideal platform for transporting heavy equipment from the Peninsula to remote North Borneo. Today it was carrying a battalion of Scorpion light tanks and truckloads of fuel and ammunition to reinforce the garrison in Brunei. The modern frigate *Lekiu* was providing escort, her Lynx helicopter probing a few miles ahead on anti-submarine patrol. The Americans had declared an "Exclusion Zone" around Brunei, but an ancient rule of international law decreed that a blockade was only legal if it was enforced. The five hundred soldiers and sailors aboard *Sri Inderapura* were the test case. The riotous abundance of marine life in these tropical waters created a cracking, hissing, cloud of confusion for *Lekiu*'s sonar operators. They knew that Chinese, Australian, American, and Indonesian submarines were lurking about, but it was almost impossible for them to pick out any definite contacts from the biological background. It would be a political disaster to attack a neutral or "friendly" sub. They could only strain to read the flickering screens, and wait.

For the sonar operators on board USS *Jefferson City* (SSN-759) 33 nm/60.4 km away, the throbbing diesels and whining turbines of the Malaysian ships rang out across the thermal layers and convergence zones like fire bells in the night. Eight weeks ago, *Jefferson City* had left Pearl Harbor on another routine peacetime patrol. A few days ago, the boat had been vectored into these shallow, treacherous waters to enforce the Brunei Exclusion Zone. And for the last six hours, the sonarmen had been tracking the enemy ships, refining the fire-control solution to enough decimal places to gladden the obsessive-compulsive heart of a nuclear submarine officer. The Weapons Control Officer spoke one last time to the Skipper. The captain replied, with a crisp, well-rehearsed and unmistakably clear order to fire.

Within a few seconds, a salvo of four RGM-84 Harpoon missiles spurted from the torpedo tubes, bored their way to the surface and emerged from their launch canisters. Even at this distance *Lekiu* must have heard the launch transient, but it was too late for the Captain of the *Sri Inderapura* to do anything except sound General Quarters and deploy damage-control teams. All he could do now was pray that the stream of 20mm slugs from the *Phalanx* weapon system atop his bridge would intersect the flight path of at least one Harpoon in the last fraction of a second before impact. It did. Another Harpoon fell to a Seawolf missile fired at the last minute by *Lekiu*. The other two Harpoons struck the LST. One penetrated into the engine room before exploding, leaving the ship dead in the water. The second struck the vehicle stowage deck, starting uncontrollable fuel and ammunition fires among the combat-loaded light tanks.

Lekiu stood by to recover survivors. By all accounts, they did a first-rate, professional job of seamanship, worthy of the traditions they had inherited from Britain's Royal Navy and their own pirate ancestors. As *Sri Inderapura* rolled over and settled into the muddy sediment of the seafloor, the overcrowded frigate turned back toward her home port. At almost the same time, the Australian submarine *Farncomb* was pumping three torpedoes into a Malaysian Ro-Ro ship, carrying vehicles and equipment for an entire brigade assigned to the defense of BSB. Malaysia would not risk any more ships to challenge the Exclusion Zone.

BSB International Airport, September 17th, 2008

Defense of an airfield against airborne assault was a typical Staff College tactical problem, and Major Dato Yasin, commanding the Malaysian Army's 9th Infantry Battalion, had graduated near the top of his class. First, block the runways to prevent surprise landings. It would inconvenience local commuters, but most of the transit buses from BSB were now parked in neat rows across every runway and taxiway of the huge airport complex. The major had wanted to block the runways with dumpsters and cargo containers filled with cement, but it might be necessary to clear the airfield rapidly to bring in supplies and reinforcements if the damned politicians could get the

American blockade lifted for even a few days. Therefore, a captain in the transport section of the major's battalion now held the buses' ignition keys.

Second, establish interlocking fields of fire across the runways to decimate parachutists in the critical few minutes after they hit the ground. The major had laid out a pattern of carefully camouflaged fighting positions for fire teams and heavy machine guns, with plenty of less carefully camouflaged dummy positions. The major had served with American troops in several UN peacekeeping missions, and while he had never seen "primary" high-resolution satellite imagery, the unclassified "secondary" imagery the Americans had shared with their UN allies was impressive enough. Three times a day (the times were carefully noted on the Major's desk calendar, thanks to a nice piece of work by Malaysian Military Intelligence) American reconnaissance satellites passed overhead, noting the smallest details of his preparations.

The third principle of defense was to maintain perimeter security, and to block any move to seize the airfield from outside. Unfortunately, the perimeter of the airport was many kilometers long, and the Major had only a reinforced battalion of a thousand men. Designed and built as a conspicuous prestige display, this vast airport was really too big for the country. Still he had managed to site his heavy weapons covering anti-tank and anti-personnel minefields along the most likely approach routes.

Fourth principle: Dispose your air-defense assets for 360° coverage and relocate them frequently. This was easy enough. The battalion's air-defense section consisted of a few man-portable Blowpipe missiles. The divisional air defense battery had emplaced a Rapier SAM launch unit and several dummy launchers on nearby hilltops, but he knew that it had little chance of surviving the first attack.

Finally, pray real hard. This was not part of the Staff College tactical solution, but as he faced west toward Mecca and knelt for the first of the five daily prayers, the major reflected that it was the most important step. He was a patriotic Malay and a good Muslim, and he had just noticed that the readout of his personal GPS receiver, programmed to indicate the exact bearing of the Holy City, was displaying gibberish. The Americans had begun "selective availability," the random garbling of the signals of the Global Positioning System. It did not matter. He knew where he was. If the Americans wanted this airfield badly enough, they would take it. Major Yasin had no illusions about his personal chances of survival. But that was in God's hands. Inshallah.

Aboard *USS Springfield* (SSN-761) in the Andaman Sea, September 17th, 2008

Naval tradition required waking the Captain whenever there was a significant event affecting the ship. The order over the Very Low Frequency broadcast was a simple code group of a few letters, but it meant "Come to periscope depth to receive a downlink of targeting data." That counted as a significant event, all right. Nobody in the communications section had ever seen that one, even in an exercise. It was a new capability to provide targeting data for the dozen BGM-109 Tomahawk cruise missiles that slumbered in vertical launch tubes just behind the boat's bow section. Now, it only required a dish antenna smaller than a dinner plate poking above the waves for a few minutes, precisely aimed at a spot in the sky. From there, information could be downloaded from the Theater Mission Planning System, which provided near-real-time targeting information.

Once the download was received and confirmed, *Springfield* silently nosed down to a comfortable, secure depth and the Captain asked his Weapons Control Officer to bring up a visual display of target coordinates and the missile flight path. The stern, unwritten rules of their nuclear fraternity required that submariners never express surprise, but none of the officers gathered around the glowing console could avoid an involuntary gasp. In two days, they were going to take out Malaysia's big air base at Kuantan on the east coast of the Malay Peninsula. The missiles would fly right across the country, skimming over the tea plantations of the Cameron Highlands, to hit sheltered F/A-18s and MiG-29s from the unexpected landward side.

Agana Harbor, Guam, September 17th, 2008

The perfumed tropical breeze carried the scent of diesel exhaust across the bay as the four big ships raised anchor and steamed out into the Pacific. You would not call them beautiful. The great boxy hulls were piled high with containers and festooned with heavy cranes. A helicopter landing pad and an awkwardly angled folding ramp were tacked onto the stern, seemingly as an afterthought. You expect ships to be named after famous admirals or powerful politicians, but these vessels carried the names of enlisted men and junior officers who had fallen in nameless rice paddies and obscure fire bases, some four decades ago: *Pfc Dewayne T. Williams,* 1st Lt. Baldomero Lopez, 1st Lt. Jack Lummus, Sgt. William R. Button.

They were no greyhounds of the sea, making 17 kt/31 kph toward their rendezvous with Marines who would fly halfway across the world to link up with the weapons, vehicles, supplies, and equipment they carried. With flat black hulls and white paint topside, they were pretty ugly ships, all things considered. But in the eyes of a logistician, the ships of Maritime Prepositioning Squadron Three (MPSRON 3) were more beautiful than any China Clipper that ever rounded Cape Horn under a full spread of canvas. Just two days behind the ships of MPSRON 3 were the ships of a similar U.S. Army unit, carrying equipment for a mountain brigade. If the U.S. could secure a lodgement ashore in Brunei, there would be a division's worth of force to back it up.

Final Confirmation Briefing, USS *Bon Homme Richard* (LHD-6), South China Sea, 2000 Hours, September 18th, 2008

Colonel Taskins plugged in her laptop and began to run though the various phases of Tropic Fury. The keys were speed and surprise. With a lot of help from the Air Force in the Philippines and on Guam and a lavish expenditure of BGM-109 Tomahawk cruise missiles, they would blind the Malaysian forces, making them unable to sense or defend against the approach of PHIBRON 11. The risks were many. The amphibious force would approach the coast of occupied Brunei with only a handful of escorts: two Aegis guided-missile cruisers and destroyers, a single *Kidd*-class (DDG-993) guided-missile destroyer, a pair of modernized *Spruance*-class

(DD-963) destroyers, and three old *Oliver Hazard Perry*-class (FFG-7) guided-missile frigates. PHIBRON 11 itself was tiny, with only *Bon Homme Richard* (LHD-6), the damaged *Germantown* (LSD-42), and the brand-new assault ship *Iwo Jima* (LPD-18). *Constellation* CVBG, which had been on a port visit in Australia, was steaming forward with the ships of MPSRON 3, and would join up with PHIBRON 11 the day after the invasion started (D+1). Meanwhile, fighter cover would be supplied by a reinforced detachment of AV-8B Plus Harrier IIs just flown in, as well as F-15C Eagle fighters of the 366th Wing's 390th Fighter Squadron deployed to Naval Air Station (NAS) Cubi Point near Subic Bay in the Philippines. The rest of the 366th, with support units, had deployed to the Western Pacific, and would work in relays to protect the amphibious force until the *Constellation* (CV-64) group arrived. The risk of attack on PHIBRON 11 was low, since it was unlikely the Malaysians would expect them so quickly. Their Navy had been driven into port, and only their Air Force was left to deal with the threat from the sea. The coming air campaign would deal with that.

Colonel Taskins continued her briefing for the assembled crowd in the officers' mess. "Folks, we're going to have to work fast, and neat. Our biggest problems are with the oil facilities on the western side of the country. This is what the Malaysians want to keep, and what we must insure that they do not destroy. North Borneo is an extremely fragile ecosystem, so a mass of burning oil wells will not do. This is why I've committed so much of the force to securing the fields. Nevertheless, we must also clear the cargo terminal in the harbor at BSB, so that follow-on forces

The invasion and liberation of Brunei.   *Jack Ryan Enterprises, Ltd., by Laura Alpher*

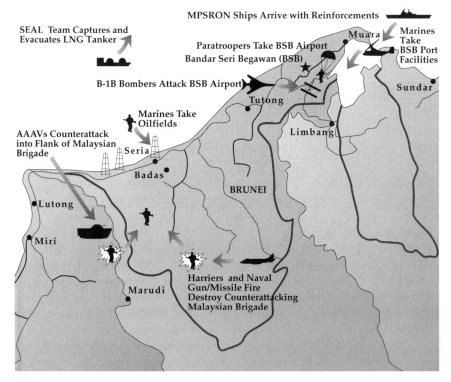

can relieve us. Finally, we must relieve our squad at the American Embassy in BSB. General Bear tells me that he wants the gunny and his detachment taken care of, and we will do this. Is that understood?"

A chorus of nods told her that it was.

"All right then," she continued, "let's get the job done, take care of each other, be Marines, and go home safe. God bless you all."

That was all they needed to hear.

Over Kota Kinabalu, Sabah (North Borneo), 0130 Hours, September 20th, 2008

Kota Kinabalu, the primary Malaysian air base in North Borneo, was taken seriously by Tropic Fury planners. Home base for two fighter squadrons and a gaggle of maritime patrol aircraft, it had to be neutralized. Since all of the submarine-launched Tomahawk cruise missiles were committed against targets on the Malay Peninsula, this one would have to be done by aircraft. The U.S. Air Force drew the assignment.

All day and most of the night, the 366th had sparred with the Malaysians, darting in and out with fighters from Cubi Point, supported by airborne tankers. It had driven the defenders at Kota Kinabalu to exhaustion, and by 0300 local time, they were near collapse. Tropic Fury's Joint Forces Air Component Commander (JFACC), the Air Force brigadier general commanding the 366th, had planned his operations to produce this result. Make them crazy, spar with them a while, and then hit them when they're too tired to notice. Now the fakes were over, and the Sunday punch was on the way. Two F-16Cs from the 389th FS equipped with targeting pods and HARM missiles dashed in to launch their weapons at the air-traffic-control and SAM radars on the field. The two F/A-18s that lifted off were rapidly dispatched by AIM-120 AMRAAMs from a pair of escorting Eagles, and that was it. Within seconds, Kota Kinabalu was blind and helpless. Now came the heavy iron.

Six B-1B Lancers of the 34th Bombardment Squadron had flown non-stop from Anderson AFB on Guam, carrying the ordnance that would shut down Kota Kinabalu for good. The first four came in from the north, very low over the China Sea at just over Mach 1, throwing up huge twin rooster tails of spray. At 10 nm/18.3 km from the coast, all four pulled up into zoom climbs. At the apex of the maneuver, each aircraft released twenty-four JDAMS guided bombs with hardened 2,000-lb/909.1-kg warheads. Within seconds, every aircraft shelter, runway, taxiway, fuel tank, and weapons bunker had been hit. The last two B-1s came from inland at medium altitude, dumping a total of sixty CBU-87/89 wind-corrected cluster bombs on the base, ensuring that Kota Kinabalu would be disabled for many weeks to come.

Around the rim of the South China Sea, similar events were taking place. On the Peninsula, every major fighter and transport air base was being hit by submarine-launched BGM-109 Tomahawk cruise missiles. Aging B-52Hs from the 2nd Bombardment Wing at Barksdale AFB, Louisiana, staged out of Diego Garcia, launched waves of cruise missiles, taking out communications and command centers. The ships of PHIBRON 11 and their escorts would be effectively invisible to the Malaysians, until they came within sight of land.

25 nm/45.7 km North of the Coast of Brunei, 0200 Hours, September 20th, 2008

The LCAC slowed to a crawl and dropped its stern ramp just long enough for six rigid raider craft to slide out onto the gentle swell. Then it turned and headed back towards *Iwo Jima* (LPD-18), its mother ship, over the horizon, as Marines of the 31st MEU (SOC)'s Force Reconnaissance Platoon started their specially silenced outboard motors and headed inshore toward the mangrove swamps along Brunei's western border. Before sunrise, the boats would be securely hidden and the Marines would be humping through coastal jungle toward a daytime hideout on the edge of the rain forest. At the same time, a single MV-22B Osprey from the MEU (SOC)'s ACE made a low-level approach to the coast east of Brunei. Hugging the hills and dodging in and out of lush valleys, it made five touch-and-go landings, dropping off four-man reconnaissance teams. With their special observation and surveillance equipment, the teams would give Colonel Taskins continuous location and status reports on Malaysian forces in Brunei. Tomorrow night, they would all be very busy Marines.

Seria LNG Terminal, Brunei, 0000 Hours, September 21st, 2008

The Brunei-Shell Tankers motor vessel *Bubuk* was one of a handful of similar merchant ships that flew Brunei's gold, black, and white flag. Extraordinary ships they were. They displaced over 51,000 tons, and their specialized cargo was liquefied natural gas, stored at frigid temperatures in huge insulated spherical tanks that filled the spacious hulls. Crewed by expatriate British officers and Pakistani hands, a fleet of these vessels shuttled between Brunei and Japan. *Bubuk* was the only one that had been caught in port by the Malaysian takeover. The ship was not just an enormously valuable asset and a symbol of national sovereign; it was a floating bomb with the potential explosive force of a tactical nuke. Accidental or deliberate detonation of over 2,648,610 ft^3/75,000 m^3 of volatile LNG would level Seria, a town of 25,000 people, along with several billion dollars worth of capital equipment. Tropic Fury planners quickly determined that *Bubuk* would have to be seized and secured, very carefully. This was exactly the kind of mission that U.S. Navy SEALs trained for, dreamed about, and salivated over. PHIBRON 11's SEAL detachment, embarked aboard *Iwo Jima* (LPD-18), drew the assignment.

Bubuk's designers had thoughtfully provided a small helipad over the stern, and this was the point of entry for the main SEAL boarding party—rappelling down a rope from a hovering CH-53. Reconnaissance had confirmed the presence of a handful of sentries on deck and around the jetty. They were taken out in just seconds after a series of stealthy bounds, followed by silenced shots from the SEALs' MP-5s. It took only a few minutes to liberate the crew from enforced captivity in the berthing areas, escort them to their stations, and get under way. Luckily, the Malaysians had allowed one engine to stay on-line to maintain the ship's electrical power, and in less than ten minutes the huge LNG ship was backing away from the pier, setting course to the north, out of harm's way.

Crossing the 12-m/22-km territorial limit, they passed a formation of fifteen AAAVs, headed ashore from *Iwo Jima* (LPD-18) at over 30 kt/55 kph. At the same time, a pair of AH-1W Cobra attack helicopters flew by, escorting the amphibious tractors to the beach. Ten minutes later, six LCACs from *Bon Homme Richard* (LHD-6) and *Germantown* (LSD-42) skimmed by, carrying M1A1 tanks and LAVs that would join the AAAVs, to form the armored task force that would take and hold western Brunei's oil production and storage facilities. It was less than thirty minutes to H-Hour.

Port of Muara, Brunei, 0100 Hours, September 21st, 2008

The patrol boats were going to be a problem. Captain Bill Schneider, commander of Golf Company, had obsessed about it for a week. His company of Marines had one of the toughest assignments of the entire operation. Dropped offshore in fragile, rigid raiding craft from *Iwo Jima*, they were to seize Muara's port facilities precisely at 0100. The sprawling cargo container port had the only wharf in the country that could accommodate the MPS ships, now standing by only 200 nm/366 km offshore. To deal with any patrol boats, he had placed Javelin teams in several of the lead boats, with orders to shoot first and count the pieces later. There was no time for such niceties as identification this evening.

Another problem was keeping the Malaysians from getting the alarm out on their arrival. The Malaysian communications net relied on almost untraceable satellite phones, registered with INMARSAT under the names of private businesses. Theoretically, the INMARSAT treaty prohibited use of its satellite channels for military operations, but the Eurocrats who controlled the system had stonewalled American attempts to impose an orbital "data embargo." International satellite telecommunications was a fiercely competitive business, and no Third World rogue state would ever trust a service provider that knuckled under to Western diplomatic pressure. But NSA technical wizards had provided the answer. One of the rigid raiders carried a compact, high-powered jammer that would disrupt cellular and satellite communications within a roughly 3-nm/5-km radius. Just enough to let the Marines establish a lodgment on the cargo wharf.

The raiders managed to get all the way to the dock before they were noticed. The two-man guard posts at the end of the pier were knocked out before they could sound an alarm. Within minutes, the Marines secured the wharf and a two-block perimeter of warehouses. They quickly set up strongpoints, anchored by a *Javelin* team and a light machine gun. This done, the young captain began to send out patrols aggressively, to determine whether the follow-on operation could start at midday. The patrols confirmed that the bulk of the Malaysian forces were dug in around the oil facilities and the international airport. Captain Schneider called Colonel Taskins in *Bon Homme Richard*'s (LHD-6) LFOC, using his own secure satellite link. He recommending committing the reserve company at the dock, where resistance was minimal. This done, he settled down to defend his position and "hold until relieved."

BSB International Airport, Brunei, 0111 Hours, September 21st, 2008

Major Yasin had been wondering when the Americans would hit him, and was surprised when they had not struck the night before. Now he was receiving scattered reports of fighting at the oil production facilities and the harbor, but nothing in his area. At the request of his brigade commander, he released one company to head west to the oil fields. He was thankful that the Malaysian command had never authorized taking hostages or holding the civil population at risk. This whole affair was economic; pure and simple. This kept the battle honorable, though theft of a whole country still bothered him.

He was still contemplating the delicate balance of national policy and personal morality when eight HARMs, launched by AV-8s from *Bon Homme Richard*, crashed into his anti-aircraft and SAM positions, followed by a rain of GBU-29 JDAMS bombs. Before the thunder of the explosions had stilled, there was another more ominous sound. He heard the engines of heavy jet transports, growing quickly louder. As a stream of big planes passed overhead, he realized what was coming, and sounded the alarm. It did him little good. The 1st Battalion of the 325th Airborne jumped from an altitude of 500 feet/152.4 meters, putting them on the ground and into action quickly. Having been dropped with surprising precision directly on their objectives, the heavy weapons positions around the field, they took most of them within seconds of hitting the ground.

This was fortunate, as the second wave, the 325th's 2nd Battalion, was only five minutes behind. These troopers had the job of clearing the runways and taxiways so they would be ready for fly-in reinforcements. Within an hour, the whole of the 325th had flown in from their staging base on Guam, and the C-17A Globemaster IIIs were headed back for another load. For Major Yasin, his Staff College problem was over. The survivors of his unit scattered, heading south into the mountains, where they would try to regroup.

Seria Oil Production Complex, Brunei, 0120 Hours, September 21st, 2008

The AAAVs hit a beach lined with petroleum storage tanks as far as the eye could see. They immediately unloaded their cargo of Marines. The vehicles then sought cover, awaiting orders to move inland. The embarked company carried only shotguns and grenade launchers, loaded with the beanbag rounds. Malaysian forces had not done much to secure the field, mainly because if it went up in flames, they lost the very reason for taking Brunei in the first place. So they had decided to cover the east and west flanks of the field, as well as the access road running along the coast. They had never expected an enemy crazy enough to come through the oil-storage facility.

The Marines were pleased to see that British Shell had managed to evacuate its security and field personnel. Word of this had come down two days earlier, so they knew any armed men in front of them were unfriendly. A handful of Malaysian soldiers patrolling the area were captured and held in a POW pen on the beach. Somehow, few Malaysian soldiers had volunteered for guard duty amid hundreds of tanks holding millions of gallons of flammable and explosive hydrocarbons. The Marines rapidly moved south to get beyond the tank farms. When the company reached the fence, they used small charges to blow holes in the chain-link-and-razor wire, and then called for their AAAVs to come and pick them up.

By this time, the LCACs carrying M1A1 tanks and LAVs had arrived, and a complete armored task force was ready to chew up any hostile force approaching the oil fields. The task force broke into platoons with the LAVs out on patrol and the tanks in reserve. Now nobody could get within 3 mi/4.8 km of the production facilities without the approval of the USMC, or a really ugly fight. Behind them came Marine combat engineers and demolitions experts to defuse any mines or booby traps the enemy might have left behind. As expected, there were none. Unlike Saddam Hussein back in 1991, Malaysia wanted Brunei intact. The Malaysians were not interested in crazed revenge. After all, business was business!

Cargo Pier, Port of Muara, Brunei, 0600 Hours, September 21st, 2008

Colonel Taskins stood on the end of the cargo pier with her counterpart from the 325th Airborne, the American ambassador, and several other officers. They were all listening to a satellite hookup to Tropic Fury Joint Task Force headquarters at NAS Cubi Point in the Philippines. General Bear was on the other end. His gruff voice came through loud and clear.

"Ambassador, were there any problems when our team arrived?"

Ambassador Jacob Arrens's voice showed his relief over his recent liberation. But he was a professional. His first order of business was report to General Bear about conditions in BSB.

"Sir, there is absolutely no damage to public utilities or facilities; and to the best of my knowledge, there have been no atrocities or other war crimes. It appears to have been a straight grab for the oil, plus leverage to negotiate for the Spratly leases next month. By the way, the gunnery sergeant of the embassy Marine detachment wants to talk with you at your convenience. He seems to feel a need to personally report in to you."

Bear smiled at the thought and replied, "Thank you Mr. Ambassador, I'd like you to put him on as soon as I can find a minute. Now let's try to get some real work done. Okay, Colonel, lay it out."

Colonel Colleen Taskins, USMC, swallowed hard, and took a second to frame her answer. In the next two or three minutes, the fate of her Marines, the prestige of the U.S., and the future of Brunei might hinge on what she was about to say. She had done her job, but now she was being asked for an on-the-scene assessment that would decide if Tropic Fury would begin its next, critical step. She remembered her first day as a plebe at Annapolis, a beautiful spring day in 1986. When she entered the service, women could not even fly in tactical squadrons. Now she was "in the loop." On another occasion, she might have frozen or been scared. But now, training and two decades of service took over, and her voice was clear and strong.

"General, we've taken all of our objectives, and casualties have been minimal. I've got less than ten wounded, and no KIAs reported as yet. The boys from the 82nd

Airborne hit their targets as planned, have made the linkup with us from the airport, and seem to be in good shape, sir. Their colonel will give you his report."

The commander of the 325th spent two minutes laying out his situation, closing with: "Sir, we've received the aerial port group that PACAF sent us, and we're ready to receive the first of your fly-in-brigade personnel. I can hold what I have, and would love to get some help to kick these bastards out of here." The paratrooper colonel's enthusiasm was infectious. It was up to Colonel Taskins now.

"Colleen, it's your call. What do you want me to do?" Taskins had never heard General Bear call her that, even when he had been one of her more terrifying instructors at Annapolis. Now he was leaving it up to her.

Her response was immediate. "General, we've got indications that the Malaysian brigade that occupied Brunei has run back over the border, to link up for a counterattack with another Malaysian brigade. Sir, send me MPSRON 3 and their fly-in brigade from the III Marine Expeditionary Force. I'll get my people out of here, and let the professionals clean this mess up." She was ready to go home with her people. In six hours, the four ships of MPSRON 3 would arrive at this pier, and begin unloading.

Prime Minister's Residence, Kuala Lumpur, Malaysia, 0900 Hours, September 21st, 2008

The Malaysian Prime Minister was running *amok*, and he had not even left his office. The reports coming in from around his country showed a series of very precise and selective strikes by aircraft and missiles, as well as some kind of counter-invasion in Brunei. Like all English-educated men, he had been raised on the stories of how Field Marshal Erwin Rommel had planned to defeat the D-Day invasion in 1944 by destroying the beachhead before the end of what he called "the longest day." There were now fifteen hours left in his "longest day," and he needed to make the most of them.

He had already called in the Chief of Staff of the Malaysian Army, and had laid out his demands. The Americans had gained two footholds in Brunei. The first, in BSB itself, seemed to be quite robust. The other, around the oil-production facilities of Seria, was smaller—and it was now holding the only thing that he had cared about in the first place. The holder of those wells and facilities might still be able to negotiate with the new Sultan, who would surely follow this invasion into BSB in the next few days. Perhaps the right to negotiate the Spratly oil leases could be traded for the survival of the existing North Borneo wells. It made sense. The two Malaysian brigades were ordered to attack the Marines defending Seria and retake the oil field. It was the last chance to salvage something from this adventure.

Cargo Pier, Port of Muara, Brunei, 1300 Hours, September 21st, 2008

There had been little time for niceties like tugboats and fenders; the captains of the MPSRON 3 ships had just driven right in. Luckily, they'd caught the tides right, and were able to moor the big Ro-Ros with a minimum of scraped paint and bent plating. The stern ramps dropped, and vehicles poured out. A few hours earlier, the first elements of the III MEF's fly-in brigade from Okinawa had arrived at BSB International Airport. Riding in the same commandeered buses that had been used to block the taxiways and runways just twelve hours earlier, the first elements of the brigade were driven directly to the wharf, where they mounted up their M1A1s, AAAVs, LAVs, and HMMWVs and began to fan out across Brunei.

It normally would take eighteen hours to finish the unloading of the combat vehicles, with another three days to off-load supplies. But now everything had to be done sooner, because the Army's AWR-3 squadron would arrive in sixty hours, and they would need to use the same port facilities. When the Army mountain brigade arrived, plus an additional fly-in brigade from the 82nd Airborne Division, there would be a division-sized task force in Brunei. The concept of the operation was to rapidly build up a force big enough to overmatch anything the Malaysians could throw against the beachhead. So far it had worked.

The biggest current worry was the armored task force holding the oil facilities on the western side of the country. Colonel Taskins knew that Major Hansen's force was stretched thin. If she were the Malaysian brigade commander, that was the place that she would attack. She strode across the pier to speak with Brigadier General Mike Newman, commander of the units off-loading from the ships.

She came to the point quickly. "Mike, I think that we have a potential problem out in the oil-fields."

Looking up from his data slate, he replied, "How so, Colleen?"

"Sir, I believe that Major Hansen's task force is overstretched out at Seria. He needs some reinforcement and support."

Newman stood up, wiped his brow for a moment, and asked, "What did you have in mind?"

Her reply was again clear and rapid, "General, I want to move another company of infantry and the heavy weapons company over to the western side this afternoon. I also want to land the 155mm battery, and get some additional surveillance assets over to them. They can probably stand up to one or two counterattacks, but anything more could cause us real problems over there, sir."

"We're scheduled to relieve them tomorrow morning with a battalion landing team," Newman replied. Then he thought for a moment, remembering that this lady had *never* given him bad advice. "Maybe you're right." He turned to his operations officer and asked, "Harry, what's the situation on the brigade ACE moving down from Cubi Point?"

The Operations Officer referred to his own data slate, and replied, "Well, sir, we've got the first squadrons of F/A-18Ds and AV-8Bs down and dispersed, as well as some tankers. Two squadrons of MV-22Bs are on the way right now. They should be ready to start flying CAP and support missions before sundown."

"Tell you what, Colleen. Why don't you send the reinforcements over this afternoon, and I'll chop your whole ACE back to you to support them. Will that do?"

"Yes, sir!" This errand done, she headed back to *Bon Homme Richard* to make arrangements.

South of the Seria Oil Production Facility, Brunei, 1400 Hours, September 21st, 2008

Bill Hansen was grateful for the news he had just received from Colonel Taskins. He was already picking up enemy activity in front of his positions, but in a couple of hours, he would have twice as much force. He was also gaining powerful fire-support assets: 155mm guns, the Harriers and Cobras of the ACE, and a couple of offshore destroyers. This was what he needed to ensure his position would hold until relief arrived in the morning. Even better, his BLT commander was on the way, to take over responsibility for the beachhead.

Headquarters, Malaysian 2nd Brigade, South of Seria, Brunei, 1415 Hours, September 21st, 2008

The two Malaysian brigade commanders had met to plan their defenses when orders to attack arrived from the Prime Minister. Both of the officers were British-trained and had no doubts about their duty. But both had severe doubts about the odds of executing this attack. The 5th Brigade, which had occupied BSB, had suffered scant losses from the Americans, but it had scattered, and most of the day had been required to bring it back together. Now they were expected to retake the oil fields, drive the Marines into the sea, and do it before dark. After the noon prayer, they spread out the maps under their camouflaged command tent and set to work, trying to organize something that might succeed.

The plan was for the 5th Brigade to attack directly north towards the sea, while the fresher 2nd Brigade would swing around to the west, to hit the Americans along the coastal road. Both attacks were coordinated to hit the Marines at 1630 hours, and would continue until the sun went down into the South China Sea. Reconnaissance indicated that they were facing two dozen armored vehicles and about six hundred Marines. All told, their two Brigades had over five thousand men, with almost a hundred light tanks and personnel carriers. The problem was that they could not use their artillery. The orders from Kuala Lumpur were explicit: No artillery could be used *anywhere* near the production facilities. Starting an uncontrollable oil fire would defeat the whole campaign. The brigade commanders drank a final cup of tea, wished each other Allah's blessing, and made ready for the last attack of this bizarre little war.

Over Western Brunei, 1500 Hours, September 21st, 2008

The Dark Star UAV had launched several days earlier from NAS Cubi Point, and was less than halfway through its five-day mission. Equipped with a television camera, infrared scanner, and synthetic aperture radar, it had been keeping track of the two Malaysian brigades in the jungle south of Seria. The heavy canopy of foliage in the foothills blocked visual sensors, but the IR and radar picked up useful imagery. On *Iwo Jima* (LHD-18), the intelligence team monitoring the data stream from the Dark Star was beginning to worry. The two Malaysian brigades were showing signs of life. The reconnaissance teams inserted into Western Brunei on D-1 were sending back a steady stream of sightings. While Colonel Taskins had been limited to committing Major Hansen's small armored force to that end of the country, she had concentrated over two thirds of the MEU (SOC)'s intelligence collection assets into the area, to avoid unpleasant surprises. After some analysis, the conclusion was reported to Colonel Taskins. There would a be a two-brigade attack to overrun the Seria beachhead and recapture the oil fields, starting about 1500 hours and running until sundown. Her intuition confirmed, she began to set a trap for the enemy units.

LFOC, USS *Bon Homme Richard* (LHD-6), 1615 Hours, September 21st, 2008

It felt good to be back in the chair of her LFOC station. Colonel Colleen Taskins felt the rush of anticipation at the start of an operation. Game time. Her workstation showed the estimated positions of the two Malaysian brigades (intelligence flagged them as the 2nd and 5th), and she was working to set up supporting fires. Right now, the enemy was under the cover of the jungle. But to attack her Marines, they would have to come out into the open. General Newman had released almost all of her units back to her for the coming operation. In addition, he had made a call back to the JFACC at NAS Cubi Point in the Philippines, who dispatched an E-8C Joint Stars radar-surveillance plane to help her out. The J-Stars bird had a large canoe-shaped radar under its belly that could detect moving vehicles in real time, sending the data directly to a terminal located here in the LFOC. A quick look at the J-Stars display confirmed her suspicions, and she shifted units around Western Brunei like a chess master. There would be just enough time.

5th Malaysian Brigade, South of Seria, 1630 Hours, September 21st, 2008

The movement of the 5th Brigade had gone well, even though the troops were tired, having been on the move continuously since midnight. The retreat out of BSB and the airport had left them angry, eager to get back at the Americans. Now they would have their chance. The line of departure was a dirt road along the lower Belait River, about 5 mi/8 km from the coast. Their plan was to drive into a gap between the production and storage facilities, then fan out along the coast to seize the objectives. Much of the route was covered by jungle, their element. They could win.

By 1635 hours, they were moving forward, infantry leading the light tanks and armored personnel carriers. Suddenly, shell fire began to drop on their heads. At first, it was just a few 155mm rounds. Then, 5-in./127mm high-explosive rounds from ships began to fall. As the infantry went for cover, the armor pushed on ahead. The troops felt safe against air attack in the jungle. The deafening shell fire kept them from hearing the arrival of the Harriers overhead. The Harrier pilots, however, seemed to know exactly where the Malaysians were. The J-Stars had given the Harrier pilots precise GPS coordinates to drop their weapons "blind" through the jungle canopy. Each aircraft delivered six CBU-87 cluster bombs. Thousands of CEM cluster munitions fell through the top of the foliage, shredding the forward battalions of the 5th Brigade. Tanks and carriers destroyed by the hollow charges of the

CEMs became small volcanoes in the darkening jungle. The guns stopped. All that was left was the sound of burning vehicles, exploding ammunition, and the low moans of the dead and dying. As the brigade commander tried to rally the remains of his unit, the desperate radio calls from his command post were identified by an ES-3A Shadow surveillance aircraft, and rapidly triangulated by several of the ships offshore. Within seconds, a fire mission was flashed over the support network, and a pair of TACMS missiles were fired by one of the offshore destroyers. These arched inland, guided by their onboard GPS systems. When the two missiles were directly over the brigade command post, they ejected a load of anti-personnel cluster munitions. In seconds, the command post and most of its vehicles were destroyed, with little left but a scar in the jungle, which would rapidly grow over. The southern prong of the Malaysian counterattack was broken.

Batang Baram River Crossing, Brunei/Malaysia Border, 1645 Hours, September 21st, 2008

The commander of the 2nd Brigade was getting increasingly frustrated trying to push across the Batang Barem River. The Americans had taken the ferry at the mouth of the river and established a series of strong-points across the river from his brigade. The firing had gotten lively, and he had already lost some vehicles to TOW and Javelin missiles. Now the Malaysians were starting to make some headway. They had forced crossings at several points along the far riverbank, and were starting to get whole platoons across. He was late at the planned start line for his attack, and communications with the 5th Brigade had been cut off. But at least his units were finally starting to move. The enemy in front of his brigade seemed to be Marines on foot, with a few tanks and LAVs. Just what his reconnaissance had told him would be there.

Mouth of the Batang Barem River, Brunei/Malaysia Border, 1700 Hours, September 21st, 2008

It had been the kind of mission that his old friends in the Army's armored cavalry would have loved. Bill Hansen had pulled his AAAVs off the line several hours earlier, handing over their defensive positions to leg Marines. He would refuel, rearm, and head back through the oil tanks to the sea. Carrying a company of Marines, his fifteen armored vehicles headed out to sea at full speed on a long, looping arc to the southwest. The sea was calm, and the fifteen vehicles were cutting through the South China Sea at over 30 kt/55 kph, trying to stay out of sight from shore-based observers. Their goal was the mouth of the Batang Barem River, where Marines from the 31st MEU (SOC) already held the north bank. It took less than an hour to reach the goal, and they barely slowed down as they entered the river.

Major Hansen and the AAAVs, moving rapidly up the Batang Barem River, were actually behind the bulk of the 2nd Malaysian Brigade. Cruising at over 20 kt/ 36 kph, they moved to crumple the Malaysians' left flank. About 3 mi/5.5 km upriver from the 2nd Brigade, the fifteen AAAVs slowed down, dropped their tracks, and retracted the bow flaps. Striking the flank of the lead battalion, they penetrated into the unit's rear area, overrunning the command post. They tore through the area and sent the battalion staff running for the hills.

At this point, Hansen broke his AAAVs into five teams, and sent them tearing through the rear of the 2nd Brigade. They shot up command vehicles and trucks with their 25mm cannons, and popped any armored vehicles that got in their way with Javelin missiles. Then, coordinating their maneuver by digital data links, they converged on the command post of the 2nd Malaysian Brigade. It had been less than an hour since they had climbed out of the Batang Barem River and started their headlong dash. They spotted the command staff of the 2nd Malaysian Brigade coming towards them with hands raised. The last effective combat unit of the Malaysian Army in Brunei had just surrendered.

BSB International Airport, 0800 Hours, September 22nd, 2008

The 7th Gurkha Rifles had flown into Brunei on chartered commercial aircraft, and were now taking control of the airport complex from the units of the 82nd Airborne Division. While it was a symbolic handover, the return of the Gurkhas meant a return of order to Brunei. Crown Prince Omar Bolkiah was arriving on one of the Royal Brunei Airlines jets that had been interned in Manila. Escorted most of the way by a pair of F-15C Eagles from the 366th at NAS Cubi Point, the Prince had insisted that the final leg be escorted only by the Marine Harriers that had done so much to liberate his country. He walked down the aircraft steps under the watchful eyes of Colonel Rai, moved to one of the grassy areas, and kneeled to kiss the soil of his liberated home.

In a few days, he would be crowned as the Sultan of Brunei. Surprisingly little damage had been done in the short liberation campaign. His half-brother, pretender to his father's throne, had fled to take up refuge in Saudi Arabia, in the same political leper colony once occupied by criminals like Idi Amin. As for himself, he would take his father's plan for the development of the South China Sea oil fields to the October UN conference. The American ambassador in Manila had returned his father's data slate, recovered from the helicopter crash several weeks earlier. It held the late Sultan's private diary, which contained his notes for the conference. There were also his father's poems and letters, which were to have been given to *him* at the time of a succession. He had them now, and knew how proud his father had been of him. He intended to make sure that his spirit always would be.

Prime Minister's Residence, Kuala Lumpur, Malaysia, 1200 Hours, September 22nd, 2008

The Malaysian Prime Minister was looking into the eyes of his Army Chief of Staff, as well as those of his other ministers. "Gentlemen," he said tiredly, "our forces have failed to retake the oil facilities in Brunei, and I must declare to the people of our Federation that this government has fallen. I just hope that we can get the word out before they come through the door to announce it to us personally. I must resign, and will retire from public life." With that, he rose and left the room. The Army Chief of Staff wondered if he would even make it home. The crowds were already in the streets, and they had a ugly history of tearing politicians who disappointed them into pieces. Small ones.

PHIBRON 11, Steaming North through the South China Sea, September 30th, 2008

It had been a frantic week, getting extracted from Brunei. Along with the unavoidable ceremonies and honors, everything that had touched the soil of North Borneo had to be meticulously cleaned for a full Japanese inspection when they returned to Okinawa the following week. But right now, everyone was getting some sleep. There is little time for rest during amphibious operations, and the ships of the ARG were unusually quiet during the transit.

For Colonel Taskins, though, there were other duties. While she was being hailed as the greatest female warrior since Joan of Arc, there still were painful tasks to deal with. One of these was to write letters home. Those letters. Casualties had been light during Tropic Fury, but there still had been five dead and thirty-four wounded. The wounded had already been evacuated through Pearl Harbor to the Balboa Navy hospital in San Diego. The bodies of the other five had been flown to Dover AFB, Delaware, then to Arlington for burial. Now her duty was to write the letters to their families. In her two command tours, these were the first fatalities her unit had suffered.

The first one was the hardest. The young Marine private from Detroit had been in the team that secured the cargo pier; he had been shot down by a sniper in the streets in BSB. She had not known him personally. But this was one command responsibility that no commander could delegate or postpone. Ahead of her was a visit to the White House and Congress, probably a general's star, and maybe even a regimental or divisional command. But now, this was where she wanted to be, with *her* Marines. After a few more minutes, she started to type on her computer, and the words came.

Conclusion:
A Corps for Five Hundred Years...

In the chapters of this book, I've tried to take you on a tour of what are undoubtedly the crown jewels of the United States Marine Corps, the MEU (SOC)s. In exploring these seven precious national assets though, I hope that you have gotten some sort of sense of what the Marines are all about. While the units themselves are wondrous and dangerous precision instruments for the national leadership to move about the chessboard of world events, it is important to remember what the basic building blocks of the MEU (SOC)s are: Marines. Marines are perhaps, as I proposed in the Introduction, the ultimate expression of America's military persona. When a person in a foreign land thinks about what the United States might send if they were angry or helpful, Marines coming ashore are often the most likely response that comes to mind.

I like to think about the Marines a lot. My own ideas about the Corps caused me to give my primary novel character, Jack Ryan, a Marine background. Many of the ethics, morals, and characteristics that I consider central to Jack, are at the core of the Marine ethos. In addition, Marines have filled the pages of my books, because they are reliable, inventive, and colorful people in real life. I like to think that those same images are the ones that an enemy would have, prior to considering a fight with them. This is perhaps their most powerful weapon against a potential opponent: the fear of what might happen if one had to face a force of American Leathernecks in battle. You see, Marines are mystical. They have *magic*.

Marines like to see themselves as firemen in a world full of pyromaniacs these days, and perhaps they are right. Busy as the various units of the Corps have been in the five decades since the end of World War II, the five years since the end of the Cold War have been positively dizzying. Liberia. Desert Shield. Somalia (the first, second, and third times). Desert Storm. Bosnia-Herzegovina. I could go on, but I think you are probably getting the idea. Marines are our sentries on the walls of the world, and they take an immense pride in the privilege of serving in that role. When they call themselves "...the 911 Force...," they really do mean it. Thank God they do, because in a world filled with nationalistic dictators, natural disasters, and other unexpected things, we really need Marines!

To this end, I perhaps owe them a few thoughts on how I view their future. Five decades ago, at the moment of the flag raising on Mount Suribachi during the Battle of Iwo Jima, there occurred the defining moment in the history of the Corps. Watching from an offshore command ship, and viewing the bravery and determination of the Marines who had made it to the top of that fire-lashed peak, was James Forrestal, then–Secretary of the Navy. As the flag went up, and all of the emotion rose in the throats of those watching, he is said to have spoken the following words to General Holland "Howlin Mad" Smith:

"Holland, the raising of that flag on Suribachi means a Marine Corps for the next five hundred years..."

At the time of the Iwo Jima invasion, the Corps had not yet celebrated its 170th birthday. Now, fifty years later with the dawn of a new millennium within sight, it is easier to see what Secretary Forrestal meant. What he was talking about was a spirit and ethos that allows the Marine Corps to attack new problems and missions with fresh ideas and perspectives. Sometimes, when the other services and our allies have decided that these things are "too hard" to accomplish within their institutional guidelines and restrictions. The U.S. Marines have been leaders in technology and tactical development in this century, and you see it in the unique mix of equipment and doctrine that they have developed. Things like precision weapons delivery (sniping and dive bombing) and over-the-horizon transport systems (air cushioned landing craft and helicopter assaults). When you want something new done, give it to a Marine!

So what does this mean about the future of the Marines in the 21st century? Well, for starters they have one. In the minds of the national command authorities, they provide a valuable contribution to maintaining America's forced entry capability. That alone should ensure their survival to their 250th birthday. Beyond that, the skies are literally the limit. Emerging technologies in systems like long-range vertical takeoff and landing transport aircraft could see a merging of the current missions of airmobile and forward deployed forces. Powered personal armor and armament systems could see the emergence of Marines looking something like Robert Heinlein's *Starship Troopers* by the middle of the next century. Beyond that, it is probably impossible to imagine just what kinds of roles and missions that they will have.

Whatever technology brings though, there will be some things that Marines will always be, even in the year 2275 when they celebrate their 500th birthday. They will, as always, be the best basically trained warriors in the world. Even in the 23rd century there will probably be drill sergeants and yellow footprints in the Palmetto groves of Parris Island. Marines will also be riflemen, or whatever kind of personal weapon is in fashion three hundred years from now. Aimed fire from a shouldered weapon will always be a vital part of the Marine ethos. Finally, they will be finding new and innovative ways to win battles, and support the execution of wars. The enemies by then may even be extraterrestrials, but I think that the Generals Krulak (father and son) would heartily approve.

When I said earlier that Marines are America, I meant it. They represent us in so many different ways, in a variety of roles. From standing watch as embassy guards to flying the President in their distinctive olive-drab helicopters, Marines are people that we trust in whatever job they are assigned to perform. That trust is born from a commitment, both institutional and personal. It means that a Marine is both a part of a finely designed machine, yet has made a commitment to stand out from the crowd, and make their own way in life. It is hard not to smile when you see them, whatever their age, rank, and assignment. Whether they are in for just a few years, or make it a lifelong commitment like some of the people we have met in this book, the Corps changes them all for life.

And whatever their reason for having originally joined the Marines, they seem to have come into a common defining experience that allows them to share something special from life.

So be proud of them, because they are proud to serve you. Respect them, because they defend the things that make our nation the finest in the world. And please love them, because they stand on the walls of freedom, and keep them safe for the rest of us to sleep soundly at night. That is the real meaning of their motto *Semper Fidelis*– "Always Faithful." They always have been, and always will be. Even if it takes five hundred years to prove it.

Bibliography

Books:

Adan, Avraham (Bren). *On the Banks of the Suez*. Presidio Press, 1980.

Albrecht, Gerhard (editor). *Weyers Flotten Taschenbuch 1992/93 (Warships of the World)*. Bernard & Graefe Verlag, Bonn, Germany, 1992.

Alexander, Joseph H., and Bartlett, Merrill L. *Sea Soldiers in the Cold War*. Naval Institute Press, 1995.

Allen, Thomas, and Polmar, Norman. *Codename Downfall: The Secret Plan to Invade Japan and Why Truman Dropped the Bomb*. Simon & Schuster, 1995.

Ambrose, Stephen E. *Pegasus Bridge: June 6, 1944*. Simon & Schuster, 1985.

   — *D-Day, June 6, 1944: The Climactic Battle of World War II*. Simon & Schuster, 1994.

Arnett, Peter, *Live from the Battlefield*. Simon & Schuster, 1994.

Atkinson, Rick. *Crusade: The Untold Story of the Persian Gulf War*. Houghton Miffen, 1993.

Baker, A.D., III. *Allied Landing Craft of World War Two*. Naval Institute Press, 1985.

   — *Japanese Naval Vessels of World War Two*. Naval Institute Press, 1987.

   — *Combat Fleets of the World 1993*. Naval Institute Press, 1993.

   — *Combat Fleets of the World, 1995*. Naval Institute Press, 1995.

Baxter, William P. *Soviet AirLand-Battle Tactics*. Presidio Press, 1986.

Beach, Captain Edward L. USN (Ret.). *The United States Navy: 200 Years*. Holt, 1986

Bin Sultan, Khaled. *Desert Warrior: A Personal View of the Gulf War by the Joint Forces Commander*. Harper Collins, 1995.

Blackwell, James. *Thunder in the Desert: The Strategy and Tactics of the Persian Gulf War*. Bantam Books, 1991.

Blair, Colonel Arthur H., U.S. Army (Ret.). *At War in the Gulf*. A&M University Press, 1992.

Blair, Clay. *The Forgotten War: America in Korea, 1950-1953*. Times Books, 1987.

Boyne, Walter J. *Clash of Wings: World War II in the Air*. Simon & Schuster, 1994.

   — *Clash of Titans: World War II at Sea*. Simon & Schuster, 1995.

Bradin, James W. *From Hot Air to Hellfire: The History of Army Attack Aviation*. Presidio Press, 1994.

Braybrook, Roy. *British Aerospace Harrier and Sea Harrier*. Osprey/ Motorbooks International, 1984.

   — *Harrier and Sea Harrier*. Osprey, 1984.

   — *Soviet Combat Aircraft*. Osprey, 1991.

Brown, Captain Eric M., RN. *Duels in the Sky: World War II Naval Aircraft in Combat*. Naval Institute Press, 1988.

Brugioni, Dino A. *Eyeball to Eyeball: The Cuban Missile Crisis*. Random House, 1991.

Burrows, William E., and Windham, Robert. *Critical Mass*. Simon & Schuster, 1989.

Bywater, Hector, C. *The Great Pacific War*. Reprint of 1925 Edition. Naval Institute Press, 1991.

Cardwell, Colonel Thomas A., III, USAF, *Airland Combat*. Air University Press, U.S. Air Force, 1992.

Chant, Christopher, *Encyclopedia of Modern Aircraft Armament*. IMP Publishing Services Ltd., 1988.

Chetty, P. R. K. *Satellite Technology and Its Applications*. 2nd Edition. McGraw Hill, 1991.

Chris, Bishop, and David, Donald. *The Encyclopedia of World Military Power*. The Military Press, 1986.

Clancy, Tom. *The Hunt for Red October*. Berkley Publishers, 1985.

   — *Red Storm Rising*. Berkley Books, 1986.

   — *The Cardinal of the Kremlin*. Putnam & Sons, 1988.

   — *The Sum of All Fears*. G.P. Putnam's Sons, 1991.

   — *Submarine: A Guided Tour Inside a Nuclear Warship*. Berkley, 1993.

   — *Armored Cav: A Guided Tour of an Armored Cavalry Regiment*. Berkley Books, 1994

   — *Debt of Honor*. G.P. Putnam Sons, 1994.

   — *Fighter Wing: A Guided Tour of an Air Force Combat Wing*. Berkley, 1995.

Cohen, Dr. Elliot A. *Gulf War Air Power Survey Summary Report.* U.S. Government Printing Office, 1993.
- *Gulf War Air Power Survey Volume I.* U.S. Government Printing Office, 1993.
- *Gulf War Air Power Survey Volume II.* U.S. Government Printing Office, 1993.
- *Gulf War Air Power Survey Volume III.* U.S. Government Printing Office, 1993.
- *Gulf War Air Power Survey Volume IV.* U.S. Government Printing Office, 1993.
- *Gulf War Air Power Survey Volume V.* U.S. Government Printing Office, 1993.

Cohen, Dr. Eliot A., and Gooch, John. *Military Misfortunes: The Anatomy of Failure in War.* Free Press, 1990.

Cooling, Benjamin F. (editor). *Case Studies in the Development of Close Air Support.* Office of Air Force History, 1990.

Coyne, James P. *Airpower in the Gulf.* Air Force Association, 1992.

Crampton, William. *The World's Flags.* Mallard Press, 1990.

Crowe, Admiral William J., Jr. The Line of Fire: *From Washington to the Gulf, the Politics and Battles of the New Military.* Simon & Schuster, 1993.

Darwish, Adel, and Alexander, Gregory. *Unholy Babylon: The Secret History of Saddam's War.* St. Martin's Press, 1991.

David, Peter. *Triumph in the Desert.* Random House, 1991.

Dawood, N.J. (editor). *The Koran.* Penguin Books, 1956.

De Jomini, Baron Antoine Henri. *The Art of War.* Green Hill Books, 1992.

Doleman, Edgar C., Jr. *The Vietnam Experience: Tools of War.* Boston Publishing Company, 1985.

Donnelly, Ralph W. *The Confederate States Marine Corps: The Rebel Leathernecks.* White Mane, 1989.

Dorr, Robert F. *The Imperial Japanese Navy.* Naval Institute Press, 1978.
- *Desert Shield – The Build Up:The Complete Story.* Motorbooks, 1991. –
- *Desert Storm Air War.* Motorbooks, 1991.

Dunnigan, James F., and Bay, Austin. *From Shield to Storm.* Morrow Books, 1992.

Dupuy, Colonel Trevor N., USA (Ret.). – *The Evolution of Weapons and Warfare.* Bobbs-Merrill, 1980.
- *Options of Command.* Hippocrene Books, Inc., 1984.
- *Numbers, Predictions & War:The Use of History to Evaluate and Predict the Outcome of Armed Conflict.* Hero Books, 1985.
- *Understanding War: History and Theory of Combat.* Paragon House, 1987.
- *Attrition: Forecasting Battle Casualties and Equipment Losses in Modern War.* Hero Books, 1990.
- *Defeat: How to Recover from Loss in Battle to Gain Victory in War.* Paragon House, 1990.
- *Saddam Hussein: Scenarios and Strategies for the Gulf War.* Warner Books, 1991.
- *Future Wars: The World's Most Dangerous Flashpoints.* Warner Books, 1993.

Edwards, Major John E., USA (Ret.). *Combat Service Support Guide.* 2nd Edition Stackpole Books, 1993.

Eliot, Joshua (editor). *Indonesia, Malaysia & Singapore Handbook.* Passport Books, 1994.

Eshel, David. *The U.S. Rapid Deployment Forces.* Arco Publishing, Inc., 1985.

Ethell, Jeffrey, and Price, Alfred. *Air War South Atlantic.* Macmillan, 1983.

Evans, Thomas J., and Moyer, James M. *Mosby's Confederacy.* White Mane Publishing Co., 1991.

Flagherty, Thomas J. *Carrier Warfare.* Time Life Books, 1991.

Flaherty, Thomas H. *Air Combat.* Time Life Books, 1990.

Flintham, Victor. *Air Wars and Aircraft: A Detailed Record of Air Combat, 1945 to Present.* Facts on File, 1990.

Foster, Simon. *Hit The Beach!* Arms and Armour Press, 1995.

Francillon, Rene, J. *Tonkin Gulf Yacht Club: U.S. Carrier Operations off Vietnam.* Naval Institute Press, 1988.
- *World Military Aviation, 1995.* Naval Institute Press, 1995.

Frank Chadwick. *Gulf War Fact Book.* Game Designers Workshop, 1992.

Frank, Richard B. *Guadalcanal: The Definitive Account of the the Landmark Battle.* Random House, 1990.

Friedman, Norman. *U.S. Naval Weapons.* Naval Institute Press, 1985.
- *Desert Victory: The War for Kuwait.* Naval Institute Press, 1991.
- *Naval Institute Guide to World Naval Weapons Systems.* Naval Institute Press, 1991.

   — *Naval Institute Guide to World Naval Weapons Systems. 1991/92.* Naval Institute Press, 1991.

   — *Naval Institute Guide to World Naval Weapons Systems. 1994. Update.* Naval Institute Press, 1994.

Gibson, James William. *The Perfect War: Technowar in Vietnam.* Atlantic Monthly Press, 1986.

Godden, John (editor). *Shield & Storm: Personal Recollections of the Air War in the Gulf.* Brassey's, 1994.

Goldstein, Donald L., et. al. *D-Day, Normandy: The Story and the Photographs.* Brassey's, 1994.

Gordon, Michael R., and Trainor, Gen. Bernard E. *The General's War: The Inside Story of the Conflict in the Gulf.* Little Brown, 1995

Gray, Colin S. *The Leverage of Sea Power.* Free Press, 1992

Grove, Eric, *Battle for the Fjords: NATO's Forward Maritime Strategy in Action.* Naval Institute Press, 1991.

   — *Sea Battles in Close Up: World War II.* Volume 2. Naval Institute Press, 1993.

Gumble, Bruce L. *The International Countermeasures Handbook.* EW Communications Inc. 1987.

Halberstadt, Hans. *Desert Storm: Ground War.* Motorbooks International, 1991.

Hallion, Dr. Richard P. *The Literature of Aeronautics, Astronautics, and Air Power.* U.S. Government Printing Office, 1984.

   — *Strike from the Sky: The History of Battlefield Air Attack, 1911-1945.* Smithsonian, 1989.

   — *Storm over Iraq: Air Power and the Gulf War.* Smithsonian Books, 1992.

Hammel, Eric. *Guadalcanal: Starvation Island.* Crown, 1987.

Hansen, Chuck. *U.S. Nuclear Weapons: The Secret History.* Orion Books, 1988

Hanson, Victor Davis. *The Western Way of War: Infantry Battle in Classical Greece.* Alfred Knopf Publishers, 1989.

Hartcup, Guy. *The Silent Revolution: Development of Conventional Weapons 1945-85.* Brassey's, 1993.

Hastings, Max. *Overlord.* Simon & Schuster, 1984.

Heatley, C. J., III. *Forged in Steel: U.S. Marine Corps Aviation.* Howell Press, Charlottesville, Virginia, 1987.

Heinlein, Robert A. *Starship Troopers.* Ace Books, 1959.

Hersh, Seymour M. *The Samson Option.* Random House Publishers, 1991.

Honan, William H. *Visions of Infamy.* St. Martin's Press, 1991.

Hudson, Heather E. *Communication Satellites: Their Development and Impact.* Free Press, 1990.

Hughes, David R. *The M16 Rifle and Its Cartridge.* Armory Publications, Oceanside, California, 1990.

Isby, David. *Weapons and Tactics of the Soviet Army.* Janes, 1981.

Isenberg, Martin T. *Shield of the Republic: The United States Navy in an Era of Cold War and Violent Peace, 1945-1962.* St. Martin's Press, 1993.

Jablonski, Edward. *America in the Air War.* Time-Life Books, 1982.

Jessup, John E., Jr., and Coakley, Robert W. *A Guide to the Study and Use of Military History,* U.S. Government Printing Office, 1991.

Keany, Thomas A., and Cohen, Eliot A. *Revolution in Warfare? Air Power in the Persian Gulf.* Naval Institute Press, 1995.

Keegan, John. *The Illustrated Face of Battle.* Viking, 1988.

   — *The Second World War.* Viking, 1989.

   — *A History of Warfare.* Alfred A. Knopf, 1993.

Kelly, Mary Pat. *"Good to Go": The Rescue of Scott O'Grady from Bosnia.* Naval Institute Press, 1996.

Kershaw, Robert J. *D-Day: Piercing the Atlantic Wall.* Naval Institute Press, 1994.

Kinzey, Bert. *U.S. Aircraft & Armament of Operation Desert Storm.* Kalmbach Books, 1993.

Knott, Captain Richard C. USN. *The Naval Aviation Guide.* 4th Edition, Naval Institute Press, 1985.

Krulak, Lt. General Victor H., USMC. *First to Fight: An Inside View of the U.S. Marine Corps.* Naval Institute Press, 1984.

Kyle, Colonel James H., USAF (Ret.). *The Guts to Try.* Orion Books, 1990.

Lake, Donald, David, and Jon (editors), *U.S. Navy and Marine Corps Air Power Directory.* Aerospace Publishing, Ltd., 1992.

Lambert, Mark (editor). *Jane's All the World's Aircraft 1991-92.* Jane's Publishing Group, 1992.

Langguth, A. J. *Patriots: The Men Who Started the American Revolution.* Simon & Schuster, 1988.

Liddell-Hart, B.H. *Strategy.* Frederick A. Praeger, Inc. 1967.

Lord, Walter. *Incredible Victory: Day of Infamy.* Holt Rinehart, 1957.

   — *The Battle of Midway.* (Reprint of 1967 Edition), Harper Collins, 1993.

Lundstrom, John B. *The First Team.* Naval Institute Press, 1984.

—   *The First Team and the Guadalcanal Campaign.* Naval Institute Press, 1994.

Luttwak, Edward, and Koehl, Stuart L. *The Dictionary of Modern War: A Guide to the Ideas, Institutions, and Weapons of Modern Military Power.* Harper Collins, 1991.

Macksey, Kenneth. *Invasion: The German Invasion of England.* July 1940, Macmillan, 1980.

Maroon, Fred J., and Beach, Edward L. *Keepers of the Sea.* Naval Institute Press, 1983.

Mason, John T., Jr. *The Pacific War Remembered: An Oral History Collection.* Naval Institute Press, 1986.

McConnell, Malcolm. *Just Cause: The Real Story of America's High-Tech Invasion of Panama.* St. Martin's Press, 1991.

McKinnon, Dan. *Bullseye – Iraq.* Berkley, 1987.

McRaven, William H. *Spec Ops.* Presidio Press, 1995.

Meisner, Arnold. *Desert Storm: Sea War.* Motorbooks International, 1991.

Melson, Charles D., and Hannon, Paul. *Marine Recon. 1940-90,* Osprey, 1994.

Middlebrook, Martin. *Task Force: The Falklands War, 1982.* Penguin Books, 1987.

Miller, Edward S. *War Plan Orange: The U.S. Strategy to Defeat Japan, 1897-1945.* Naval Institute Press, 1991.

Mills, Anastasia R. *Fodor's 96, Spain.* Fodor's, 1996.

Moore, Captain John, RN. *Janes's American Fighting Ships of the 20th Century.* Modern Publishing, 1995.

Morrocco, Jon. *The Vietnam Experience: Thunder from Above.* Boston Publishing Company, 1984.

Morse, Stan. *Gulf Air War Debrief.* Aerospace Publishing Limited, 1991.

Moskin, J. Robert. *The U.S. Marine Corps Story.* McGraw-Hill, 1987.

Nalty, Bernard C. *The United States Air Force Special Studies: Air Power and the Fight for Khe Sanh.* U.S. Government Printing Office, 1986.

Newhouse, John. *War and Peace in the Nuclear Age.* Alfred Knopf Publications, 1989.

Nichols, Commander John B., USN (Ret.), and Tillman, Barrett. *On Yankee Station: The Naval Air War Over Vietnam.* Naval Institute Press, 1987.

Nordeen, Lon O., Jr. *Air Warfare in the Missile Age.* Smithsonian , 1985.

O'Ballance, Edgar. *No Victor, No Vanquished.* Presidio Press, 1978.

O'Grady, Captain Scott, USAF. *Return with Honor.* Doubleday, 1995.

Pagonis, Lt. General William G., USA, with Cruikshank, Jeffrey L. *Moving Mountains: Lessons in Leadership and Logistics from the Gulf War.* Harvard Business School Press, 1992.

Peebles, Curtis. *Guardians: Strategic Reconnaissance Satellites.* Presidio Press, 1987.

Pocock, Chris. *Dragon Lady: The History of the U-2 Spyplane.* Motorbooks International, 1989.

Polmar, Norman. *Naval Institute Guide to the Ships and Aircraft of the U.S. Fleet.* 15th Edition, Naval Institute Press, 1993.

Potter, Michael C. *Electronic Greyhounds: The Spruance-Class Destroyers.* Naval Institute Press, 1995.

Poyer, David. *The Med: A Novel of the Navy.* St. Martin's Press, 1988.

Pretty, Ronald T. *Jane's Weapon Systems 1981-82.* Jane's Publishing Company Limited, 1981.

Price, Alfred, *Harrier at War.* Ian Allen, 1984.

—   *Air Battle Central Europe.* Warner Books, 1986.

—   *Instrument of Darkness: The History of Electronic Warfare.* Peninsula Publishing, 1987.

—   *The History of U.S. Electronic Warfare.* Association of Old Crows, 1989.

Rapoport, Anatol (editor). *Carl Von Clausewitz on War.* Penguin Books, 1968.

Reynolds, Clark G. *The Carrier War.* Time-Life Books, 1982.

Rhodes, Richard. *The Making of the Atomic Bomb.* Simon & Schuster, 1986.

Richelson, Jeffrey. *The U.S. Intelligence Community.* Ballinger Publishing Company, 1985.

—   *Sword and Shield: Soviet Intelligence and Security Apparatus.* Ballinger Publishing Company, 1986.

—   *American Espionage and the Soviet Target.* William Morrow and Company, 1987.

—   *America's Secret Eyes in Space.* Harper & Row Publishers, 1990.

Rommel, Erwin. *Infantry Attacks.* Presidio, 1990.

Santoli, Al. *Leading the Way: How Vietnam Veterans Rebuilt the U.S. Military.* Ballantine Books, 1993.

Scales, Brig. General Robert H., Jr. USA. *Certain Victory: The U.S. Army in the Gulf War.* Brassey's, 1994.

Schmitt, Gary. *Silent Warfare: Understanding the World of Intelligence.* Brassey's (U.S.), 1993.

Schneider, Wolfgang (editor). *Taschenbuch der Panzer (Tanks of the World).* 7th edition, Bernard & Graefe Verlag, Bonn, Germany, 1990.

Serber, Robert. *The Los Alamos Primer: The First Lectures on How to Build an Atomic Bomb.* University of California Press, 1992.

Sharp, Admiral U.S.G., USN (Ret.). *Strategy for Defeat.* Presidio Press, 1978.

Sharpe, Captain Richard, RN. *Jane's Fighting Ships 1989-90.* Jane's Publishing Company Limited, 1990.

Shaw, Robert L. *Fighter Combat: Tactics and Maneuvering.* Naval Institute Press, 1985.

Sherrod, Robert. *History of Marine Corps Aviation in World War II.* Nautical and Aviation Publishing, 1987.

Smith, Gordon. *Battles of the Falklands War.* Ian Allen, 1989.

Smith, Peter C. *Close Air Support: An Illustrated History, 1914 to the Present.* Orion Books, 1990.

Spector, Ronald, H. *Eagle against the Sun: The American War with Japan.* Free Press, 1985.

St. Vincent, David. *Iran: A Travel Survival Kit,* Lonely Planet, 1992.

Staff, U.S. News and World Report. *Triumph without Victory: The Unreported History of the Persian Gulf War.* Random House, 1992.

Stephen, Martin. *Sea Battles in Close Up: World War II.* Naval Institute Press, 1991.

Stevens, Paul D. (editor), *The Navy Cross: Vietnam.* Sharp & Dunnigan, 1987.

Stevenson, William. *90 Minutes at Entebbe.* Bantam Books, 1976.

Summers, Colonel Harry G., Jr., USA (Ret.). *A Critical Analysis of the Gulf War.* Dell Publishing, 1992.
    — *The New World Strategy.* Simon & Schuster, 1995.

Swanborough, Gordon, and Bowers, Peter. *United States Military Aircraft since 1909.* Smithsonian, 1989.
    — *United States Navy Aircraft since 1911.* Naval Institute Press, 1990.

Thornborough, Anthony. *Sky Spies: The Decades of Airborne Reconnaissance.* Arms and Armour, 1993.

Toffler, Alvin and Heidi. *War and Anti-War-Survival at the Dawn of the 21st Century.* Little Brown, 1993.

Toscano, Louis. *Triple Cross: Israel, the Atomic Bomb & the Man Who Spilled the Secrets.* Birch Lane Press, 1990.

Valenzi, Kathleen D. *Forged in Steel: U.S. Marine Corps Aviations.* Howell Press, 1987.

Van der Vat, Dan. *The Pacific Campaign, World War II.* Simon & Schuster, 1991.

Von Hassell, Agostino. *Strike Force: U.S. Marine Special Operations.* Howell Press, 1991.

Wagner, William. *Lightning Bugs and other Reconnaissance Drones.* Aero Publishers, 1982.
    — *Fireflies and other UAV's.* Midland Publishing Limited, 1992.

Walker, Bryce. *Fighting Jets.* Time-Life Books, 1983.

Waller, Douglas C. *The Commandos: The Inside Story of America's Secret Soldiers.* Simon & Schuster, 1994.

Ward, Commander Nigel "Sharkey," DSC, AFC, RN. *Sea Harrier Over the Falklands: A Maverick at War.* Naval Institute Press, 1992.

Warden, Colonel John A., III, USAF. *The Air Campaign: Planning for Combat.* Brassey's, 1989.

Ware, Lewis B. *Low Intensity Conflict in the Third World.* U.S. Government Printing Office, 1988.

Watson, Bruce W.;George, Bruce; Tsouras, Peter; and Cyr, B.L. *Military Lessons of the Gulf War.* Greenhill Books, 1991.

Wedertz, Bill. *Dictionary of Naval Abbreviations.* Naval Institute Press, 1977.

Weinberg, Gerhard. *A World at Arms: A Global History of World War II.* Cambridge, 1994.

Weinberger, Caspar. *Fighting for Peace: Seven Critical Years in the Pentagon.* Warner Books, 1990.

Weisgall, Jonathan M. *Operation Crossroads: The Atomic Tests at Bikini Atoll.* Naval Institute Press, 1994.

Weissman, Steve, and Krosney, Herbert. *The Islamic Bomb.* Times Books, 1981

Whipple, A. B. *To the Shores of Tripoli: The Birth of the U.S. Navy and Marines.* Morrow, 1991.

Winnefeld, James A. and Johnson, Dana J. *Joint Air Operations: Pursuit of Unity in Command and Control 1942-1991.* Naval Institute Press, 1993.

Winnefeld, James A., Niblack, Preston, and Johnson, Dana J. *A League of Airmen: U.S. Air Power in the Gulf War.* Rand Project Air Force, 1994

Wood, Derek. *Jane's World Aircraft Recognition Handbook.* Fifth Edition. Jane's Information Group, 1992.

Woodward, Robert. *The Commanders.* Simon & Schuster, 1991.

Woodward, Admiral Sandy, RN. *One Hundred Days: The Memoirs of the Falklands Battle Group Commander.* Naval Institute Press, 1992.

Zaloga, Steven J. *Red Trust: Attack on the Central Front, Soviet Tactics and Capabilities in the 1990's.* Presidio Press, 1989
    — *Target America: The Soviet Union and the Strategic Arms Race,* 1945-1964. Presidio Press, 1993.

Zumwalt, Admiral Elmo, USN, (Ret.). *On Watch.* Admiral Zumwalt Associates, Arlington, Virginia, 1976.

Bosnia: Country Handbook. U.S. Department of Defense, 1995.

Conduct of the Persian Gulf War. U.S. Government Printing Office, 1992.

GPS-A Guide to the Next Utility. Trimble Navigation, 1989.

Japan at War. Time-Life Books, 1980.

Space Log-1993. TRW, 1994

Special Forces and Missions. Time Life Books, 1990.

The World's Missile Systems. General Dynamics, 1988.

TRW Space Data. 4th Edition, TRW, 1992.

Wings at War Series, No. 2: Pacific Counterblow. Headquarters, Army Air Forces, 1992.

Magazines:

Air & Space

Air Force

Aviation Week and Space Technology

Command: Military History, Strategy, & Analysis

Leatherneck

Marine

Marine, Almanac 95

Marine, Almanac 96

Marine Corps Gazette

Naval History

The Economist

The Hook

U.S. Naval Institute Proceedings

U.S. News & World Report

World Airpower Journal

Videotapes:

AAAV: *Leading the Way.* General Dynamics Land Systems, 1995.

AAV7A1: *First to Fight.* United Defense, 1995.

Advanced Amphibious Assault Vehicle (AAAV). United Defense, 1995.

AH-1W Supercobra, August 1994. Bell Textron, 1994.

Army TACMS, Loral Vought Systems, 1994.

Barrett M82A1 and M90. Barrett Firearms Mfg., Inc., 1995.

Behind Enemy Lines: USMC Reconnaissance. Headquarters, U.S. Marine Corps, 1995.

Bell Helicopters in the Gulf War. Bell Textron, 1991.

Beyond the Horizon. Litton Ingalls Shipbuilding, 1988.

BLU-109B: Penetrate and Destroy. Lockheed Missiles and Space Company.

Chaos in the Littorals — March 21st, 1995. Major General James Myatt, USMC (Ret.). 1995.

Christening Kearsarge (LHD-3). Litton Ingalls Shipbuilding, 1992.

Customer Satisfaction through Total Quality. Texas Instruments, 1996.

Forward from the Sea. Bell Textron, 1994.

Harrier II Plus Remanufacture Program. McDonnell Douglas, 1994.

Hellfire — The Difference. Rockwell International, 1992.

Hercules and Beyond. Lockheed Aeronautical Systems Company. 1994.

Hercules Multi-Mission Aircraft. Lockheed Aeronautical Systems Company. 1994.

History of the Advanced Amphibious Assault Vehicle (AAAV). AAAV Program Manager. 1995.

It's About Performance. Sight & Sound Media. 1995.

Javelin 94. TI/Martin Joint Venture, 1994.

Joint Stars. Grumman. 1994.

Joint Stars One System Multiple Missions. Grumman. 1993.

JSOW Update 1994. Texas Instruments, 1994.

LHD Mission Conference. Litton Ingalls Shipbuilding, 1994.

M1A1 for the USMC. General Dynamics Land Systems, 1995.

MAG-13 Music Video – Long Version. McDonnell Douglas, 1991.

Maritime Preposition Force. Headquarters, U.S. Marine Corps, 1994.

MLRS: In the Storm. Loral Vought Systems, 1993.

Navy ATD Narration. Lockheed Martin Loral Vought Systems. 1993.

Navy League Music Loop '94. Newport News Shipbuilding, 1994.

Night of the Cobra. Bell Textron, 1993.

Nite Hawk F/A-18 Targeting FLIR Video. Lockheed Martin Loral, 1994.

On the Road Again. McDonnell Douglas, Northrop, General Electric, Hughes. 1994.

Operation Desert Storm Nite Hawk and Pave Tack FLIR Video for IRIS. Loral Aeronutronic. 1991.

Paveway Stock Footage. Defense Systems & Electronics Group. 1991.

Predator Presentation. Lockheed Martin Loral, 1993.

Putting Technology to Sea. Litton Ingalls Shipbuilding, 1994.

Storm from the Sea. Naval Institute, 1991.

Systems Integration: Forging New Frontiers. Bell Textron, 1993.

Tiltrotor Technology: Taking 21st Century Flight to a New Dimension. Bell Helicopter Textron, Visual Communications Center. 1995.

Uncooled FLIR. Texas Instruments, 1995.

United Defense AAAV. Texas Instruments, 1995.

V-22s Are Coming. Bell Textron, 1994.

War in the Gulf Video Series,1-4. Video Ordnance Inc., 1991.

Warriors From the Sea. Headquarters, U.S. Marine Corps, 1993.

Wings of the Red Star, Volume 1,2, and 3. The Discovery Channel, 1993.

CD-ROMs/Software:

Academic Year 1994 Curriculum: Multimedia CD-ROM. Air Command and Staff College, USAF, 1994.

Academic Year 1995 Curriculum: Multimedia CD-ROM (2 CDs). Air Command and Staff College, USAF, 1995.

Atomic Age. Softkey, 1994.

Desert Storm: The War in the Persian Gulf. Warner New Media, 1991.

Distance Learning Course, Multimedia Edition. Air Command and Staff College, USAF, 1995.

Encarta 96 Encyclopedia. Microsoft, 1996.

Infopedia. Future Vision Multimedia, 1995.

Warplanes: Modern Fighting Aircraft. Maris, 1994.

Wings (4 CD set). Discovery Communications, 1995.

World Factbook 1995 Edition. Wayzata, 1995.

Games:

Flight Commander 2. Avalon Hill Company, 1994.

Flying Nightmares. Domark Software, 1994.

Harpoon (3rd Ed.). Game Designers Workshop, 1987.

Harpoon Classic (Version 1.5). Alliance Interactive Software, 1994.

Harpoon II. Three Sixty, 1995.

Phase Line Smash. Game Designers Workshop. 1993.

TAC OPS: Modern Tactical Combat 1994-2000. Arsenal Publishing, 1994.